Dejan Zivkovic

Foundations of Algorithm Design and Analysis

Dejan Zivkovic

Foundations of Algorithm Design and Analysis

VDM Verlag Dr. Müller

Impressum/Imprint (nur für Deutschland/ only for Germany)
Bibliografische Information der Deutschen Nationalbibliothek: Die Deutsche Nationalbibliothek
verzeichnet diese Publikation in der Deutschen Nationalbibliografie; detaillierte bibliografische
Daten sind im Internet über http://dnb.d-nb.de abrufbar.

Coverbild: www.purestockx.com

Verlag: VDM Verlag Dr. Müller Aktiengesellschaft & Co. KG
Dudweiler Landstr. 99, 66123 Saarbrücken, Deutschland
Telefon +49 681 9100-698, Telefax +49 681 9100-988, Email: info@vdm-verlag.de

Herstellung in Deutschland:
Schaltungsdienst Lange o.H.G., Berlin
Books on Demand GmbH, Norderstedt
Reha GmbH, Saarbrücken
Amazon Distribution GmbH, Leipzig
ISBN: 978-3-639-13504-6

Imprint (only for USA, GB)
Bibliographic information published by the Deutsche Nationalbibliothek: The Deutsche
Nationalbibliothek lists this publication in the Deutsche Nationalbibliografie; detailed
bibliographic data are available in the Internet at http://dnb.d-nb.de.

Cover image: www.purestockx.com

Publisher:
VDM Verlag Dr. Müller Aktiengesellschaft & Co. KG
Dudweiler Landstr. 99, 66123 Saarbrücken, Germany
Phone +49 681 9100-698, Fax +49 681 9100-988, Email: info@vdm-verlag.de

Printed in the U.S.A.
Printed in the U.K. by (see last page)
ISBN: 978-3-639-13504-6

Contents

Preface

The field of algorithms is now a well-established part of computer science and mathematics due to its theoretical elegance and practical importance. On the one hand, computer practitioners can directly use good algorithms in complex programming projects. Moreover, for computer researchers algorithms are an indispensable tool in getting insight on hardness of problems. On the other hand, efficient algorithms often pose a challenging mathematical problem because their analysis uses deep mathematical results from different fields. Mathematicians have also recognized the computational significance per se and even in existential proofs make every effort to prove that objects exist by providing an algorithmic way to construct them. Thus the modern study of algorithms is based on an interplay of their computational and mathematical aspects.

Algorithms are basic ingredients of all technologies used in contemporary computers. Even if on the surface it does not appear that an advanced technology requires algorithmic content at the application level, under the hood it probably relies heavily on sophisticated algorithms. The hardware design, graphical user interfaces, and routing in networks are just a few examples of computer technologies that make extensive use of algorithms.

The goal of this book is to lay down foundations of the fascinating area of algorithms. In keeping with this goal, we have attempted to present some of the most fundamental methods for algorithm construction. In particular, the book introduces a number of paradigms and techniques useful for designing and analyzing data structures and algorithms.

The content of the book touches upon various themes related to algorithms. The following is an informal overview of the topics that are covered in the book.

- Data structures: to appreciate importance of data representation, besides the basic structuring techniques we discuss more advanced solutions including balanced (AVL) trees, hash tables, and binomial heaps.
- Ubiquitous algorithms: to develop basic skills, we study the classic tree and (weighted) graph algorithms.
- Algorithm design: to recognize similar problems and apply standard solutions, we study algorithm paradigms such as divide-and-conquer, dynamic programming, greedy algorithms, and randomized algorithms.
- Algorithm analysis: to rank algorithms on efficiency, we discuss how to analyze their time (or space) complexity.
- NP-completeness: to understand why some problems are particularly hard, we study a class of such problems that most likely do not admit of an efficient algorithm.

We have tried in the book to put emphasis on ideas and ease of understanding rather than implementation details. Another equally important feature of the book is the treatment of algorithms as mathematical abstractions that can be reasoned about formally. Thus the book aims to teach students how to precisely describe properties of an algorithm and to strive for proving claims about them.

Each chapter contains a few representative problems of techniques or topics discussed. These are explained in great detail, which is helpful to initially grasp the concepts. Furthermore, the end of each chapter contains a number of exercises whose difficulty ranges from reinforcement exercises to creativity problems. To get the concepts and techniques well in hand though, students are strongly encouraged to actually implement their algorithmic solutions in whatever programming language they are most familiar with. These implementation projects are not mentioned explicitly, with a few exceptions, but will be obvious when encountered.

This book is intended primarily as a textbook for a sophomore/junior first course on the design and analysis of algorithms and data structures. Reading of this book requires a certain level of maturity in programming and mathematics. Instead of providing the necessary background in the form of an appendix or a preliminary chapter, we have chosen to assume that the reader has already completed necessary computer science and mathematics courses.

In particular, the reader is expected to have completed programming-based introductory computer science courses. For example, the readers should have the experience of writing programs that use arrays and pointers, as well as programs that are nontrivially recursive. The reader's exposure to a higher-level programming language, such as C or Java, is needed so as to have a feeling of how algorithms are actually run on the computer and what is typical execution time of various programming constructs. Background in discrete mathematics and elementary calculus is necessary because we assume that the reader is able to handle basic mathematical concepts such as sets, functions, proofs, series, (recurrence) equations, and induction. Some knowledge of discrete probability is also needed to follow probabilistic analysis in certain sections, and it is especially important for full understanding of Chapter 11.

This book is structured to allow the instructor a great deal of freedom on how to organize and present the material. That's why the dependence between chapters is rather flexible, allowing the instructor to customize a course by highlighting those topics that one feels are most important. Moreover, numerous exercises after each chapter provide the instructor with an ample source of homework problems.

The text is accompanied by more than 250 pictures illustrating the explanations. Algorithms in the text are specified in plain English and clear pseudocode, so the reader is relieved of the burden to learn a new programming language in order to understand how these algorithms work.

The text has grown out of class notes that we have used to successfully teach many classes of computer science and mathematics majors. Based on our experience, the rule of thumb is that a one-semester introductory course should cover most of Chapters 1–9, along with selected topics from the remaining chapters.

In addition, the book is also well suited for self-study by technical professionals because we have tried to convey to the reader an understanding of how to implement the algorithms discussed.

Acknowledgments. I am indebted to many people who helped me in preparing this book. My sincere expression of thanks goes to all colleagues and students who used incomplete versions of the book in their courses and responded to my request to make constructive remarks and report errors.

Finally, I would appreciate receiving feedback on the book from the readers. Any comment or error report can be emailed to dzivkovic@singidunum.ac.rs.

DEJAN ŽIVKOVIĆ

Belgrade, Serbia
January, 2009

1

Introduction

Implementation of large computer systems is a very complex task and there are many reasons why this is so. One is that large programming projects require co-ordinated effort of many people with different background and expertise. Different software engineering techniques can help us cope with this aspect of the complexity of large computer systems. However, this is only part of the story.

Another part is that the large systems often involve solving algorithmic problems for which straightforward solutions may not be efficient enough. Roughly speaking, we use computers because we need to process large amounts of data. When we run a program on a large amount of input data, we must be certain that it uses a reasonable amount of memory and terminates in a reasonable amount of time. The program's memory and execution time are almost always independent of the software technology we use (for example, procedural or object-oriented approach). They mostly depend on the program's underlying algorithm for basic operations that the computer is supposed to carry out when the program is run. This means that efficient algorithms by which large systems should perform their basic tasks are an important aspect of the large systems' complexity.

In this book we restrict ourselves to smaller components of large systems and consider methods for the design and analysis of their algorithms. Although the algorithms that we study often amount to only a tiny fraction of the code for a large computer system, this small part can make a big difference for the success of the whole project. A naive approach to solving an algorithmic problem by a simplistic algorithm frequently leads to poor performance on large inputs. Moreover, no amount of fine-tuning of a badly designed algorithm is going to make a substantial difference. Thus, the design of algorithms that perform well on large input data is one of the most fundamental aspects of good programming.

1. Algorithms

To shed some light on how to design efficient algorithms, we will first try to define basic notions used throughout the book. An **abstract algorithmic problem** is described informally in common language that usually gives a short hint on the underlaying algorithmic problem. An **algorithmic problem** is a clear specification of an abstract problem whose solution is to be obtained computationally on the computer. Since in the book we consider only algorithmic problems, for simplicity we often drop the redundant modifier algorithmic and refer to them simply as problems.

For example, the abstract *sorting problem* can be stated as follows: given an array of numbers in arbitrary order, sort them in increasing order. The corresponding algorithmic sorting problem requires more precise specification of the desired input-output relationship: the input of the sorting problem is an array of numbers in arbitrary order, and the output is (possible the same) array of the same numbers reordered from smallest to largest. Another example is the *change-making problem*: given a set of coin denominations and a dollar amount, find the minimum number of coins needed to make change for the amount. In this case, we obviously need a more precise specification for the problem: given a list of $k \geq 1$ positive integers $c_1, \ldots, c_k$ ("coin denominations") and a positive integer a ("dollar amount"), find a list of k nonnegative integers $b_1, \ldots, b_k$ ("individual coin numbers") such that $a = \sum_{i=1}^{k} b_i \cdot c_i$ and $\sum_{i=1}^{k} b_i$ is minimal.

In general therefore, an algorithmic problem is specified by a set of **input values** and a set of **output values**, both sets satisfying some constraints given in the problem statement. An **instance of a problem** consists of the valid input for the problem, and a **solution to the problem instance** consists of the corresponding output for the input instance. The standard format for specifying a problem consists of two parts defining its input and output values. Both parts are expressed in terms of various mathematical objects, such as numbers, sets, functions and so on, with the input part representing a generic instance of the problem and the output part representing a generic solution to the generic instance.

For example, we can formally specify the change-making problem as having the following input-output relationship:

Input: A positive integer k, a k-element list of positive integers $\langle c_1, \ldots, c_k \rangle$, and a positive integer a.

Output: A k-element list of nonnegative integers $\langle b_1, \ldots, b_k \rangle$ which satisfies two conditions: $a = \sum_{i=1}^{k} b_i \cdot c_i$ and the sum $\sum_{i=1}^{k} b_i$ is minimal.

An instance of the change-making problem is a specification of concrete values for the input parameters of the problem: k, k-element list, and a. For example, $k = 4$, the 4-element list $\langle 1, 5, 10, 25 \rangle$, and $a = 17$ represents an instance of the change-making problem. The solution to this problem instance is the 4-element list $\langle 2, 1, 1, 0 \rangle$.

An **algorithm** is a well-defined computational procedure that takes some values as **input** and produces some values as **output**. Like a cooking recipe, an algorithm is a precise and unambiguous specification of a sequence of steps that can be carried out mechanically. An algorithm for a problem provides a step-by-step instructions that transform the problem's input into its output, achieving the desired input-output relationship. A *correct* algorithm halts with the correct output for every input instance. We then say that the algorithm *solves* the problem. In other words, an algorithm for a problem is a clearly specified sequence of instructions the computer is to execute to solve the problem.

Algorithm design refers to the process of specifying those unambiguous instructions that make up an algorithm. This is a creative endeavor though, so there is no prescribed formula for getting algorithms in general case. However, we will see that there are standard techniques that we can use to algorithmically solve many types of problems. These paradigms include divide-and-conquer algorithms, greedy algorithms, dynamic-programming algorithms, and randomized algorithms.

Algorithm analysis tries to measure resource usage of the algorithms. The most expensive resources that we usually wish to minimize are execution time and memory consumption of an algorithm, and accordingly we talk about its time and space complexity. It should be noted though, that a measure on the amount of resources needed by an algorithm really depends on the computer model on which the algorithm is executed. In this book we exclusively work under the limits of sequential machine models—parallel computational models and parallel algorithms deserve their own book.

2. Computational model

Generally speaking, computational models are simplified descriptions of real computers that strip away all of their complexity except for a few features deemed to be critical to the understanding of the phenomenon under study. Since measuring computational resources used by algorithms is one of our principal subject, we should first specify a precise model of a computing device for executing algorithms and define possible primitive operations in computations. However, we will refrain from a detailed discussion on the formal model of computation and instead later describe informally how we determine the algorithm complexity. This suffices for our purpose, since we will concentrate on asymptotic performance of algorithms, that is, their efficiency on large input. Moreover, formal computational models are really important for proving lower bounds on the algorithm complexity. Since we have more to say about this aspect of the subject only in the last chapter, there is no reason to bother with low-level details in the rest of the book.

Nevertheless, to be more precise, we now briefly outline the main features of the computing model we are going to work with. In most of the book we assume that our principal model of computation is an idealized "real" computer with single processor. We place no upper bound on its memory size, but assume that its programs cannot modify themselves. The exact repertoire of its instructions is not too important, as long as the instruction set resemble the one found in real computers. We assume there are arithmetic (add, subtract, multiply, divide, remainder, floor, ceiling), logical (and, or, not), input-output (read, write), data movement (load, store, copy), shift (by a small number of bit positions to the left and the right), and control (conditional and unconditional branch, subroutine call and return) instructions. These instructions operate on their operands that are stored in memory words whose addresses are determined using direct or indirect addressing scheme. We do not distinguish between memory hierarchy, such as cache or virtual memory, that exists in real computers. This model is a generic one-processor, *random-access machine* (*RAM*) model and should be quite familiar to anyone who has programmed in an assembly language.

The data types in the RAM model are integer and floating-point. We also assume a limit on the size of each memory word so that stored numbers cannot grow arbitrarily. For example, if the size of the memory word is w, then we can represent integers less than or equal to 2^{w-1} in absolute value. The floating-point numbers are represented in the IEEE standard format.

In the sequential RAM model, instructions are executed one after another with no concurrent operations. We postulate that our algorithms will be implemented as computer programs that are run on the RAM machine. Before we can analyze an algorithm though, we must specify the time required to execute each instruction and the space used by each operand. We choose to measure the time and space complexity of a program running on our idealized computer under the *unit-cost model*. In it, each instruction requires one unit of time and each operand requires one unit of memory space.

This model is appropriate if it is reasonably to assume that each number encountered in a program can be stored in one memory word. If a program computes extremely large numbers, which cannot be realistically used as single operands, and the unit-cost model is used without care or abused for that matter, it can give flawed impression of efficient programs.[1] However, if the unit-cost model is used as intended (for example, we break computation with large numbers into simpler steps and do not use unrealistically powerful instructions), the complexity analysis under this model is greatly simplified and accurately reflects experimental observations.

In brief, our guide is how real computers are designed. That is, we don't want to provide our idealized computer with unrealistic power. On the other hand, we would like it to allow us to work comfortably with at least random access data arrays and unit-cost operations on moderately large numbers.

3. Algorithm notation

The next question we need to answer about algorithms is what notation we should use to clearly specify the sequence of instructions comprising an algorithm? Essentially, the only requirement is that the specification must provide a precise description of the computational procedure to be followed. To express the steps comprising an algorithm, we can use, in the order of increasing precision, the natural language, pseudocode, and real programming language. Unfortunately, ease of expression goes in the reverse order.

---

[1]That's why some authors use the *log-cost model*, in which each instruction takes time proportional to the logarithmic length of the operands of the instruction, and each operand requires space proportional to the logarithm of its magnitude.

In describing and communicating algorithms we would like a notation that is more natural and easier to understand than an underlying program for the RAM machine, or even a computer program in a high-level programming language. Sometimes, the most expressive and concise way to specify an algorithm is the natural language. In more complicated cases, however, we need programming constructs to describe an algorithm unambiguously.

Notice that the notion of an algorithm is not the same as the more concrete notion of a computer program. A computer program is run on a real computer, and so a program must be written in a programming language according to strict rules of the language. Thus, every letter, comma, or symbol is important so that we have zero margin of error if we want to be able to run the program on the computer. We do not need such an absolute formalism with algorithms—after all, algorithms are not run on the real, "stupid" computers. Algorithms can be thought of as running on some sort of idealized computer with unlimited memory. They are really mathematical objects that humans can reason about in order to focus on the most essential aspects of presumably hard problems. Consequently, it is very desirable to be able to abstract away unnecessary details of a programming language. Another difference between algorithms and real programs is that we are not concerned with software engineering issues in algorithms. For example, error handling is often ignored in order to convey the essence of the algorithm more clearly and succinctly.

Generally, algorithms are small in size or they consist of short sections of code intermingled with text that explains crucial ideas or main points. This usually represents the first step in solving problems on the computer since picky, nonessential details of programming languages can only blur the big picture. This does not mean that an algorithm is allowed to contain any ambiguity with respect to the type of instructions executed or their exact order of execution. This simply means that we can be sloppy with a semicolon or coma here and there, if that does not cause any confusion to the reader. Also, some formally required elements of computer programs, such as data types, often are simply understood from the context.

For all these reasons, algorithms are most often expressed in an informal style as a pseudocode. Actually, we frequently describe the ideas of an algorithm in English, moving to pseudocode to clarify certain tricky details of the algorithm. The pseudocode we use in this book is similar to C, Pascal, or Java, so the readers who have done programming in any of these languages should have no trouble reading our algorithms. Our pseudocode uses ordinary programming language concepts such as variables, expressions, conditions, statements and procedures. We shall make no attempt to give a precise definition of the pseudocode syntax and semantics, since that is neither within the scope of the book nor is necessary. Instead, we provide a brief synopsis of conventions that are used in our pseudocode:

1. Primitive data types are integers, floating-point numbers, characters, and pointers. Additional data types such as strings, lists, queues and graphs are introduced as needed.

2. Variables and (compound) data types are not formally declared. Rather, their meaning should be evident either from their names or from the context.

3. Comments are enclosed in between symbols ⟦ and ⟧.

4. The loop statements **for**, **while**, and **until**, as well as and the conditional statement **if-then-else**, have their usual interpretations. The loop-counter variable in the **for** loop retains its value after exiting the loop. The **else** part in the conditional statement is optional.

5. Every statement ends with a semicolon (;). One line may contain multiple short statements.

6. The assignment operator is denoted by the symbol =. A multiple assignment of the form $x = y = e$ assigns to both variables x and y the value of expression e. It is equivalent to the assignments $x = e; y = e;$.

7. The boolean operators are denoted by the mathematical symbols $\land$, $\lor$, and $\neg$. Boolean expressions that use these operators are always fully parenthesized. They are short-circuit evaluated from left to right. That is, to evaluate the expression $p \land q$, we first evaluate p. If p evaluates to FALSE, we do not evaluate q since $p \land q$ must evaluate to FALSE. Similarly, for the expression $p \lor q$ we evaluate q only if p evaluates to FALSE, since otherwise $p \lor q$ must evaluate to TRUE. Boolean expressions evaluated in this way allow us to correctly use conditions such as $(i \leq n) \land (A(i) = 0)$ even in the case when the index i exceeds the size n of the array A.

8. Block structure of compound statements is indicated solely by indentation. Using indentation in small algorithms instead of conventional bracket symbols, such as { and } or **begin** and **end**, actually enhances clarity.

9. Variables are local to the procedure in which they are used. We do not use the same name for two different variables. Global variables are always explicitly mentioned and used only when needed. They are assumed to exist in an appropriate outer environment.

10. Array elements are accessed by writing the array name followed by the index in parenthesis. For example, $A(i)$ denotes the i-th element of the array A. The ellipsis (...) are used to indicate a range of array elements. Thus, $A(i \ldots j)$ denotes the subarray of A whose elements are $A(i), \ldots, A(j)$.

11. Procedures are used in one of two ways. One way is as a function with one result. In this case the last statement executed in the procedure must be a **return** statement followed by an expression. The **return** statement causes the expression to be evaluated and the execution of the procedure to terminate. The result of the procedure is the value of this expression. The second way of using a procedure is as a subroutine. In this case, the **return** statement is not required, and execution of the last statement in the procedure completes the

execution of the procedure itself.

A procedure may communicate with other procedures by means of the global variables and through parameters. Input (or input/output) parameters are specified in the procedure heading, and the output parameters are listed in the **return** statement. Arguments are passed to a procedure in a transparent way that correctly treats formal input, input/output, and output parameters of the procedure. Arguments of compound data types are passed only by reference; that is, the called procedure receives a pointer to the actual parameter.

12. We use descriptive, high-level statements in the natural language whenever such statements make an algorithm more understandable than an equivalent sequence of programming, low-level statements. For example, the statement

$$\textbf{for } \text{each vertex } v \text{ of graph } G \textbf{ do}$$

is more expressive than the details of its irrelevant implementation.

4. Algorithm design

The primary tool in designing good algorithms are algorithmic paradigms. They represent general approaches to the construction of *correct* and *efficient* solutions to problems. Such methods are of interest because

- they provide templates suited to solving a broad range of diverse problems;
- they can be translated into common control and data structures provided by most high-level languages;
- they lead to algorithms whose time and space requirements can be precisely analyzed.

Later in the book we will examine a number of design paradigms including divide-and-conquer, greedy method, dynamic programming, and randomization. Although more than one technique may be applicable to a specific problem, it is often the case that an algorithm constructed by one approach is clearly superior to equivalent solutions built using alternative techniques. The choice of a design paradigm is an important aspect of algorithm design.

Two fundamental issues must be addressed when designing good algorithms: correctness and efficiency. Since algorithms are mathematical abstractions, we typically need to establish an algorithm's correctness using mathematical methods and reasoning. This is especially important for complex algorithms, where correctness is not obvious at all and requires a careful mathematical proof. For simple algorithms, a short justification of the algorithm's basic properties is usually sufficient.

Efficiency of an algorithm is a measure of the computational resources, typically time and memory, that the algorithm requires for execution. Intuitively, the amount of computational resources depends heavily on the size and structure of the input values of the algorithm. For example, taking the analogy of an algorithm with a cooking recipe, it is clear that carrying out a pizza recipe to

bake pizzas for 10 persons requires more time and cookware than to bake pizzas for two persons. Thus, to estimate an algorithm's efficiency is to determine
the amount of resources it requires for execution as a function of the input size.
However, this function is usually very complicated, so to simplify its analysis we
may consider the maximum possible running time (or memory consumption),
among all the algorithm's inputs of the same size. We will usually focus on this
worst-case estimate in the book, although other types of performance measures
are also discussed when appropriate.

Other aspects of good algorithms are also important, particularly simplicity and clarity. Simple algorithms are desirable because such an algorithm is
easier to implement as a working program than a complex one. The resulting
program is less likely to have subtle bugs that get exposed in a production environment. Also, an algorithm should be understandable by a person reading the
pseudocode that describes the algorithm. Alas, our quest for a fast and simple
algorithm is often contradictory, and we must strike a balance between these
two desired characteristics.

4.1. Correctness of algorithms. No matter how we manage to come up with
an algorithm, the first and foremost requirement is that the algorithm correctly
solves a given problem. There are many difficulties, both practical and theoretical, in proving algorithms correct.

Practical obstacles have to do with the fact that most programs are written
to satisfy some informal specification, which itself may be incomplete or incon-
sistent. On the theoretical side, despite many attempts to formally define the
meaning of a program, there is no commonly accepted formalism one can use
to show nontrivial algorithms correct.

Anyway, it is beneficial to state and prove properties about algorithms. Such
properties are often the most useful short explanations that address the question about the way of how a particular algorithm works. The designer of an algorithm should be able to at least envision these properties, even though it may
be impractical to write out their proofs in detail. These properties should also
serve as a guide in the creative process of constructing an algorithm.

An indispensable tool in proving properties about algorithms is mathematical induction. Inductive proofs are essential in arguing that an algorithm does
what it is claimed to do. This applies to both iterative and recursive algorithms.

4.2. Proving iterative algorithms correct. The key to proving iterative algorithms correct is to show that loops satisfy certain properties. Necessary property for every loop is that its exit condition will be satisfied at some point so
that the loop will eventually terminate. Another kind of loop property is the one
whose validity is not changed with loop execution. Such a property is called the
loop invariant and represents a formal statement that holds true prior to every
iteration of the loop. This means that the loop invariant will be true after execution of the last iteration (or before the first unexecuted iteration), that is, it will
be true when the loop as a whole terminates. Recognizing the right loop invariant is important because we can usually make use of the loop invariant and the

logical value of the loop exit condition to conclude something useful about the correctness of an algorithm.

To show that any loop execution does not change the loop invariant, we must show two facts about it:

1. It is true before the first iteration of the loop.
2. If it is true before an iteration of the loop, it remains true after the iteration (that is, before the next iteration) of the loop.

Notice the similarity to mathematical induction, where we prove a base case and an inductive step. Here, showing that the invariant holds before the first iteration is like the base case, and showing that each iteration maintains the loop invariant is like the inductive step.

As an example, consider the problem of finding the greatest common divisor of two positive integers x and y. We denote the greatest common divisor of x and y by $\gcd(x, y)$. The **greatest common divisor** (or **gcd** for short) of the pair of positive integers (x, y) is defined as the largest integer dividing both x and y evenly. That is, $d = \gcd(x, y)$ divides both x and y evenly, so $x = md$ and $y = nd$ for some $m, n \in \mathbb{N}$,[2] and any other such common divisor of x and y is less than d.

Does this definition make sense? In other words, does every pair of positive integers have the greatest common divisor? Any pair of positive integers clearly has 1 as a common divisor, and the largest number that could possibly be a common divisor of x and y is the minimum of x and y. Thus, $\gcd(x, y)$ always exists and lies somewhere between 1 and $\min\{x, y\}$, inclusive.

An iterative algorithm searches for $\gcd(x, y)$ starting with the minimum of x and y, and then checks each smaller integer in turn until the first common divisor is found. Since we are repeating tests from the largest possibility down to the smallest one, the first common divisor found will be the greatest.

```
GCD1(x, y)
    d = min{x, y};
    while (x ≠ 0 (mod d)) ∨ (y ≠ 0 (mod d)) do
        d = d - 1;
    return d;
```

Let's argue more formally that the algorithm GCD1(x, y) correctly computes $\gcd(x, y)$. To begin, let us show that the **while** loop must terminate. Notice that the exit condition checks whether d does not divide x or y evenly. If that is the case, the body of the loop is repeated; otherwise, the loop terminates. To see why the exit condition will eventually become false, observe that initially $d = \min\{x, y\}$, and d decreases by 1 each time around the loop. In the worst case therefore d will eventually become 1, the exit condition will be false because $x = 0$ (mod 1) *and* $y = 0$ (mod 1), hence the loop will terminate.

[2]Throughout the book we let $\mathbb{N}$ denote the set $\{0, 1, 2, \dots\}$ of nonnegative integers.

Now we must prove correctness of the GCD1 algorithm. The appropriate invariant of the **while** loop is the property $d \geq \gcd(x, y)$. This property clearly holds prior to the first iteration of the loop, since initially d gets the value of $\min\{x, y\}$ and $\min\{x, y\} \geq \gcd(x, y)$. Next, we need to show that if the property holds before an iteration, then it will hold after execution of the iteration. Well, suppose that the value of d is greater than or equal to $\gcd(x, y)$ before an iteration, and suppose that the iteration of the **while** loop has been executed. This means that the exit condition evaluated to true and the value of d was decreased by 1. Since the exit condition was true, the old value of d was *not* a divisor of both x and y, and so the old value of d could not be equal to $\gcd(x, y)$; rather, it must have been strictly larger than $\gcd(x, y)$. Since the new value of d is one smaller than the old value, it follows that the new value of d is again greater than or equal to $\gcd(x, y)$ before the next iteration. This completes the argument that $d \geq \gcd(x, y)$ is an invariant property of the **while** loop in the GCD1 algorithm.

Finally, we examine that happens when the **while** loop terminates. The loop ends when d is a divisor of both x and y. Hence, $d \leq \gcd(x, y)$ when the loop terminates. Since the loop invariant $d \geq \gcd(x, y)$ is also true, it follows that $d = \gcd(x, y)$. But this value is returned as the result of the GCD1 algorithm, which shows that the algorithm is correct.

4.3. Proving recursive algorithms correct. Given a problem, we implicitly use induction in deriving a recursive definition of the solution to the problem. Then it is usually a simple matter to transform the solution into a recursive algorithm that implements it.

In a ***recursive definition***, we define some concept in terms of the concept itself. Of course, such a definition must not be meaningless or paradoxical; rather, it must define a class of objects by the way of closely related "smaller" objects. More precisely, a recursive definition involves a basis, in which one or more of the simplest objects in the class are defined, as well as an inductive step, in which more complex objects are defined in terms of simpler objects in the class.

As an example, consider again the problem of finding the greatest common divisor of two positive integers x and y. Let's suppose that we have somehow obtained an equivalent recursive definition of $\gcd(x, y)$ that states that if $x \geq y$ then

$$(1) \qquad \gcd(x, y) = \begin{cases} y, & x = 0 \pmod{y} \\ \gcd(y, x \pmod{y}), & x \neq 0 \pmod{y}. \end{cases}$$

We will prove in a moment that this recursive definition is equivalent to the non-recursive definition of $\gcd(x, y)$. For now, we want to demonstrate how trivial it is to turn the recursive definition into a recursive algorithm whose correctness is obvious. The following recursive algorithm GCD2(x, y) closely mirrors the recursive definition (1). In the algorithm we assume that $x \geq y$ holds upfront, which we can easily ensure by swapping x and y if necessary before making the call to the GCD2 procedure.

$$GCD2(x, y)$$
$$[\![\text{Assumes } x \geq y]\!]$$
$$z = x \pmod{y};$$
if $z = 0$ **then**
    return y;
else
    return $GCD2(y, z)$;

The harder part is to show the equivalence of the nonrecursive and recursive definitions of the gcd. That is, for any two positive integers x and y such that $x \geq y$, the recursive computation according to the equation (1) produces $\gcd(x, y)$. To prove this, we use induction on the number of applications k of the recursive equation (1) in the computation.

If $k = 1$, it must be the case that $x = 0 \pmod{y}$, that is, y divides x evenly, and so $\gcd(x, y) = y$. Since y is also the result of the one-step recursive computation of the gcd, the base case of the induction is true. As the inductive hypothesis, suppose that the claim is true for some $k \geq 1$. That is, given any two positive integers u and v such that $u \geq v$, if the recursive computation of the gcd for the pair (u, v) takes k steps, then we obtain $\gcd(u, v)$ as the result. Now, take two positive integers x and y such that $x \geq y$, and suppose the recursive computation of the gcd for the pair (x, y) takes $k + 1$ steps. We can write the sequence of $k + 1$ applications of the equation (1) as the following $(k + 1)$-step recursive computation:

$$\text{the gcd for } (x, y) \rightarrow \text{the gcd for } \left(y, x \pmod{y}\right) \rightarrow \cdots \rightarrow d_2,$$

where "$\rightarrow$" denotes one application of the equation (1) on a pair of positive integers. Let $d = \gcd(x, y)$ and $d_1 = \gcd\left(y, x \pmod{y}\right)$. Since $0 < x \pmod{y} < y$ and the recursive computation of the gcd for the pair $\left(y, x \pmod{y}\right)$ takes k steps, it follows by the inductive hypothesis that $d_2 = d_1$. Thus, it suffices to show that $d_1 = d$, since then for the $(k + 1)$-step recursive computation of the gcd for the pair (x, y) giving d_2 we will have $d_2 = d_1 = d$ as required. To show $d = d_1$, we argue that the sets of common divisors of the integer pairs (x, y) and $\left(y, x \pmod{y}\right)$ are the same. Well, if a evenly divides both x and y, then a evenly divides both y and $x \pmod{y} = x - iy$, $(i \in \mathbb{N})$. Conversely, if b evenly divides both y and $x \pmod{y} = x - iy$, $(i \in \mathbb{N})$, then b evenly divides both $x = x \pmod{y} + iy$, $(i \in \mathbb{N})$, and y. This completes the inductive proof of the equivalence of the nonrecursive and the recursive definition of the gcd.

5. Algorithm analysis

Once an algorithm is given for a problem and the algorithm is demonstrated to be correct, an important next step is to determine the amount of resources the algorithm will require. This task of estimating the efficiency of an algorithm is known as algorithm analysis. The design and analysis of algorithms are rarely independent and separate processes though. They mutually influence each other and equally contribute to the quest for correct, efficient, and elegant algorithms.

Two most precious computing resources that an algorithm must conserve are time and memory space. An algorithm that requires years to finish or several gigabytes of main memory is not much useful even if it is completely correct. Other possible criteria for algorithm efficiency are, for example, the amount of network traffic it generates, or the amount of data it moves to and from disks. However, in the book, we will almost exclusively deal with time complexity— space will be mentioned only if it significantly deviates from "normal" amounts. In other words, we identify performance of an algorithm with its running time.

The amount of time that any algorithm takes to run almost always depends on the amount of input data it must process. We expect, for example, that sorting 10,000 numbers requires more time than sorting 10 numbers. The running time of an algorithm is thus a function of the size of input data. Algorithms for different problems can have different kinds of input sizes. For example, for the problem of sorting an array of numbers, the input size is the number of array elements to be sorted; for the multiplication of two large integers x and y, the input size is the number of digits in x plus the number of digits in y; for graph problems, the input size is the number of vertices and/or the number of edges in a graph. Fortunately, input size of a particular problem is usually clear from the nature of the problem itself.

To analyze an algorithm, we therefore begin by grouping input data according to its size given by a nonnegative integer n. Then we would like to use a function $T(n)$ to represent the running time of the algorithm on the input of size n. In other words, for a nonnegative integer n, the value $T(n)$ should give the number of time units taken by the algorithm on any input of size n. Figure 1 illustrates these ideas.

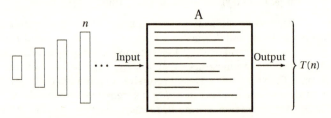

FIGURE 1. The running time $T(n)$ of the algorithm A as a function of the input size n.

However, given an input of size n, the exact running time of an algorithm on that particular input is dependent on many factors, such as the speed of the host machine or the structure of the input data itself, and not just on the input size. So, to simplify the analysis and to make the definition of $T(n)$ meaningful, we assume that all *basic operations* require one unit of time to execute and we look at the worst-case of input data. To obtain the running-time function $T(n)$ of an algorithm, we then simply count the number of basic operations performed by the algorithm in the *worst-case scenario*. More precisely, we define the **worst-case running-time** function $T(n)$ of an algorithm to be the algorithm's maximum running time over all inputs of size n. This function is also called the **worst-case time complexity**. Following the tradition, we almost exclusively refer to the worst-case running time and the worst-case time complexity simply as the *running time* and *time complexity*.

A *basic operation* is a computer operation whose execution time can be bounded above by a constant depending only on the particular implementation used (machine, programming language, compiler, and so on). Later we'll see that we are only interested in asymptotic behavior of the running time, that is, in its growth to within a constant multiplied by some simple function of the input size. Thus, it is only the total number of basic operations that matters in the algorithm analysis, not the exact time required by all of them. Equivalently, we say that any of the basic operations can be executed at unit cost. Typical basic operation are:

- an assignment;
- a comparison of two values;
- an arithmetic operation;
- a logical operation;
- an input/output operation.

A word of caution: a mathematical operation as apparently innocent as addition cannot be sometimes considered basic if the size of its operands is large. In this case, the time needed to execute it increases with the size of the operands, hence cannot be bounded above by a constant. Anyway, with this in mind, in what follows we consider previous operations as basic unless explicitly stated otherwise.

The *worst-case scenario* is a course of execution in which the *most* basic operations are performed. As a simple example, consider these two algorithm fragments A1 and A2:

A1	A2
$n = 5$;	**read**(n);
repeat	**repeat**
read(m);	**read**(m);
$n = n - 1$;	$n = n - 1$;
until $(m = 0) \vee (n = 0)$;	**until** $(m = 0) \vee (n = 0)$;

In A1 we have 5 iterations of the loop in the worst case, while in A2 we have n iterations of the same loop in the worst case.

Counting the number of basic operations in the worst case to determine the running time of an algorithm is the process that does not have to start over from scratch every time. Instead, it can be greatly simplified if we know the running times of typical program constructs. Namely, we can derive the running time of a compound construct from the running times of its building blocks. Table 1 summarizes the running times of basic constructs, where T_B denotes the (worst-case) running time of a block B. However, the table does not include recursion whose detailed discussion is put off until Chapter 5.

The most important rule here is that the running time of a loop is at most the running time of the statements inside the loop body (including tests) times the number of iterations. (Notice though that in Table 1 we ignored, for simplicity,

Construct	Time complexity
Sequence S: P; Q;	$T_S = T_P + T_Q$
Conditional S: **if** C **then** P; **else** Q;	$T_S = \max\{T_P, T_Q\}$
Loop S: (1) **while** C **do** P; (2) **repeat** P; **until** C; (3) **for** $i = j$ **to** k **do** P;	$T_S = T_P \cdot n$, where n is the worst-case number of iterations

TABLE 1. Time complexity of basic algorithm constructs.

the cost of performing tests in conditional and loop statements.) This in turn applies to nested loops: the running time of a group of simply nested loops is the running time of the statements in the innermost loop body multiplied by the product of the number of iterations of all loops. For example, for this fragment

```
for i = 1 to n do
  for j = 1 to n do
    if i < j then
      swap(A(i, j), A(j, i));   ⟦ basic operation ⟧
```

the worst-case running time is $T(n) = n \cdot n \cdot 2 = 2n^2$, because each of the two nested loops iterates n times and the innermost loop body is the **if** statement taking 2 basic operations.

Besides the worst-case running time, there are other approaches in measuring efficiency of (deterministic) algorithms. One is probabilistic analysis of the average running time of an algorithm. The ***average running time*** is a function $T_a(n)$ that gives the average value of the running time of an algorithm on random inputs of size n. The average running time is a more realistic measure of the algorithm performance in practice, but it is quite often much harder to calculate than the worst-case running time. This is so because we must know the probability distribution of inputs, which is hardly plausible in practice. Anyway, to make calculations more mathematically tractable, we usually assume that all inputs of size n are equally likely, which might not be appropriate in the given case.

Another approach is to measure efficiency of an algorithm with respect to efficiency of other algorithms that solve the same problem. This technique consists of developing a small collection of typical inputs that serve as ***benchmarks***. That is, we take the benchmark inputs as representative of all inputs, and agree to consider an algorithm that performs well on the benchmark inputs as if it performs well on *all* inputs.

Our primary emphasis in this book is on the worst-case running time, and we will not pursue the benchmarking in algorithm analysis at all. This point of view is changed in Chapter 11, where we study algorithms that are "probabilistic" in nature. Such randomized algorithms make random decisions during the course of execution so that their running time depends on the outcomes of these decisions. For randomized algorithms we measure the *expected running time*, where the expectation is calculated with respect to the random decisions.

Once we have found the running time of an algorithm, we may also try to answer three additional questions:

1. How can we tell whether an algorithm is a good algorithm?
2. How do we know if an algorithm is optimal?
3. Why do we need an efficient algorithm in the first place—why not just buy a supercomputer?

To answer the first question, we can simply compare the running time of the algorithm in question with those of existing algorithms solving the same problem. If it favorably compares to the other algorithms, we can quantitatively conclude that it is a good algorithm for some problem.

The second question is generally hard. For, although we can tell which of any two given algorithms is better by comparing their running times, we still don't know if a better algorithm is the best. We need to demonstrate a lower bound on the running times of all algorithms for a specific problem. This means that not only known algorithms, but also unknown ones, necessary must take time for execution greater than some bound. Finding such tight bounds is not an easy task, and we shall not generally undertake it in this book.

The third question is usually brought up by ignorant people too rich to design efficient algorithms. Besides being more intellectually challenging, a faster algorithm running on a slower computer will always win for sufficiently large input instances, as we shall see. In fact, problems don't have to get that large before the faster algorithm becomes superior, even when running on a personal computer.

Exercises

1. Consider the *searching problem*:

   Input: An integer n, an array of n numbers $A = [a_1, a_2, \ldots, a_n]$, and a number x.
   Output: The first index i such that $x = A(i)$ or $i = 0$ if x does not appear in A.

   (a) Give several instances of the searching problem and the corresponding solutions.
   (b) Write an iterative algorithm which looks for x by scanning through the array from its beginning. Using a loop invariant, prove correctness of your algorithm.

2. Consider the problem of calculating the size of monthly payments to a loan. Let L be the amount of the loan, I be the yearly interest rate of the loan (with $I = 0.08$ indicating an 8% interest rate), and P be the monthly payment of the loan. What is the amount of the loan L_n outstanding after the n-th month?

   Each month, the loan amount increases because of the monthly interest and decreases because of the monthly payment. Thus, if $M = 1 + I/12$, then $L_0 = L$ is the original loan amount, $L_1 = ML_0 - P$ is the loan amount after one month, $L_2 = ML_1 - P$ is the loan amount after two months, and so on.

   (a) Write an iterative algorithm for calculating L_n, given the values for L, I, P, and n. Using a loop invariant, prove correctness of your algorithm. (*Hint:* The loan outstanding after i iterations of a loop is L_i.)
   (b) Derive a recursive formula for L_n and write a recursive algorithm for calculating L_n.
   (c) Prove by induction that for each $n \geq 0$,

   $$L_n = L \cdot M^n - P \cdot \left(\frac{M^n - 1}{M - 1} \right).$$

   (d) Derive a formula for the actual monthly payment P of a loan L, yearly interest rate I, and the number of payments n. (*Hint:* Assume that, after n payments, the loan amount is reduced to 0.)

3. Exactly determine the (worst-case) running time for the following algorithm fragments:

   (a) **if** $x = 0$ **then**
       for $i = 1$ **to** n **do**
           $A(i) = i$;

   (b) $i = 1$;
       repeat
           $A(i) = B(i) + C(i)$;
           $i = i + 1$;
       until $i = n$;

   (c) **if** $x = 0$ **then**
       for $i = 1$ **to** n **do**
           for $j = 1$ **to** n **do**
               $A(i, j) = 0$;
       else
       for $i = 1$ **to** n **do**
           $A(i, i) = 1$;

   (d) **for** $i = 1$ **to** n **do**
       for $j = 1$ **to** n **do**
           for $k = 1$ **to** j **do**
               if $i = j$ **then** $A(k, k) = 0$;

2

Algorithmics

The design and analysis of algorithms is at the very heart of computer science. This subject, which is sometimes referred to as *algorithmics*, requires a certain amount of mathematical and programming maturity. The first part of this chapter contains a sorted list of introductory examples in order to establish the basic framework we will use throughout the book to think about algorithms. This gentle introduction will also clarify the fundamental ideas introduced in Chapter 1. Second part of the chapter introduces asymptotics as a way to cope with often complicated algorithm analysis.

There is no magic formula in getting good algorithms. It is largely a matter of judgment, intuition, and experience. Nevertheless, for many problems this process includes the following steps:

1. We first eliminate all nonessential and distracting complications of a practical problem so as to get the algorithmic problem as clean as possible. This typically means that the problem is abstracted away in terms of the simple mathematical concepts such as sets, functions, graphs, etc.
2. Having specified the problem with enough mathematical rigor, we construct an algorithm for the problem and prove its correctness. This step for the most part is carried out with the help of algorithmic paradigms and mathematical tools.
3. Finally, we analyze the algorithm so as to establish its efficiency. Since an initial analysis may give rise to a complicated-looking running-time function, involving summations or recurrences, we try next to simplify this function using the asymptotic notation and other techniques that we are going to learn in the second part of the chapter.

These steps are not completely independent of each other and often interlace. Moreover, not every aspect of this process is needed in full generality for simpler problems.

1. Examples of algorithm design and analysis

In this section we will work out several representative examples to get at the essence of the process of algorithm design and analysis. We present the examples without motivation, skipping first of the above steps, but the examples do have practical importance. They also have great pedagogical value as the later parts of this book will build upon this base.

1.1. Exchanging values of two variables. Given two variables, the problem is to exchange their values so that the new value of each variable is equal to the old value of the other variable. Not assuming anything about the type of the variable values, we need an auxiliary variable to hold the old value of one of the variables before we overwrite it with the other variable's value.

$\text{SWAP}(x, y)$
$z = x;$
$x = y;$
$y = z;$
return $x, y;$

Since each operation involved here is a basic one, the running time of the SWAP algorithm is clearly

$$T = 1 + 1 + 1 + 1 = 4.$$

We say that the SWAP algorithm runs in ***constant*** time.

1.2. Maximum element of an array. Given an array A of n numbers, the problem is to find the index of the (first) largest element of the array. A natural solution is to assume initially that the first element is the largest. Then we sequentially check each other element to see if it is greater than the largest element found so far, and if so, the current element becomes the largest. This is justified by inductive reasoning: given that we know $A(j)$ is the largest element in the subarray $A(1 \ldots i)$, to determine the largest element in the one-element longer subarray $A(1 \ldots i + 1)$, it is necessary and sufficient to compare the elements $A(j)$ and $A(i + 1)$.

$\text{FINDMAX}(A)$
   $m = A(1);$ [the largest element found so far]
   $j = 1;$ [index of the largest element]
   $i = 2;$ [check other elements ...]
  while $i \leq n$ **do**
    if $m < A(i)$ **then** [found a larger element]
      $m = A(i);$
      $j = i;$
    $i = i + 1;$
  return $j;$

Since the number of iterations of the **while** loop is $n - 1$, the running time of the algorithm FINDMAX is

$$T(n) = 1 + 1 + 1 + (n-1)(3+2) + 1$$
$$= 4 + 5(n-1)$$
$$= 5n - 1.$$

We say that the FINDMAX algorithm runs in **_linear_** time.

1.3. Sorting an array. Given an array A of n numbers, the problem is to re-arrange the array elements so that they form a nondecreasing sequence. That is, given an array A with n elements $A(1), A(2), \ldots, A(n)$, we have to produce a permutation of the array indices $i_1, i_2, \ldots, i_n$ such that the new first element $A(i_1)$, the new second element $A(i_2)$, and so on, the new n-th element $A(i_n)$ satisfy $A(i_1) \leq A(i_2) \leq \cdots \leq A(i_n)$.

In this section we present three simple algorithms for solving the sorting problem. They all require asymptotically the same time for execution, which is **_quadratic_** in the size of input. Later we will discuss more efficient sorting methods.

1.3.1. *Bubble sort.* The best way to understand the basic idea behind probably the simplest sorting method called the **_bubble sort_** is to imagine that the array to be sorted is kept vertically. Then the array elements with low values are "light" and bubble up to the top. Figure 1 shows the action of bubble sort on an array of size 6.

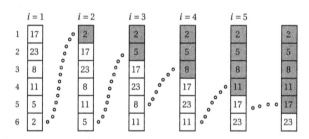

FIGURE 1. Bubble sort.

We make repeated passes over the array from bottom to top. As we go, if two adjacent elements are out of order, that is, if the "lighter" element is below, we reverse them. The effect of this operation is that on the first pass the "lightest" element, that is, the element with the lowest value, rises all the way to the top. On the second pass, the second lowest element rises to the second position, and so on. We need not, on the second pass, try to bubble up to position one, because we know that the lowest element is already there. Similarly for the third pass, we need not try to bubble up to position one and two, because we know that they are already occupied by the smallest and second smallest element. Thus in general, on pass i we need not try to bubble up to a position less than i.

The following is pseudocode for the bubble sort.

BUBBLE-SORT(A)
1. **for** $i = 1$ **to** $n - 1$ **do**
2.   **for** $j = n$ **downto** $i + 1$ **do**
3.   **if** $A(j) < A(j - 1)$ **then**
4.   SWAP($A(j)$, $A(j - 1)$);
5. **return** A;

If we take the *swap* operation as a basic operation (which is a reasonable approach, since its execution time is bounded by a constant), the body of the loop of lines 2–4 takes at most 4 time units. Here we account for testing and decrementation of j, the test $A(j) < A(j - 1)$ of line 3, and the *swap* operation. For a fixed value of i, the body of the inner loop of lines 2–4 is executed exactly $n - i$ times. The body of the outer loop consists of the inner loop, as well as three basic operations: testing and incrementation of i and initialization of j. Consequently, if we take into account the initialization of i at the beginning, as well as the line 5 at the end, the running time of the entire algorithm is

$$
\begin{aligned}
T(n) &= 2 + \sum_{i=1}^{n-1} (3 + 4(n - i)) \\
&= 2 + \sum_{i=1}^{n-1} 3 + 4 \sum_{i=1}^{n-1} (n - i) \\
&= 2 + 3(n - 1) + 4 \sum_{i=1}^{n-1} i \\
&= 2 + 3n - 3 + 4 \cdot \frac{(n - 1)n}{2} \\
&= 2n^2 + n - 1.
\end{aligned}
$$

Notice that in the best case when the array A is initially already sorted, hence no swaps are ever needed, the BUBBLE-SORT algorithm requires similar time because the test of line 3 is executed $n(n - 1)/2$ times.

1.3.2. *Insertion sort.* The second sorting method is called ***insertion sort*** because on the i-th pass we insert the i-th element $A(i)$ into its correct place among the elements $A(1), A(2), \ldots, A(i-1)$, which were previously already placed in sorted order. This is illustrated in Figure 2.

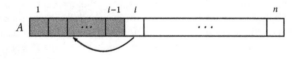

FIGURE 2. Insertion sort.

To make the algorithm cleaner, it helps to introduce a ***sentinel*** element $A(0)$ whose value is smaller than any value among $A(1), A(2), \ldots, A(n)$.

INSERT-SORT (A)
1. $A(0) = -\infty$;
2. **for** $i = 2$ **to** n **do**
3. $j = i$;
4.   **while** $A(j) < A(j-1)$ **do**
5.   SWAP$(A(j), A(j-1))$;
6. $j = j - 1$;
7. **return** A;

We refrain from detailed analysis of the running time of the INSERT-SORT algorithm, because it's similar to that of the BUBBLE-SORT algorithm. Instead, we proceed in a way that is usually applied when we need to analyze complexity of algorithms. That is, the total time required for a group of basic operations like initializations, incrementations, decrementations, assignments, tests, and so on, will be encompassed by some constant. However, even this is to much detail, and we will later see how to further get rid of these constants when we discuss the asymptotic notation.

Now, the **while** loop of lines 4–6 cannot take more than $i - 1$ steps, since j is initialized to i at the beginning of the loop at line 3, and decreases each time around the loop. The loop always terminates when $j = 1$ since then $A(j - 1) = A(0) = -\infty$, which forces the test of line 4 to be false. Consequently, in the worst case, the INSERT-SORT algorithm performs $n - 1$ iterations of the outer **for** loop varying the counter i from 2 to n, and each of these iterations for $i = 2, 3, \ldots, n$ executes at most $i - 1$ iterations of the inner **while** loop. Let's represent by constant c_1 the time taken by the initializations at lines 1 and 2 and by the **return** statement at line 7. Also, represent by constant c_2 the time taken by the test at line 2 and the initialization at line 3. Similarly, let c_3 be the constant time taken by the test at line 4, the *swap* operation at line 5, and the decrement operation at line 6. The running time of the entire INSERT-SORT algorithm is then

$$T(n) = c_1 + \sum_{i=2}^{n} \bigl(c_2 + c_3(i - 1)\bigr)$$

$$= c_1 + c_2(n - 1) + c_3 \sum_{i=2}^{n} (i - 1)$$

$$= c_1 + c_2(n - 1) + c_3 \frac{(n - 1)n}{2}$$

$$\leq c_4(n - 1)n,$$

for some other constant $c_4 = c_1 + c_2 + c_3/2$ and $n > 1$. The reader may check that if the array is initially sorted in reverse order, then the **while** loop of lines 4–6 is executed exactly $i - 1$ times. Therefore, in the worst case the INSERT-SORT algorithm really runs in the above $T(n)$ time.

1.3.3. *Selection sort.* In the **selection sort** we also make repeated passes over the array A. However, this time in the i-th pass we select the element with minimum value among the elements $A(i), A(i + 1), \ldots, A(n)$, and then swap it with $A(i)$. As a result, after i passes, the i lowest values will occupy the elements $A(1), A(2), \ldots, A(i)$ in sorted order. This is illustrated in Figure 3.

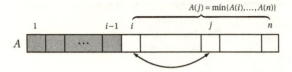

FIGURE 3. Selection sort.

The following is pseudocode for the selection sort algorithm, which uses FINDMIN as a subprocedure.

```
SELECT-SORT(A)
   for i = 1 to n - 1 do
      j = FINDMIN(A, i, n);
      SWAP(A(i), A(j));
   return A;
```

The algorithm FINDMIN(A, i, n) returns the index of the smallest element among elements of the subarray $A(i \dots n)$. This algorithm is an easy modification of the previous FINDMAX algorithm and is left as an exercise. The reader can also check that the running time of the FINDMIN algorithm is proportional to $n - i$. Since *swap* is a constant-time operation, this implies that the body of the **for** loop in the selection sort takes $c_1(n - i) + c_2$ time, for some constants c_1 and c_2. If we ignore the constant time needed for the initialization of the **for** loop and for execution of the **return** statement (as we would be doing in asymptotic analysis), the total time taken by the SELECT-SORT algorithm is

$$T(n) = \sum_{i=1}^{n-1} \left(c_1(n - i) + c_2\right)$$
$$= c_1 \frac{(n-1)n}{2} + c_2(n-1)$$
$$\leq c_3(n-1)n,$$

for some other constant $c_3 = c_1/2 + c_2$.

1.4. Searching a sorted array. Given an array A of n elements sorted in nondecreasing order, and given some item x, the searching problem is to determine whether the item x appears among the n array elements.

The searching problem, in essence, models the problem of efficiently looking up a word in a dictionary or a name in a telephone directory. First we give obvious and not so efficient solution that does not exploit the fact that the array is sorted. Then we present an algorithm that uses this fact to decide much faster whether the given item is indeed in the array.

1.4.1. *Sequential search.* The obvious approach to the searching problem is to look sequentially at each element of A until we either come to the end of the array or encounter an element equal to the given item x. The algorithm SEQ-SEARCH(A, x) implements this approach, and returns 0 if x is not found in the array A of size n; otherwise, it returns the index of the first array element equal to x.

$$\text{SEQ-SEARCH}(A, x)$$
$$j = 0; \; [\![\text{ index of } x \text{ in the array }]\!]$$
$$i = 1;$$
$$\textbf{while } (i \le n) \wedge (j = 0) \textbf{ do}$$
$$\textbf{if } A(i) = x \textbf{ then } j = i;$$
$$i = i + 1;$$
$$\textbf{return } j;$$

Carefully counting basic operations in the worst-case, we see that the running time of the SEQ-SEARCH algorithm is

$$T(n) = 4n + 4.$$

1.4.2. *Binary search.* To speed up the search considerably, we use the fact that the array is sorted. The idea is to start search by isolating x either in the first half or in the second half of the array. We must first find the index m of the middle element, which is $m = \lfloor (1+n)/2 \rfloor$.[1] Then we compare the lookup item x with $A(m)$. If they are equal, we are done since we have found x in the array A. If $x < A(m)$, we recursively repeat the search in the first half of the array. If $x > A(m)$, we recursively repeat the search in the second half of the array. If we are trying to search an "empty" array, we know that x is not in the array A. This approach is illustrated in Figure 4.

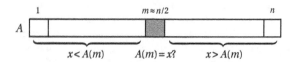

FIGURE 4. Binary search.

The algorithm BIN-SEARCH(A, i, j, x) below implements the binary search technique. It searches for x in the portion of the array A indexed by the lower bound i and the upper bound j. These parameters are necessary because of the recursive nature of the algorithm. However, to search for x in the original array, we simply need to make the initial call BIN-SEARCH$(A, 1, n, x)$.

---

[1]If x is any real number, then $\lfloor x \rfloor$, called the ***floor*** of x, denotes the integer part of x, or the largest integer less than or equal to x. Thus, $\lfloor 7.51 \rfloor = 7$ and $\lfloor 7 \rfloor = 7$. Also, $\lceil x \rceil$, called the ***ceiling*** of x, denotes the smallest integer greater than or equal to x. Thus, $\lceil 7.51 \rceil = 8$ and $\lceil 7 \rceil = 7$.

$\text{BIN-SEARCH}(A, i, j, x)$
   if $j < i$ **then** $[\![$ array is "empty" $]\!]$
     return 0;
   $m = \lfloor (i + j)/2 \rfloor$;
   if $A(m) = x$ **then**
     return m;
   else if $A(m) > x$ **then**
     return $\text{BIN-SEARCH}(A, i, m - 1, x)$;
   else $[\![$ $A(m) < x$ $]\!]$
     return $\text{BIN-SEARCH}(A, m + 1, j, x)$;

The BIN-SEARCH algorithm first checks whether the search range bounded by i and j has the length greater than zero. If at any time the lower bound exceeds the upper, we are trying to search for x in a range of the array of zero length. This means that we have failed to find x, and we return 0. If the length of the range is not zero, the algorithm calculates the index m of the element in the middle of the range and examines the middle element $A(m)$. If this element is equal to x, the search succeeds and we return m. If not, the algorithm continues the search in the lower or upper half of the range, depending on whether x is less than or greater than $A(m)$.

The running time $T(n)$ of a recursive algorithm is analyzed quite differently from a nonrecursive algorithm. The line of reasoning usually considers two cases. One is the base case when the input size n is equal to, say, 1, and the other is the general case when the input size n is strictly greater than 1. Depending on the logic of the algorithm, the running time $T(n)$ in the general case is expressed in terms of the value $T(m)$ representing the running time for some smaller input size m $(m < n)$. In this way, we obtain a recurrence equation that can be solved for $T(n)$ using standard mathematical techniques.

For example, the BIN-SEARCH algorithm for $n = 1$ clearly takes constant amount of time. Thus, $T(1) = c$ for some constant $c > 0$. If $n > 1$, the algorithm in the worst-case spends some time to test whether the middle element is equal to the lookup item, and then repeats itself on the first or second half of the array. Thus, $T(n) = T(\max\{m_1, m_2\}) + d$, where $d > 0$ is a constant that represents the fixed amount of time the algorithm takes exclusive of the recursive call, and m_1 and m_2 are the lengths of the first and second half of the array, respectively. Note that, since we assume that n is an integer, hence n can be even or odd, m_1 and m_2 are not equal in general. The reader can easily check that actually $m_1 = \lceil n/2 \rceil - 1$ and $m_2 = \lfloor n/2 \rfloor$. Since always $\lceil n/2 \rceil - 1 \le \lfloor n/2 \rfloor$, this implies $T(n) = T(\lfloor n/2 \rfloor) + d$. (Here we reasonably assume that $T(n)$ is an increasing function and, for simplicity, write = rather than more precisely $\le$.) In other words, we obtain the following recurrence equation:

$$T(1) = c,$$
$$T(n) = T(\lfloor n/2 \rfloor) + d, \quad n > 1,$$

where c and d are positive constants. To solve this recurrence equation for $T(n)$, we use the method known as *repeated substitution*. This and other techniques are discussed in more detail in Chapter 5.

Now, to get rid of the floor function, first suppose n is a power of 2, that is, $n = 2^k$ for some $k \geq 1$. If $T(m)$ with $m < n$ is repeatedly replaced using the general recurrence for $T(n)$, then

$$T(n) = T(n/2) + d$$
$$= T(n/4) + d + d = T(n/4) + 2d$$
$$= T(n/8) + d + 2d = T(n/8) + 3d$$
$$= T(n/16) + d + 3d = T(n/16) + 4d$$
$$\vdots$$
$$= T(n/2^k) + k \cdot d.$$

Since $n = 2^k$, or equivalently $k = \log_2 n$, we further have[2]

$$T(n) = T(n/2^k) + k \cdot d$$
$$= T(1) + \log n \cdot d$$
$$= c + d \log n.$$

In the case n is not a power of 2, we can use a similar argument with more careful analysis of n to show that $T(n) = c + d \lfloor \log n \rfloor$ (see Exercise 3).[3] Therefore, binary search is an order of magnitude faster than sequential search.

1.5. The greatest common divisor. Consider again the problem of finding the greatest common divisor of two positive integers x and y. In Section 4.1 we proved correctness of two algorithms for this problem. Let us analyze first the iterative version, which we repeat here.

```
GCD1(x, y)
    d = min{x, y};
    while (x ≠ 0 (mod d)) ∨ (y ≠ 0 (mod d)) do
        d = d − 1;
    return d;
```

It is clear that the running time of this algorithm is proportional to the number of iterations of the **while** loop, which in turn depends on the input numbers x and y. For example, if $x = y$, then the number of iterations is 0. If we apply the GCD1 algorithm to two consecutive positive integers, say, x and $y = x + 1$, then the **while** loop (including tests) is executed x times. Therefore, the worst-case running time of the GCD1 algorithm is proportional to the $\min\{x, y\}$.

Since the worst-case time complexity function of the GCD1 algorithm is linear in the smaller of the two input numbers, does this mean that it is a fast algorithm? Not really. A subtle question here is the size of the problem. If we apply

---

[2]In computer science, logarithm is almost always taken for base 2, rather than for base 10 or base e. Following the tradition, throughout the book we omit the notation of the base of logarithm and understand it to be 2 unless specified otherwise.

[3]In fact, in the rest of the book, we'll ignore the floors and ceilings in recurrences altogether. This can be always done without loss of generality, and so for simplicity we choose to be sloppy rather than formally correct.

the GCD1 algorithm to two numbers with, say, 100 digits, then we cannot expect that the arithmetic operations involved in the computation may be done in one unit of time. In this case, we must implement "extended precision" integer arithmetic ourselves, and use it to perform the assignment, modulus, and decrement operations. Thus, a more realistic measure of the size of the problem is the sum of the number of digits (or bits in the binary notation) of the two input numbers.

What we have demonstrated is that, even if we can do basic arithmetic operations efficiently,[4] and if the sum of the number of digits of the two input numbers is n, then the worst-case running time of the GCD1 algorithm is at least, roughly, $10^{n/2}$. This follows from the case when the input numbers are consecutive positive integers, hence they have roughly $n/2$ digits, and so the running time of the GCD1 algorithm is proportional to the smaller of the two numbers, which we can always choose to be roughly $10^{n/2}$. This means that the GCD1 algorithm actually takes exponential time in the worst-case and is therefore useless for large numbers.

Let us now see if we are more lucky with the recursive algorithm for finding the greatest common divisor of two positive integers x and y. This version is called Euclid's algorithm, and we repeat it here as well.

GCD2(x, y)
$[\![$ Assumes $x \geq y$ $]\!]$
$z = x$ (mod y);
if $z = 0$ **then**
    return y;
else
    return GCD2(y, z);

To get started with the analysis of the GCD2 algorithm, let's experiment with some examples to see how many steps the algorithm takes. We represent each recursive call by the symbol $\rightarrow$. Some number pairs require a few recursive calls:

$$\text{GCD2}(1055, 15) \rightarrow \text{GCD2}(15, 5) = 5,$$
$$\text{GCD2}(400, 24) \rightarrow \text{GCD2}(24, 16) \rightarrow \text{GCD2}(16, 8) = 8,$$
$$\text{GCD2}(101, 100) \rightarrow \text{GCD2}(100, 1) = 1,$$
$$\text{GCD2}(34, 17) = 17.$$

Yet the computation for some other pairs is more labor intensive:

$$\text{GCD2}(89, 55) \rightarrow \text{GCD2}(55, 34) \rightarrow \text{GCD2}(34, 21) \rightarrow \text{GCD2}(21, 13) \rightarrow$$
$$\rightarrow \text{GCD2}(13, 8) \rightarrow \text{GCD2}(8, 5) \rightarrow \text{GCD2}(5, 3) \rightarrow$$
$$\rightarrow \text{GCD2}(3, 2) \rightarrow \text{GCD2}(2, 1) = 1.$$

These examples suggest that the question of the running time of the GCD2 algorithm is very delicate. We see that the number of recursive calls the algorithm makes may be quite different, depending on the numbers in question. However, we can bound the number of recursive calls, and since each recursive call takes

---

[4]Fast computation of basic arithmetic operations (addition, subtraction, multiplication, division with remainder) is discussed in Chapter 5.

some constant time,[5] we can get a time bound for the Euclid's algorithm by multiplying this constant and the upper bound on the number of recursive calls. The
key to this approach is the following observation about the Euclid's algorithm.

LEMMA *For any pair* (x, y) *of positive integers such that* $x > y$, *if in the call*
GCD2(x, y) *is made another recursive call* GCD2(y, z) *with* $z = x$ (mod y), *then*
$yz < \frac{1}{2}xy$.

PROOF: To prove that in each recursive call the product of two input numbers drops by a factor of at least two, consider the recursive call when the pair
(x, y) with $x > y$ is replaced by the pair (y, z) with $z = x$ (mod y). Then we have
$x = ky + z$ for some positive integer k and $0 < z < y$. Hence $x \geq y + z > 2z$, that is,
$xy > 2yz$. Thus, $yz < \frac{1}{2}xy$ as claimed. ∎

Suppose that computation of the Euclid's algorithm on two positive integers
x and y makes m ($m \geq 0$) recursive calls. In other words, the call GCD2(x, y)
generates a sequence of m recursive calls, and the last nonrecursive call returns
$g = \gcd(x, y)$:
$$\text{GCD2}(x, y) \rightarrow \cdots \rightarrow \text{GCD2}(h, g) = g.$$
It follows by the previous lemma applied m times that $hg \leq xy/2^m$. Since certainly $hg \geq 1$, we get $xy \geq 2^m$, that is,
$$m \leq \log(xy) = \log x + \log y.$$
Thus, we have shown that the number of recursive calls of the Euclid's algorithm,
applied to two positive integers x and y, is at most $\log x + \log y$. Let n_1 be the
number of bits of x written in base 2, n_2 be the number of bits of y written in
base 2, and $n = n_1 + n_2$. Notice that $\log x < n_1$ and $\log y < n_2$, hence $\log x + \log y <
n_1 + n_2 = n$. Now, since each of the m recursive calls of Euclid's algorithm, as well
as the last nonrecursive call, takes a constant time, say c, the algorithm's running
time $T(n)$ can be estimated by
$$T(n) = c(m + 1) \leq c(\log x + \log y + 1) \leq c(n + 1) \leq dn,$$
where $d = 2c$ is a constant independent of n. This means that the running time
of the GCD2 algorithm is at most linear in the problem size and is therefore very
fast algorithm.

1.6. The maximum contiguous subsequence sum problem. Suppose we
are given an array A of n numbers $a_1, a_2, \ldots, a_n$ (positive and negative). A *contiguous subsequence* of the array A is a subsequence made up of consecutive
elements of A. For example, if
$$A = [2, -1, 1, 3, -4, -6, 7, -2, 3, -5],$$
then $1, 3, -4, -6, 7$ and -1 are contiguous subsequences, but $2, 1, 3$ is not. Note
that the empty subsequence of length zero is considered to be a contiguous subsequence. For simplicity, we also use the term *subsequence* to mean *contiguous
subsequence* throughout this section.

---

[5]This constant time is the time needed for four basic operations: calculating remainder of an
integer division, assigning the remainder to a variable, testing the variable for zero, and making a
procedure call.

Given a contiguous subsequence $a_i, \ldots, a_j$, its sum, denoted by $S_{i,j}$, is simply the sum of its elements: $S_{i,j} = \sum_{k=i}^{j} a_k$. We postulate that the sum of the empty subsequence is zero. **The maximum contiguous subsequence sum (mcss) problem** is to determine the maximum sum of all contiguous subsequence of a given array. (In full generality, we also need to identify the corresponding maximum-sum contiguous subsequence.)

For the preceding example, the answer would be 8 corresponding to the contiguous subsequence $7, -2, 3$. If an array contains only positive numbers, the answer is clear and would be the sum of all the array elements. If an array contains only negative numbers, the contiguous subsequence of maximum sum is the empty subsequence and the answer would be zero. The interesting case is therefore when we have both positive and negative numbers in a given array.

Given an array A of n numbers, let M be the sum of its maximum-sum contiguous subsequence. Formally,

$$M = \max\{0, \max_{1 \le i \le j \le n} S_{i,j}\}.$$

For notational convenience, in our discussion that follows we assume that we implicitly always take 0 as maximum in computing various maximum sums if the "real" maximum is less than zero. (By contrast, our algorithms will correctly compute the maximum sums in all cases.) For example, we may succinctly write $M = \max_{1 \le i \le j \le n} S_{i,j}$.

The maximum contiguous subsequence sum problem arises in several contexts in computational biology in the analysis of DNA or protein sequences. The problem is computationally interesting because there are many algorithms of various efficiency to solve it. In this section we present two algorithms. The first algorithm is a natural quadratic algorithm that serves as a warm-up to an improvement. The second algorithm is a not-so-obvious linear algorithm that makes use of the structure of the input array to improve performance. Two additional algorithms are discussed as exercises in Chapter 5 (Exercise 13) and Chapter 7 (Exercise 12).

1.6.1. *A quadratic algorithm.* To compute M by the definition, we need to compute the sums of all subsequences that begin at position 1, then the sums of all subsequences that begin at position 2, and so on, then the sums of all subsequences that begin at position n. Finally, we need to compute the maximum of these $n + (n-1) + \cdots + 1 = n(n+1)/2$ sums. We can slightly speed-up this approach if in each step i we additionally compute the maximum of the sums of all subsequences that begin at position i, and in the end we compute the maximum of these n maximum sums. In other words, we can imagine that we are computing first another array of n elements $b_1, b_2, \ldots, b_n$ such that

$$b_i = \max_{j=i,\ldots,n} S_{i,j},$$

and then M as

$$M = \max_{i=1,\ldots,n} b_i.$$

The pseudocode implementation of this idea is broken down into two algorithms: the first one computes b_i, and the second one uses it to solve the mcss problem.

The algorithm $B(i)$ below that computes b_i goes along the lines of the FIND-MAX algorithm from Section 1.2 that finds (the index of) the largest element of an array.

$B(i)$
    $m = 0;$ [the max subsequence-sum]
    $s = 0;$ [$S_{i,j}$ for $j = i, \ldots, n$]
    for $j = i$ **to** n **do**   [compute $S_{i,j}$ for $j = i, \ldots, n$]
        $s = s + A(j);$ [$S_{i,j} = S_{i,j-1} + a_j$]
        if $m < s$ **then**   [a larger sum found]
            $m = s;$ [update the running max sum]
    return $m;$

The algorithm $B(i)$ works as follows. We keep in m the running maximum sum of the subsequences beginning at position i, as well as their sums in s. (These variables are properly initialized to 0). In each step j from i through n, we compute $S_{i,j}$ in s by adding a_j to the previously computed sum $S_{i,j-1}$, and immediately check whether $m < s$. If so, we update the running maximum sum since we encountered a larger-sum subsequence beginning at position i.

Clearly, the running time of this algorithm is dominated by the **for** loop whose counter j ranges from i to n. So, the simple body of the loop is repeated $n - i + 1$ times, implying that the running time of $B(i)$ is proportional to $n - i + 1$.

The procedure that solves the mcss problem by first computing all b_i's and then taking their maximum is encompassed by the following MCSS1 algorithm.

MCSS1 (A)
    for $i = 1$ **to** n **do** $b_i = B(i);$
    $M = \max_{i=1,\ldots,n} b_i;$
    return $M;$

To estimate the running time of MCSS1 algorithm, we observe that the i-th iteration of its **for** loop takes time proportional to $n - i + 1$, so the time that the entire loop takes is proportional to $\sum_{i=1}^{n} (n - i + 1) = n(n+1)/2$. Computing $M = \max_{i=1,\ldots,n} b_i$ can be done by a simpler version of the FINDMAX algorithm, and so it certainly takes linear time in the size n of the array $b_1, b_2, \ldots, b_n$. Thus, the total time of the MCSS1 algorithm is proportional to $n(n+1)/2 + n$. The expression $n(n+1)/2 + n = n^2/2 + 3n/2$ is of order n^2 for large n, so the MCSS1 algorithm is a quadratic algorithm.

We can further improve the algorithm if we compute "on the fly" the overall maximum M of the b_i's. That is, instead of finding the maximum of b_i's after we compute all of them, we can maintain a running maximum of them as we compute them in order. Moreover, this way we don't need a separate array to hold b_i's, so we also save on space. In fact, we can make still one more improvement by maintaining the running overall maximum not of b_i's but of the sums $S_{i,j}$ as we compute them. The following is a pseudocode for this "streamlined" version of the algorithm.

```
MCSS1(A)
    M = 0;                    【 the overall max subsequence-sum 】
    for i = 1 to n do
        s = 0;
        for j = i to n do
            s = s + A(j);
            if M < s then
                M = s;
    return M;
```

1.6.2. *A linear algorithm.* To greatly improve performance of the MCSS1 algorithm, we need to look more closely at the computation of b_i's and bring it down to linear time. We will see that we can basically skip computation of many b_i's, which will enable us to determine M in a single pass scanning the input array from left to right.

Recall our general strategy for determining M. For each $i = 1, \ldots, n$, we compute b_i by computing the sums $S_{i,i}, S_{i,i+1}, \ldots, S_{i,n}$ and maintaining their running maximum. The sums that we compute in order satisfy one of the following two conditions: either all of them are nonnegative or some of them are negative. We will show that in both cases we can skip many subsequent computations. This is formalized in the next two observations.

The first observation deals with the case when the sums $S_{i,i}, S_{i,i+1}, \ldots, S_{i,n}$ are all nonnegative. Namely, if during computation of any b_i we detect such a case, then the remaining $b_{i+1}, \ldots, b_n$ are all at most b_i and so we don't have to consider them in computing M. This is so because $\max\{b_i, b_{i+1}, \ldots, b_n\} = b_i$, hence $M = \max\{b_1, \ldots, b_i, b_{i+1}, \ldots, b_n\} = \max\{b_1, \ldots, b_i\}$.

OBSERVATION 1 *For any $i \in \{1, 2, \ldots, n\}$, if $S_{i,j} \geq 0$ for every $j = i, i+1, \ldots, n$, then $b_k \leq b_i$ for every $k = i+1, \ldots, n$.*

PROOF: For $i < k \leq n$, suppose $b_k = S_{k,r}$ for some $r \geq k$, as illustrated in the following picture.

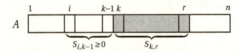

Then

$$b_k = S_{k,r}$$
$$\leq S_{i,k-1} + S_{k,r}, \quad \text{since } S_{i,k-1} \geq 0 \text{ by assumption}$$
$$= S_{i,r}$$
$$\leq b_i, \quad \text{since } b_i = \max_{j=i,\ldots,n} S_{i,j}. \blacksquare$$

The second observation deals with the case when some of the sums $S_{i,i}, S_{i,i+1}, \ldots, S_{i,n}$ are negative during computation of any b_i. The observation informally says that if j is the first index encountered such that $S_{i,j} < 0$, then we can ignore the remaining sums $S_{i,j+1}, \ldots, S_{i,n}$, skip completely computation of $b_{i+1}, \ldots, b_j$, and proceed with computation of $b_{j+1}, \ldots, b_n$.

OBSERVATION 2 *For any $i \in \{1, 2, \ldots, n\}$, if some of $S_{i,i}, S_{i,i+1}, \ldots, S_{i,n}$ are negative, let j be the lowest index such that $S_{i,j} < 0$. Then either*

Case 1. $b_k \leq b_{j+1}$ for every $k = i, i+1, \ldots, j$, or
Case 2. $b_k \leq b_i$ for every $k = i, i+1, \ldots, j$ and $b_i = \max\{S_{i,i}, S_{i,i+1}, \ldots, S_{i,j-1}\}$.

PROOF: Fix $i \in \{1, 2, \ldots, n\}$. First of all, we claim that always $b_k \leq b_i$ for every $k = i, i+1, \ldots, j$. This is certainly true for $k = i$, and for $i < k \leq j$ suppose $b_k = S_{k,r}$ for some $r \geq k$. Since j is the lowest index such that $S_{i,j} < 0$, it implies that $S_{i,k-1} \geq 0$, and so

$$b_k = S_{k,r} \leq S_{i,k-1} + S_{k,r} = S_{i,r} \leq b_i.$$

Now suppose the second case of the observation does not hold. This implies

$$b_i \neq \max\{S_{i,i}, S_{i,i+1}, \ldots, S_{i,j-1}\},$$

that is, $b_i = S_{i,r}$ for some $r > j$. (Note that $r \neq j$ since $S_{i,j} < 0$.) This is illustrated in the following picture.

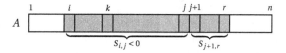

Then

$$b_i = S_{i,r}$$
$$= S_{i,j} + S_{j+1,r}$$
$$\leq S_{j+1,r}, \qquad \text{since } S_{i,j} < 0 \text{ by assumption}$$
$$\leq b_{j+1}, \qquad \text{since } b_{j+1} = \max\{S_{j+1,j+1}, \ldots, S_{j+1,n}\}.$$

In this case therefore $b_i \leq b_{j+1}$. But since the inequality $b_k \leq b_i$ for every $k = i, i+1, \ldots, j$ is always true, we can conclude $b_k \leq b_i \leq b_{j+1}$ for every $k = i, i+1, \ldots, j$. That is, the first case of the observation holds as required. ∎

In plain English, this observation implies that during computation of any b_i when we first detect $S_{i,j} < 0$, we can safely jump and continue with computation of $b_{j+1}, \ldots, b_n$. This is possible because at that moment holds either Case 1 or Case 2. Case 2 means that our running maximum already contains correctly computed b_i, and $b_{i+1}, \ldots, b_j$ are all at most b_i and thus irrelevant for M. In Case 1 we haven't yet computed b_i as the running maximum, but b_i and $b_{i+1}, \ldots, b_j$ are all at most b_{j+1} and so $M = \max\{b_1, \ldots, b_{i-1}, b_{j+1}, \ldots, b_n\}$. Thus, assuming we have correctly determined $\max\{b_1, \ldots, b_{i-1}\}$, we can ignore $b_i, b_{i+1}, \ldots, b_j$ and continue updating the running maximum with b_{j+1} through b_n.

The following is the MCSS2 algorithm that incorporates previous observations.

```
MCSS2(A)
   M = 0; i = 1;
   while i < n do              [[ compute b_i ]]
      s = 0;
      for j = i to n do
         s = s + A(j);         [[ S_{i,j} = S_{i,j-1} + a_j ]]
         if s < 0 then         [[ check S_{i,j} < 0 ? ]]
            break;             [[ if so, break from b_i, b_{i+1}, ..., b_j ]]
         else if M < s then    [[ else, update the max sum as usual ]]
            M = s;
      i = j + 1;               [[ continue with b_{j+1}, ..., b_n ]]
   return M;
```

Even though the MCSS2 algorithm contains two nested loops, the running time of this algorithm is linear. Observe that at each step in the outer **while** loop we advance its counter i to the array position that is one more than the position where the inner **for** loop stopped. This implies that the counter j of the inner loop gets each value from 1 to n exactly once. Thus the body of the inner loop is executed exactly n times, which means that the running time of the MCSS2 algorithm is proportional to n.

1.7. The stable marriage problem. In this section we investigate a problem that arises in modeling social-economic systems with deep ramifications. The problem in its most basic form was introduced by Gale and Shapley in 1962 in relation to the problem of college admissions, but since then other variants have been studied in computer science, economics, game theory, and operations research. An algorithm for solving the problem reveals the problem's rich structure and serves as an example of the algorithm analysis where correctness is more complex than running time.

Formally, we are given two disjoint finite sets M and W each of n elements, and we are to produce a bijection f of M onto W subject to certain constraints. We follow the tradition however, which uses more coloristic description of the problem by identifying the elements of the two sets with men and women. So, imagine that n men and n women seek a lifelong spouses. We assume a monogamous and heterosexual society, implying that each person wants exactly one individual of the opposite gender. Thus the bijection $f: M \to W$ from the formal formulation corresponds to pairing (marrying) each man off to exactly one woman (and vice versa). Such a bijection f is called a *(perfect) matching*. To bring the analogy further, if $f(m) = w$ in a perfect matching for some $m \in M$ and $w \in W$, this is denoted as the pair (m, w).

The constraint that a perfect matching has to satisfy is defined in the spirit of making all the men and women maintain a longterm relationship. To this end, we suppose that each person has a preference list of all the persons of the opposite sex, which reflects the person's preference for the partners he or she would like to marry. This means that every man ranks all the women from the most preferred down to the least preferred that he wants to marry, and so does every woman. These rankings are assumed strict without ties, and so the preference list of each person may be represented by a sorted array in decreasing order of all n persons of the opposite sex. Note also that these rankings do not change over time.

For example, if $n = 4$ and $M = \{A, B, C, D\}$ and $W = \{a, b, c, d\}$, the preference lists of the men in M and the women in W may be

$$A: c, b, a, d \quad B: b, a, c, d \quad C: a, d, c, b \quad D: d, c, a, b$$
$$a: A, B, C, D \quad b: D, C, B, A \quad c: B, C, A, D \quad d: C, D, A, B$$

Representing the men's and women's lists more conveniently in a tabular form, the same information may be provided by the following two tables:

men	1	2	3	4
A	c	b	a	d
B	b	a	c	d
C	a	d	c	b
D	d	c	a	b

women	1	2	3	4
a	A	B	C	D
b	D	C	B	A
c	B	C	A	D
d	C	D	A	B

Given the preference list of a person p, we say that the person p **prefers** a person q over a person r of the opposite sex if q comes before r on the p's preference list. The fact that p prefers q to r is denoted by $q <_p r$. Thus in the preceding example, the woman c prefers B to D and the man B prefers a to c, which is denoted $B <_c D$ and $a <_B c$.

Given all the preference lists, the goal is to find a matching that avoids any potential consensual adultery. That is, a matching that avoids creating a dissatisfied pair of opposite-sex persons because they both prefer each other to their assigned spouses in the matching. Formally, a pair m–w with $m \in M$ and $w \in W$ is called a **dissatisfied pair** in a matching if m and w are not matched together, i.e., in the matching we have (m, w') and (m', w) for some different $w' \in W$ and $m' \in M$, but m prefers w to w' and w prefers m to m' (the opinions of m' and w' are irrelevant).

Since nothing prevents a dissatisfied pair to split from their mates and happily remarry each other, the problem asks to find a matching with no dissatisfied pair. (We assume no one would go for someone he or she liked even less than the assigned spouse.) A matching that has no dissatisfied pair is called a **stable matching**.

Consider the matching $(A, a), (B, b), (C, c), (D, d)$ in the previous example of preference lists. The pair C–d is dissatisfied in the matching because C prefers d to his wife c and d prefers C to her husband D. Thus the matching is not stable.

If we try to remove instability by matching this dissatisfied pair (imagine the couples (C, c) and (D, d) divorce), and also match D and c as the only remaining choice for them, we obtain the new matching $(A, a), (B, b), (C, d), (D, c)$. However, this matching is also unstable because of A and c. Now, A prefers c over a, and c prefers A over D. We may proceed in this fashion of successive divorces and remarriages, but it is not hard to come up with an example in which that approach would lead us to a cycle never ending in a stable solution.

This immediately brings up the question of existence of a stable matching in every instance of the stable marriage problem. That is, given arbitrary preference lists of all the men and women, does there always exist at least one stable matching? The answer to this question is affirmative and, somewhat surprisingly, the proof is quite constructive. In fact, the proof consists of an algorithm developed by Gale and Shapley that produces a stable matching for the input preference lists.

The main idea of the Gale-Shapley algorithm can be described succinctly as "men propose, women dispose." This means that during the course of the algorithm various couples get engaged due to men's proposals, and sometimes engagements get broken by women if they receive better proposals.

The following is an overview of the algorithm:

- Initially, all men and women are free (not engaged), and then the algorithm proceeds in rounds.
- At the beginning of each round a free man is selected who proposes to the best (first) women on his list among those to whom he has never previously proposed (so no man ever proposes to the same woman twice).
- When a *free* woman receives a proposal she always accepts, and the pair becomes engaged (and not free). When an *engaged* woman receives a proposal, she checks her preference list and accepts the new proposal only if she prefers the man who proposes to her over her current fiancé. If so, the proposer and the proposed woman become a new engaged pair and the woman's old mate becomes free. If an engaged woman prefers her current fiancé over the proposer, she rejects the proposer and nothing else changes.
- The algorithm goes to the next round until we have every man engaged (not free), and then all engagements are declared final as marriages. That is, the final engagements are returned as the resulting stable matching.

This informal description can be converted more precisely in the following pseudocode.

```
SM(men's preference lists, women's preference lists)
   Initialize each person to be free;
   while there is a free man m do
       w = first woman on m's list to whom m has not yet proposed;
       [[ m proposes to w ]]
       if w is free then                         [[ w accepts m ]]
           Assign m and w to be engaged;         [[ m and w are not free ]]
       else if w prefers m to her mate m' then   [[ w accepts m ]]
           Assign m and w to be engaged;         [[ m and w are not free ]]
           Assign m' to be free;                 [[ m' becomes free ]]
       else                                      [[ w rejects m ]]
           Do nothing;                           [[ m is free, w is not free ]]
   return engagements as a stable matching;      [[ celebrate the weddings ]]
```

Ignoring for the moment the actual implementation of the SM algorithm, its pseudocode specification has three potential deficiencies. First of all, it is not clear whether the algorithm ever terminates at all, let alone whether it returns a stable matching. Namely, in each iteration of the **while** loop we have a free proposer that can remain free if his first-choice already engaged woman rejects him. Even if she accepts him, her old mate becomes free, so the number of free men won't decrease. Thus it is conceivable that there will always be a free man at the beginning of each iteration, hence the algorithm will loop forever.

The second potential error of the algorithm is that it might crash because a selected free man in the very beginning of some iteration does not have a woman to choose from his list and to propose to. Namely, it might be the case that a free man m already proposed to all the women. Equivalently, there is no woman on m's preference list to whom m has not yet proposed, hence m cannot take the first such a woman to propose to.

Finally, the description of the algorithm is not completely deterministic because in the test of the **while** loop we only check whether there is a free man. If so, we haven't specified how we pick one if there are many of them as candidates for the current iteration. Thus we may wonder if the order in which they are selected is going to make any difference in the resulting matching—perhaps using a "wrong" order could affect the possibility of getting a stable matching.

In other words, we need to show more carefully the correctness of the SM algorithm. We settle all of the above questions through a series of formal arguments about the properties of the SM algorithm.

OBSERVATION 1 *Every man proposes to a given woman at most once.*

PROOF: This is clear because a man proposes only if he is free and then always proposes to some woman to whom he has not yet proposed. ∎

OBSERVATION 2 *Every woman is engaged to at most one man, and every man is engaged to at most one woman.*

PROOF: This is also clear because a woman can only be engaged to his proposer, and if she is already engaged, she breaks the old engagement. The symmetric claim is immediate, since a man can be engaged only if he is free. ∎

OBSERVATION 3 *Once engaged, every woman stays engaged, possibly changing mates but every new engagement is an improvement from her perspective.*

PROOF: This is also clear because from the moment a free woman accepts a proposal, she never becomes free again—she can only change her mate if she gets a better proposal in her view, i.e., she prefers the proposer to her current mate. ∎

OBSERVATION 4 *The* SM *algorithm terminates in at most n^2 rounds, where n is the number of men (women).*

PROOF: To show termination, we need to find a good measure of progress of the **while** loop. (We've seen that this is not the number of free men.) We argue that the number of proposals that men have made works. Indeed, no man proposes to the same woman twice by Observation 1, so each man can make at most n proposals. Thus, all of the n men together can make at most n^2 proposals in total. In each round exactly one new proposal is made, and so there can be at most n^2 rounds. ∎

OBSERVATION 5 *Every free man has a woman to whom he has not yet proposed.*

PROOF: If m is a free man, then by Observation 2 there is also a free woman w. This implies that no man has ever proposed to free w, since otherwise she would get engaged and stay engaged by Observation 3. In particular, m has not yet proposed to w. ∎

OBSERVATION 6 *The SM algorithm constructs a stable matching.*

PROOF: We first argue that the SM algorithm constructs a perfect matching, that is, a bijection from the set of men onto the set of women. When the **while** loop terminates, there is no free man. This implies that all the men are engaged to at least one woman. On the other hand, by Observation 2 each man is engaged to at most one woman. This together means that each man is engaged to exactly one woman. Since there are exactly as many women as men, we have a perfect matching.

To show that the resulting matching is stable, we argue by contradiction. Suppose there is a dissatisfied pair m–w in the matching. That is, in the matching we have (m, w') and (m', w), but m prefers w to his wife w' ($w <_m w'$) and w prefers m to her husband m' ($m <_w m'$). Since (m, w') is a marriage, m must have proposed to w', because a man can get engaged only by proposing. But then m must have proposed to all of the women that are before his wife w' on his preference list, because a man proposes to women in the order in which they appear on his preference list. Since $w <_m w'$, it implies that during the course of the algorithm m must have proposed to w as well (and moreover m proposed to w before he proposed to his wife w').

On the other hand, since (m', w) is a marriage, Observation 3 implies that m' is the best (w's highest-ranking) man who ever proposed to w. In particular, since m proposed to w, it means $m' <_w m$. But this contradicts our assumption that m–w is a dissatisfied pair because w prefers m to her husband m' (i.e., $m <_w m'$). Therefore our assumption that there is a dissatisfied pair is untenable, that is, no dissatisfied pair can exist in the matching. ∎

OBSERVATION 7 *The SM algorithm constructs the same stable matching no matter how we select a free man in each round.*

PROOF: No free man's proposal is affected in any way by what another man does. Likewise, no woman's choice is affected by what other women do. So, in each round it does not matter in what order the proposals are made or in what order the women make their choices. ∎

Let's now analyze the running time of the SM algorithm. Given the preference lists of n men and n women, we know that the number of iterations of the **while** loop is at most n^2. Thus, the algorithm's worst-case running time is going to be proportional to n^2 if we can choose an implementation such that each iteration takes constant time. Without going into much detail (see Exercise 12 in the next chapter), in the list below we only hint at possible data structures that allow for constant-time implementation of each step in the body of the **while** loop.

- To represent n men and n women, we can arbitrary call them $1, 2, \ldots, n$.
- To represent their engagements, we can maintain two n-element arrays m and w. Naturally, $m(i) \in \{1, \ldots, n\}$ denotes the woman to whom man i is engaged, and $m(i) = 0$ if man i is free. Likewise, $w(i) \in \{1, \ldots, n\}$ denotes the man to whom woman i is engaged, and $w(i) = 0$ if woman i is free.
- To represent free men, we can store them in an n-element array F and maintain its actual length in a variable. A free man is taken from the end of the array, and a new free man is put in the end of the array. (The reader familiar with data structures may recognize that we simulate a stack here, although a queue would do as well.)
- To represent the men's preference lists, we can use an $n \times n$ table M so that the i-th row of the table stores the preference list of man i, sorted from the best to the worst woman for man i. Thus $M(i, j) = k$ means that woman k is in the j-th place on the man i's preference list.
- To represent the rank of the woman to whom a man proposed last, we can use an n-element array L so that $L(i) = j$ means that the man i proposed last to the woman in the j-th place on his preference list (i.e., that woman is $M(i, j)$). This way we can easily determine the first woman to whom a man should propose next.
- To represent the women's preference lists, we can use an $n \times n$ "ranking table" W so that $W(i, j)$ denotes woman i's ranking of man j on her preference list. Thus $W(i, j) = k$ means that man j is in the k-th place on the woman i's preference list. The reason for asymmetry compared to the men's preference lists is because this way we can easily test whether woman i prefers man j' to man j''.

2. Growth rate of functions

When analyzing running time of algorithms, we may obtain very complicated functions. For instance, a problem of input size n may have time complexity function $f_1(n) = 10n^3 + n^2 + 40n + 80$. For another problem we may get $f_2(n) = 17n \log n - 23n - 10$. Since we want fast algorithms particularly when the input size n is large, even these exact expressions can get into our way of understanding the time complexity of a problem. Namely, if n is large (imagine n having value one million or one billion), we know that functions $f_1(n) = 10n^3 + n^2 + 40n + 80$ and $g_1(n) = n^3$, or $f_2(n) = 17n \log n - 23n - 10$ and $g_2(n) = n \log n$, are pretty much the same, i.e., they have very similar values. So, to further simplify our analysis, we will say that f_1 is of order g_1 and f_2 is of order g_2. In other words, we want to measure the growth rate of the functions f_1 and g_1.

This is justified for three reasons. First, for sufficiently large arguments, the value of a function is largely determined by its dominant term. Figure 5 illustrates this by a plot of functions f_1, g_1, f_2, g_2, where we can see that curves of f_1 and g_1, as well as f_2 and g_2, are close together for large values of n.

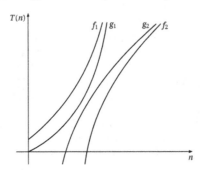

FIGURE 5. Functions f_1, g_1, f_2, g_2.

Second, the exact value of the leading constant of the dominant term is not meaningful across different machines (although the relative values of the leading constant for identically growing functions might be). The leading constant depends on the execution time of basic operations, which we assumed represents a unit time, and these times are certainly very different for a PC and a Cray computer. To better appreciate this, suppose $A1$, $A2$, and $A3$ are three algorithms with the worst-case running-time functions 2^n, $5n^2$, and $100n$, respectively, where n is the input size. The values of these functions for different n are given in the left table of Figure 6. On the other hand, if each algorithm is run on a machine that executes one million (10^6) operations per second, then each algorithm will approximately take the real time to finish according to the values in the right table of Figure 6.

n	$A1$	$A2$	$A3$
1	2	5	100
10	1024	500	1000
100	2^{100}	50000	10000
1000	2^{1000}	$5 \cdot 10^6$	100000

n	$A1$	$A2$	$A3$
1	$1\,\mu sec$	$5\,\mu sec$	$100\,\mu sec$
10	$1\,msec$	$0.5\,msec$	$1\,msec$
100	$2^{70}\,years$	$0.05\,sec$	$0.01\,sec$
1000	$2^{970}\,years$	$5\,sec$	$0.1\,sec$

FIGURE 6. Running times of three algorithms.

The real-time figures in the right table clearly show that for large n (say, $n > 100$) the algorithm $A1$ is the slowest, then $A2$, and $A3$ is the fastest. Thus, the growth of running time in terms of n (i.e., 2^n, n^2, and n) is more significant than the exact constant factors (i.e., 1, 5, and 100).

Third reason to measure only the growth rate of functions is that small values of the input size are generally not important. This is so because any algorithm is usually good enough for small input. That is, it is mostly irrelevant

whether some problem for small input is solved for, say, 1 second or 10 millisec-
onds. In other words, we are only concerned with asymptotic performance of an
algorithm, i.e., with the growth of its running time as the input size increases.

3. Asymptotic notation

We use asymptotic notation (big-oh, big-omega, big-theta, and little-oh) to
represent the growth rate of functions. Asymptotic notation allows us to establish a relative order among functions by comparing dominant terms. Although
we use big-oh notation almost exclusively throughout the book, we also define
three other types of asymptotic notation that are related to big-oh and used oc-
casionally.

DEFINITION (big-oh)   *For two nonnegative functions $f, g: \mathbb{N} \to \mathbb{R}$ we say that
$f(n) = O(g(n))$ if there are positive constants c and n_0 such that $f(n) \le c \cdot g(n)$ for
all $n \ge n_0$.*

DEFINITION (big-omega)   *For two nonnegative functions $f, g: \mathbb{N} \to \mathbb{R}$ we say
that $f(n) = \Omega(g(n))$ if there are positive constants c and n_0 such that $f(n) \ge c \cdot
g(n)$ for all $n \ge n_0$.*

DEFINITION (big-theta)   *For two nonnegative functions $f, g: \mathbb{N} \to \mathbb{R}$ we say
that $f(n) = \Theta(g(n))$ if $f(n) = O(g(n))$ and $f(n) = \Omega(g(n))$.*

DEFINITION (little-oh)   *For two nonnegative functions $f, g: \mathbb{N} \to \mathbb{R}$ we say
that $f(n) = o(g(n))$ if for any positive constant c there exists a constant n_0 such
that $f(n) \le c \cdot g(n)$ for all $n \ge n_0$.*

For example, in the case of functions f_1, g_1, f_2, g_2 from the beginning of the
previous section, we can write $f_1(n) = O(g_1(n))$ or $f_1(n) = O(n^3)$, and $f_2(n) =
O(g_2(n))$ or $f_2(n) = O(n \log n)$.

The definition for big-oh states that there is a point n_0 such that for all values
of n past this point, the function $f(n)$ is bounded by some constant multiple of
the function $g(n)$. For example, if the running time of an algorithm is $O(n^2)$,
then, ignoring constants, we are guaranteeing that at some point we can bound
the running time by a quadratic function. Notice that if the true running time
is linear, then the statement that the running time is $O(n^2)$ is technically correct
because the inequality holds. However, $O(n)$ would be a more precise claim.

If we use traditional inequality operators to compare growth rates, than the
big-oh definition says that the growth rate of $f(n)$ is less than or equal to that of
$g(n)$. In other words, the function $g(n)$ represents *asymptotic upper bound* for
$f(n)$.

The big-omega definition says that the growth rate of $f(n)$ is greater than or
equal to that of $g(n)$. Equivalently, the function $g(n)$ represents *asymptotic lower
bound* for $f(n)$. This form is usually used in lower-bound arguments when we
want to show that an algorithm must take at least some specific time to execute.

The big-theta definition says that the growth rate of $f(n)$ is equal to that of
$g(n)$, i.e., the function $g(n)$ is *asymptotically tight bound* for $f(n)$. When we use

this form in algorithm analysis, we are not only providing an upper bound on the running time of an algorithm, but also providing assurances that the analysis that leads to the upper bound is as good (tight) as possible. In spite of the additional precision of the big-theta notation, big-oh is more commonly used because it is usually easier to come up with only an upper bound.

Finally, the little-oh definition says that the growth rate of $f(n)$ is an order of magnitude less than that of $g(n)$. Intuitively, the function $f(n)$ becomes insignificant relative to $g(n)$ as n approaches infinity, i.e.,

$$\lim_{n \to \infty} \frac{f(n)}{g(n)} = 0.$$

Table 1 illustrates the meaning of these four definitions.

Asymptotic notation	Relative rate of growth
$f(n) = O(g(n))$	growth of $f(n) \leq$ growth of $g(n)$
$f(n) = \Omega(g(n))$	growth of $f(n) \geq$ growth of $g(n)$
$f(n) = \Theta(g(n))$	growth of $f(n) \approx$ growth of $g(n)$
$f(n) = o(g(n))$	growth of $f(n) \ll$ growth of $g(n)$

TABLE 1. Meaning of various asymptotic notations.

To briefly reinforce the asymptotic definitions, let's show that $n^3 + n^2 - 1 = \Theta(n^3)$. To do so, we must determine positive constants c_1, c_2, and n_0 such that

$$c_1 n^3 \leq n^3 + n^2 - 1 \leq c_2 n^3,$$

for all $n \geq n_0$. Dividing this by n^3 yields

$$c_1 \leq 1 + \frac{1}{n} - \frac{1}{n^2} \leq c_2.$$

Now it's clear that constants $c_1 = 1$, $c_2 = 2$, and $n_0 = 1$ will do.

As another illustration, observe that the running times of all three sorting algorithms we discussed in Section 1.3 of this chapter can be more succinctly expressed using the asymptotic notation as $O(n^2)$. Also, the running time of the binary search algorithm is $O(\log n)$.

There is an alternative way of showing that a given function satisfies some asymptotic bound. Namely, it is often easier to use limits instead of formal definitions for such purposes. The following limit rule, which is easy to prove (see Exercise 17), gives sufficient conditions for asymptotic bounds.

THEOREM (LIMIT RULE)   *Given nonnegative functions $f(n)$ and $g(n)$,*

$$\lim_{n \to \infty} \frac{f(n)}{g(n)} = c > 0 \Rightarrow f(n) = \Theta(g(n))$$

$$\lim_{n \to \infty} \frac{f(n)}{g(n)} = c \geq 0 \Rightarrow f(n) = O(g(n))$$

$$\lim_{n \to \infty} \frac{f(n)}{g(n)} \neq 0 \Rightarrow f(n) = \Omega(g(n))$$

$$\lim_{n \to \infty} \frac{f(n)}{g(n)} = 0 \Rightarrow f(n) = o(g(n)),$$

where c is a constant, and in the case of Ω-notation the limit can be infinity.

This Limit Rule is almost always easier to apply than the formal definitions. Moreover, although it apparently requires showing stronger conditions, it can be applied in almost every instance of the running-time functions. The only exception might be cases when the limit does not exist (e.g., $f(n) = n^{\sin n}$), but since most running times are well-behaved functions, this is rarely a problem.

For example, if we need to compare two functions $\log^2 n$ and $n/\log n$, we can evaluate the following limit (by using, say, L'Hôpital's rule):

$$\lim_{n \to \infty} \frac{\log^2 n}{n/\log n} = 0.$$

By the Limit Rule then $\log^2 n = O(n/\log n)$.

It should also be mentioned that the asymptotic notation does not introduce a total order among functions. This means that there are pairs of functions which are not comparable by, say, big-oh relation. That is, there are functions $f(n)$ and $g(n)$ such that $f(n)$ is neither $O(g(n))$ nor $g(n)$ is $O(f(n))$. For example,

$$f(n) = \begin{cases} n^3, & \text{if } n \text{ is even} \\ n, & \text{if } n \text{ is odd} \end{cases} \quad \text{and} \quad g(n) = \begin{cases} n, & \text{if } n \text{ is even} \\ n^2, & \text{if } n \text{ is odd} \end{cases}$$

are such ***incommensurate*** functions, neither of which is big-oh of the other.

Another note worth mentioning is that the generalization of asymptotic notation from one variable to many variables is surprisingly tricky. For example, one proper generalization of the O-notation for two-variable functions is the following:

DEFINITION *For two nonnegative functions $f, g: \mathbb{N} \times \mathbb{N} \to \mathbb{R}$ we say $f(m, n) = O(g(m, n))$ if there are positive constants $m_0, n_0,$ and c such that $f(m, n) \leq c \cdot g(m, n)$ for all $m \geq m_0$ or $n \geq n_0$.*

Observe that this definition requires that the inequality holds for $m \geq m_0$ or $n \geq n_0$, rather than for $m \geq m_0$ and $n \geq n_0$. The later condition would allow that the function $f(m, n)$ is not dominated by $c \cdot g(m, n)$ on infinitely many points (m, n). But this is contrary to the spirit of the O-notation for one-variable functions: there exists a constant $c > 0$ such that $f(n) \leq c \cdot g(n)$ for all but finitely many values of n (see Exercise 20).

Finally, notice that in all definitions we are misusing the symbol =, because in the asymptotic notation the equality sign is not symmetric (reflexive). This means that if $f_1(n) = O(g(n))$ and $f_2(n) = O(g(n))$, then it does not imply that $f_1(n) = f_2(n)$. Strictly speaking, for a nonnegative function $g: \mathbb{N} \to \mathbb{R}$ we should define $O(g(n))$ as a class of nonnegative functions:

$$O(g(n)) = \left\{ f: \mathbb{N} \to \mathbb{R} \mid (\exists c > 0)(\exists n_0 \in \mathbb{N})(\forall n \geq n_0) \, f(n) \leq cg(n) \right\}.$$

However, for simplicity reasons, we prefer traditional slight abuse of the notation to being more precise. We hope it will not cause any confusion to more
"mathematically correct" readers.

4. Asymptotic running time

Let's now apply the general mathematical theory of asymptotic notation in
context of the running-time function of an algorithm. We consider here only
the big-oh notation, although similar observations apply to other types of the
asymptotic notation as well.

Let $T(n)$ be the running time of some algorithm, represented as a function
of the input size n. By the nature of the function $T(n)$, we can make some reasonable assumptions about it. Namely, the argument n is a nonnegative integer,
and $T(n)$ is nonnegative for all values of n. Given a function $f(n)$ defined on
the nonnegative integers, formally we say that $T(n) = O(f(n))$ if there exist an
integer n_0 and a constant $c > 0$ such that $T(n) \leq cf(n)$ for all integers $n \geq n_0$. In
other words, $T(n)$ is at most some constant multiplied by $f(n)$, except possibly
for some small values of n. For particular functions $T(n)$ and $f(n)$, we can use
this definition to prove that $T(n) = O(f(n))$. We do so by producing a concrete
choice for n_0 and $c > 0$, and showing that $T(n) \leq cf(n)$ if n is least as large as the
chosen value n_0.

For example, let us prove that the running time of the BUBBLE-SORT algorithm is $O(n^2)$. Since in this case we calculated that $T(n) = 2n^2 + n - 1$, for $n \geq 1$
we have

$$2n^2 + n - 1 \leq 2n^2 + n \leq 2n^2 + n^2 = 3n^2.$$

Thus, choosing $n_0 = 1$ and $c = 3$ we can conclude that by the definition $T(n) =
O(n^2)$.

Alternatively, we could chose $c = 2.2$ and $n_0 = 4$, because $2n^2 + n - 1 \leq 2.2n^2$
for all $n \geq 4$, as the reader may check. In fact, we can also claim that $T(n) =
2n^2 + n - 1$ is big-oh of any fraction of n^2, say, $O(n^2/100)$. To see why, let $n_0 = 1$
and $c = 300$. Then for $n \geq 1$ we know from the previous discussion that

$$2n^2 + n - 1 \leq 3n^2 = 300 \cdot \frac{n^2}{100},$$

that is, $T(n) = O(n^2/100)$.

Moreover, we can derive a general principle that constant factors do not
matter: for any positive constant d and any function $T(n)$, $T(n) = O(dT(n))$. This
can be seen if we let $n_0 = 1$ and $c = 1/d$, and so $T(n) \leq c(dT(n))$ since $c \cdot d = 1$.

The second underlying general principle is that low-order terms do not matter in a polynomial sum. Suppose that $T(n)$ is a polynomial of the form

$$T(n) = a_k n^k + a_{k-1} n^{k-1} + \cdots + a_2 n^2 + a_1 n + a_0,$$

where the leading coefficient $a_k > 0$. Then we can throw away all terms but the
first, ignore the a_k by the first principle, and conclude that $T(n) = O(n^k)$. To
prove this, let $n_0 = 1$ and let c be the sum of all the positive coefficients among

$a_0, a_1, \ldots, a_k$. If some coefficient a_i is 0 or negative, then surely $a_i n^i \le 0$. If some $a_i > 0$, then $a_i n^i \le a_i n^k$ for every $i \le k$ and $n \ge 1$. Therefore,

$$T(n) = \sum_{i=0}^{k} a_i n^i = \sum_{a_i>0} a_i n^i + \sum_{a_i<0} a_i n^i \le \sum_{a_i>0} a_i n^i \le \sum_{a_i>0} a_i n^k = n^k \cdot \sum_{a_i>0} a_i = cn^k,$$

as required.

The third general principle generalizes the second one in that it states that only dominant term matters in a sum of terms. For example, if $T(n) = 3n + 17\log n$, i.e., $T(n) = O(n) + O(\log n)$, then $T(n) = O(n)$. In proof, suppose that $T(n) = T_1(n) + T_2(n)$, where $T_1(n) = O(f(n))$, $T_2(n) = O(g(n))$, and $g(n) = O(f(n))$. In other words, there are constants $n_1, n_2, n_3, c_1, c_2, c_3$ such that

(i) if $n \ge n_1$, then $T_1(n) \le c_1 f(n)$;
(ii) if $n \ge n_2$, then $T_2(n) \le c_2 g(n)$;
(iii) if $n \ge n_3$, then $g(n) \le c_3 f(n)$.

Let $n_0 = \max\{n_1, n_2, n_3\}$ so that (i), (ii), and (iii) hold when $n \ge n_0$. Now, if $c = c_1 + c_2 c_3$, then

$$\begin{aligned} T(n) &= T_1(n) + T_2(n) \\ &\le c_1 f(n) + c_2 g(n) \\ &\le c_1 f(n) + c_2 c_3 f(n) \\ &= (c_1 + c_2 c_3) f(n) \\ &= cf(n). \end{aligned}$$

This means $T(n) = O(f(n))$.

As an example, consider the following fragment of an algorithm that initializes an $n \times n$ matrix A to the identity matrix:

```
for i = 1 to n do
    for j = 1 to n do
        A(i, j) = 0;
for i = 1 to n do
    A(i, i) = 1;
```

The running time $T(n)$ of this fragment is the sum of the running times of the first outer **for** loop and the second **for** loop that follows behind the outer loop. This means $T(n) = O(n^2) + O(n)$, from which we can immediately conclude that $T(n) = O(n^2)$, because the function n^2 dominates the function n.

The reader can now easily check another property of the big-oh relation: product of big-oh's is big-oh of the product. More precisely, for any two functions $f(n)$ and $g(n)$ defined on nonnegative integers,

$$O(f(n)) \cdot O(g(n)) = O(f(n)g(n)).$$

For example, we can use another way to initialize an $n \times n$ matrix A to the identity matrix:

```
for i = 1 to n do
  for j = 1 to n do
    if i = j then
      A(i, j) = 1;
    else
      A(i, j) = 0;
```

However, we see that the running time $T(n)$ of this fragment is asymptotically the same as for the former one, because $T(n) = n \cdot n \cdot O(1) = O(n^2) \cdot O(1) = O(n^2)$.

Finally, the last principle that we consider regarding the O-notation is the transitivity law. That is, if $f(n) = O(g(n))$ and $g(n) = O(h(n))$, then $f(n) = O(h(n))$. To see why, let n_1 and n_2 be nonnegative integers and c_1 and c_2 be positive constants such that

(i) if $n \geq n_1$, then $f(n) \leq c_1 g(n)$;
(ii) if $n \geq n_2$, then $g(n) \leq c_2 h(n)$.

Let $n_0 = \max\{n_1, n_2\}$ and $c = c_1 c_2$. Then for all $n \geq n_0$,

$$f(n) \leq c_1 g(n) \leq c_1 c_2 h(n) = c \cdot h(n),$$

proving that $f(n) = O(h(n))$.

For example, if $T(n) = 2n^2 + n - 1$, then we know $T(n) = O(n^2)$. Since clearly $n^2 = O(n^3)$, by transitivity we can say that $T(n) = O(n^3)$.

The last example raises the question as to how we actually describe the running time of an algorithm using the asymptotic notation. On one hand, by the transitivity law, we can specify many functions with ever larger growth rate. For instance, in the last example we could further say that $T(n) = O(n^{10})$, or even $T(n) = O(2^n)$.

On the other hand, we generally want the "tightest" big-oh upper bound we can prove. That is, if $T(n) = 2n^2 + n - 1$, we want to say $T(n) = O(n^2)$, rather than make technically true but weaker statement that $T(n) = O(n^3)$. But then, by the principle that constants do not matter, we can specify an infinite number of tighter bounds, like $T(n) = O(0.5n^2)$ or $T(n) = O(0.01n^2)$.

To resolve this seemingly contradictory requirement, we choose to use, whenever possible, a big-oh expression that is as simple as possible. This means that it is given by a single term and the constant factor of that term is 1. Be aware though, that there are situations where tightness and simplicity are conflicting goals. Fortunately, such cases are rare in practice.

4.1. Typical asymptotic running-time functions. In algorithm analysis we often work with a couple of standard functions that commonly describe algorithm running times. Table 2 arranges these functions in order of increasing growth rate from top to bottom.

Notice that in Table 2 each informal function name identifies actually a class of functions of the same growth rate. For example, the quadratic function refers to all functions that are $\Theta(n^2)$. Also, the constant function is a shorthand for "some constant" and we will frequently use $\Theta(1)$, or even $O(1)$, for this purpose.

Function	Informal name
1	constant
$\log n$	logarithmic
n	linear
$n \log n$	$n \log n$
n^2	quadratic
n^3	cubic
2^n	exponential

TABLE 2. Typical functions in the order of increasing growth rate.

To better grasp the growth rate as a function of input size, Figure 7 displays a rough plot of the linear, $n \log n$, quadratic, and cubic functions.

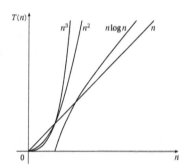

FIGURE 7. Four typical running-time functions.

In Figure 7 we see that for small values of n there are points where one curve is initially better than another, even though eventually this does not prove to be true. For example, initially the quadratic curve is better than the linear curve, but as n gets sufficiently large, the quadratic function loses its advantage. For small amounts of input, it is difficult to make comparisons between functions because leading constants become very significant. Thus, the function $7n + 120$ is larger than n^2 when n is less than 15. Eventually though, the linear function will always be less than the quadratic function. Most importantly, for small input sizes the running times are generally a split of a second, and so we need not worry about them.

Figure 7 clearly demonstrates the difference between the various running-time functions for large input sizes. A linear algorithm solves a problem of size, say, 10,000 in a small fraction of a second. An $n \log n$ algorithm uses roughly ten times as much time. Note that the actual time difference would depend on the constants involved and thus might be more or less. Depending on these constants, it is possible that an $n \log n$ algorithm might be faster than a linear algorithm for fairly large input sizes. However, for similar constants (i.e., for log-ically equally complex algorithms), linear algorithms tend to win over $n\log n$ algorithms in practice.

The most striking feature of these curves is that the quadratic and cubic al-gorithms are not competitive with the others for reasonably large inputs. We can code the quadratic algorithm in highly efficient machine language, do a poor job coding the linear algorithm, and the quadratic algorithm still loses badly. Even the most clever programming tricks or shiny machines cannot make an ineffi-cient algorithm fast. Thus, before we waste effort attempting to optimize code or waste money investing in new machines, we need to come up with an efficient algorithm.

Does all this mean that quadratic and cubic algorithms are useless? The answer is no. In some cases the most efficient known algorithms are quadratic or cubic, and in others the most efficient algorithm is even worse (exponential). Moreover, the algorithms that are not asymptotically efficient are nonetheless easy to program and so for small inputs, because any algorithm will do, that's the way to go.

The bottom line is therefore this: out of the common functions encountered in algorithm analysis, the constant function is the best, although hardly achiev-able, while the exponential function is really inferior and the corresponding al-gorithms are impractical for large inputs.

Exercises

1. Give the asymptotic running time for the following algorithm fragments:

(a) **if** $x = 0$ **then**
　　for $i = 1$ **to** n **do**
　　　$A(i) = i;$

(b) $i = 1;$
　　repeat
　　　$A(i) = B(i) + C(i);$
　　　$i = i + 1;$
　　until $i = n;$

(c) **if** $x = 0$ **then**
　　for $i = 1$ **to** n **do**
　　　for $j = 1$ **to** n **do**
　　　　$A(i, j) = 0;$
　　else
　　for $i = 1$ **to** n **do**
　　　$A(i, i) = 1;$

(d) **for** $i = 1$ **to** n **do**
　　for $j = 1$ **to** n **do**
　　　for $k = 1$ **to** j **do**
　　　　if $i = j$ **then** $A(k, k) = 0;$

(e) $i = 0; k = 1;$
　　while $k \le n$ **do**
　　　$k = 2 \cdot k;$
　　　$i = i + 1;$

(f) **for** $i = 1$ **to** $\lfloor n/2 \rfloor$ **do**
　　for $j = i$ **downto** 1 **do**
　　　if $j \bmod 2 = 0$ **then**
　　　　$A(i, j) = B(i) + C(j);$

2. Give an efficient algorithm for finding the largest and the second largest element of an array A of n elements. How many comparisons of the array elements does your algorithm do?

3. Suppose that the constants c and d in the analysis of the running time $T(n)$ of the BIN-SEARCH algorithm are equal to 1. Thus, assume the exact recurrence relation for $T(n)$ is given by

$$T(1) = 1,$$
$$T(n) = T(\lfloor n/2 \rfloor) + 1, \quad n > 1.$$

Prove by induction on n that $T(n) = 1 + \lfloor \log n \rfloor$ for every $n \geq 1$.

4. **Ternary search.** Given a range indexed from i to j in a sorted array A, we first compute the approximate 1/3-point k of the range, i.e., $k = \lfloor (2 \cdot i + j)/3 \rfloor$, and compare the lookup item x with $A(k)$. If they are not equal, we proceed as follows. If $x < A(k)$, we continue search in the range from i to $k - 1$. If $x > A(k)$, we compute the approximate 2/3-point m of the original range, i.e., $m = \lceil (i + 2 \cdot j)/3 \rceil$, and compare x with $A(m)$. In other words, we want to isolate x to within exactly one of the three subranges of length one third of the range from i to j. Write a recursive algorithm that implements the ternary search and give its running time $T(n)$ in the form of a recurrence relation. Can you solve the recurrence relation for $T(n)$?

5. **Sport fan.** Consider an array A of size n that contains only red and white elements in an arbitrary order.[6] Assuming that testing the color of an element is a basic operation, the goal is to rearrange elements so that all red elements are in the beginning and all white elements are in the end of the array A. Figure 8 illustrates the final result.

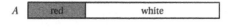

A red white

FIGURE 8

Give at least two algorithms that solve this problem and estimate the running time of each algorithm.

6. **Dutch national flag.** Consider now an array A of size n that contains red, blue, and white elements in an arbitrary order. Assuming that testing the color of an element is a basic operation, the goal is to rearrange elements so that all red elements are in the beginning, all white elements are in the middle, and all blue elements are in the end of the array A. Figure 9 illustrates the final result.
Give at least two algorithms that solve this problem and estimate the running time of each algorithm.

---

[6]Red and white are supposedly official colors of a sport team.

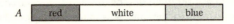

$$A$$

FIGURE 9

7. Assume each row and each column of an $n \times n$ table A is sorted in the increasing order. (Rows are sorted in the left-right, and columns in the top-down direction.) Give at least four algorithms to search for a given key x in the table A and estimate the running time of each algorithm.

8. Let x and y be positive integers. Prove the following statements about their greatest common divisor.

    (a) If $x = 2a$ and $y = 2b$, then $\gcd(x, y) = 2 \cdot \gcd(a, b)$.
    (b) For any $k \geq 0$, if $x = 2^k a$ and $y = 2^k b$, then $\gcd(x, y) = 2^k \cdot \gcd(a, b)$.
    (c) If x is even and y is odd, then $\gcd(x, y) = \gcd(x/2, y)$.
    (d) If $x \geq y$, then $\gcd(x, y) = \gcd(x - y, y)$.

9. Consider another version of Euclid's algorithm for computing the greatest common divisor of positive integers x and y. Given x and y, it starts with computing the largest power of 2 dividing both x and y. If this is 2^k, then it replaces x by the quotient of x divided by 2^k and y by the quotient of y divided by 2^k. After this "preprocessing" step, it iterates through the following:

    1. Swap the numbers if necessary to have $x \geq y$.
    2. If $y > 1$, then check the parities of x and y. If one of x or y is even and the other is odd, then replace the even number by its half, that is, x by $x/2$ if x is even or y by $y/2$ if y is even. If both x and y are odd, replace x by $x - y$. Then in both cases, repeat the same from step 1.
    3. If $y = 1$, then return 2^k as the result.
    4. If $y = 0$, then return $2^k x$ as the result.

For this version of Euclid's algorithm, do the following:

    (a) Carry out this algorithm to compute $\gcd(18, 8)$.
    (b) It seems that in step 2 we ignored the case when both x and y are even. Show that this never occurs.
    (c) Show that the modified Euclid's algorithm always terminates with the right answer.
    (d) Show that the product of two running numbers in x and y drops by a factor of 2 in at least one of any two successive iterations. In other words, if (x_0, y_0) is the pair in step 1 of any iteration and (x_2, y_2) is the pair obtained in step 2 after two successive iterations, then $x_2 \cdot y_2 \leq x_0 \cdot y_0 / 2$.
    (e) Derive an upper bound on the number of steps performed by this algorithm.

10. **Extended Euclidean algorithm.** Euclid's algorithm produces much more than just the gcd of two positive integers x and y. Namely, if we carry out

Euclid's algorithm to compute $\gcd(x, y)$, all the numbers obtained during the computation can be written as the sum of an integer multiple of x and an integer multiple of y. Using this observation, prove the following fact:

THEOREM *If $d = \gcd(x, y)$, then d can be written as a linear combination of x and y in the form $d = mx + ny$, where m and n are integers (not necessarily positive).*

Write the extended Euclidean algorithm that computes not only the greatest common divisor of x and y, but also finds integers m and n such that $\gcd(x, y) = mx + ny$. Prove carefully the correctness of your algorithm.

11. Given the following preference lists of men and women, find all stable matchings. (*Hint:* There are exactly five of them.)

men	1	2	3	4	5
A	a	b	c	d	e
B	b	c	d	e	a
C	c	d	e	a	b
D	d	e	a	b	c
E	e	a	b	c	d

women	1	2	3	4	5
a	B	C	D	E	A
b	C	D	E	A	B
c	D	E	A	B	C
d	E	A	B	C	D
e	A	B	C	D	E

12. Show that the SM algorithm indeed performs approximately n^2 rounds for some instances of the stable marriage problem. (*Hint:* Take the preference list of each of n men to be $1, 2, \ldots, n-1, n$, take the preference list of each of n women to be $n, n-1, \ldots, 2, 1$, and suppose free men are selected in a circular order $1, 2, \ldots, n$ and then back from 1, as if they are stored in a queue.)

13. Argue that the Gale-Shapley algorithm is "unfair," favoring men (and the opposite is true if women propose). Specifically, give an example of the men's and women's preferences so that the resulting stable matching of the algorithm is as good as possible for the men and as bad as possible for the women.

14. Prove that the Gale-Shapley algorithm constructs *man-optimal* stable matching! This means that no man could be better off in any other stable matching than he is in the stable matching produced by the Gale-Shapley algorithm. Put it another way, there is no stable matching in which any single man individually does better than in the Gale-Shapley matching.

15. Assuming the property of the Gale-Shapley algorithm from Exercise 14, show that the algorithm constructs *woman-pessimal* stable matching! This means that each woman winds up in the Gale-Shapley matching as badly as she possibly could in any other stable matching. Put it another way, there is no stable matching in which any single woman individually does worse than in the Gale-Shapley matching.

16. Argue that the following one-loop version of the linear MCSS2 algorithm is correct.

```
MCSS2(A)
    M = 0; s = 0;
    for j = 1 to n do
        s = s + A(j);
        if s < 0 then
            s = 0;
        else if M < s then
            M = s;
    return M;
```

17. Prove the Limit Rule.

18. Formally prove the following asymptotic relations:

    (a) $10n^3 + n^2 + 40n + 80 = \Theta(n^3)$
    (b) $17n\log n - 23n - 10 = \Theta(n\log n)$
    (c) $n\log n = O(n^2)$
    (d) $\log^2 n = O(n)$
    (e) $\log n = O(\sqrt{n})$
    (f) $n^{\log n} = O(2^n)$
    (g) $\sin n = O(1)$

19. Order the following functions by their growth rate for large values of the argument n:

$$n, \log n, \log\log n, \log^2 n, \frac{n}{\log n}, n!, \sqrt{n}\log^2 n, (\sqrt{2})^{\log n}, (1/3)^n, (3/2)^n, 17$$

20. The definition of the O-notation for two-variable functions requires that the inequality holds for $m \geq m_0$ *or* $n \geq n_0$, rather than for $m \geq m_0$ *and* $n \geq n_0$. Illustrate the observation that the former condition is more natural than the latter one with a diagram of relevant regions of the $m \times n$ plane.

3

Basic Data Structures

Many algorithms require that we use proper representation of data to achieve efficiency. Thus, in fast algorithms we need a good way to store and organize data in order to facilitate access and modifications. This data representation, together with operations that are allowed on the data, is called the *data structure*. In fact, people often use other terms for the data representation, such as "data type" (or just "type"), "abstract data type", "abstract data structure," that all sound alike, but their meaning calls for a clarification.

1. Abstract data structures

An ***abstract data structure*** (or ***ADS*** for short) is a mathematical model of a set of objects, together with a collection of operations defined on that model. For example, the abstract data structure *stack* models a dynamic set of objects in which the object removed from the set is the one most recently added to the set. Thus, the stack implements a *last-in, first-out (LIFO)* policy on the organization of data. In contrast, the abstract data structure *queue* models a dynamic set of objects in which the removed object is the oldest, i.e., the one that has been in the set for the longest time. The queue implements a *first-in, first-out (FIFO)* policy on the organization of data.

In a programming language, the ***data structure***, ***data type***, or simply ***type***, are all synonyms and refer to the representation of data and all operations allowed on that data. More precisely, they denote the set of values that a variable of some type may assume, together with a set of operations allowed on the variables of that type. For example, a variable of boolean type can assume either the value TRUE or the value FALSE, and no other value. Allowed operations are usual logical operations such as negation, conjunction, disjunction, and so on. Each programming language provides some basic data types. They vary from language to language, but most of them commonly provide integer, real, boolean, and character data types. Each programming language also contains the rules for constructing composite data types out of basic ones, which vary from language to language as well.

Abstract data structures, as the name implies, are an abstraction of data structures, and since algorithms are abstractions of programs, we usually design algorithms in terms of abstract data structures. This is not always possible though, because sometimes we need to know low-level details of the representation of the mathematical model underlying an abstract data structure. For example, some graph algorithms depend on the fact whether a graph is represented by an adjacency matrix or by a list of edges.

This is the reason why in the rest of the book we will often omit the adjective abstract and interchangeably use the terms data structure and abstract data structure. In spite of this slight abuse of terminology, you should keep in mind the distinction between the two notions and understand that, strictly speaking, the abstract data structure is an algorithmic notion, whereas the data structure is a programming notion.

Abstract data structures are generalizations of primitive data types (integer, real, and so on), just as procedures are generalizations of primitive operations (+, −, and so on). An abstract data structure encapsulates a data type in the sense that the definition of the type and all operations on that type can be localized to one section of the program. This is of great practical importance, because if we wish to change the implementation of an abstract data structure, first we know where to look at. Second, we are assured that revising one small section of code we won't cause subtle errors elsewhere in the program not related to this data type. Moreover, outside the section in which the abstract data structure is implemented, we can treat it as a primitive type without concern about the underlying implementation.

Abstract data structures have two desired properties, generalization and encapsulation, that allow us to achieve an important software engineering goal, reusability. Once we have implemented each abstract data structure, it can be used over and over again in various applications. Practically, an abstract data structure defines an interface between the user and the data.

2. Basic abstract data structures

The simplest aggregating mechanisms in most modern programming languages are the *array, record,* and *pointer* data types.[1] They represent basic building blocks for construction of complex data structures. We assume that the reader already has a good working knowledge of such basic composition rules for grouping data out of simple ones.

In this chapter, we study some of the most fundamental abstract data structures. We consider lists, which are sequences of elements, and two special cases of lists: stacks, where elements are inserted and deleted at one end only, and queues, where elements are inserted at one end and deleted from the other end.

---

[1]Another grouping method found in almost all programming languages is the *file*data type, which we won't consider in the book.

We also discuss closely related mathematical notion of sets. For each of these abstract data structures we consider several implementations and compare their relative merits.

Operations defined on abstract data structures differ wildly from one data structure to another, but all of them have at least two operations in common: one for inserting new elements into, as well as one for removing old elements from, the underlying dynamic set of elements of an abstract data structure. Here we should mention also the operation for making a brand new, empty data structure. However, this depends very much on the particular programming language and the details are rather trivial, although somewhat involved. Moreover, this is usually done once as part of the general setup of the program. Since we are mainly interested in the running times of more complex operations, not in their exact implementations, we will usually skip this "constructor" operation in our considerations. We also take an idealized viewpoint and skip other implementation details, such as error handling and freeing unused elements.

Other typical operations include finding whether a given element exists in a data structure, rearranging the elements in a particular order, determining the size (number of elements) of a data structure. In general, operations on data structures can be grouped into two categories: *queries*, which simply return information about data structures, and *modifying operations*, which change data structures.

Time for execution of an operation on an abstract data structure is usually measured in terms of the current size of the underlying set of elements. For example, if a list contains n elements, an operation that finds a given element in the list may take $O(n)$ time.

3. The abstract data structure LIST

A *linked list* (or *list* for short) has a particularly flexible structure because, in addition to the property that a list is able to grow and shrink on demand, its elements can be accessed, inserted, and deleted at any position within the list. Lists can also be concatenated together or split into sublists. They have numerous applications in all areas of computer science including programming language translation, memory management, and information retrieval.

The abstract data structure LIST represents the (linked) list in which the elements of a given type are arranged in a linear order.[2] Unlike an array, in which the linear order is implicit from the array indices, the order in a list is determined by a pointer in each element. It is convenient to postulate the existence of a special pointer NIL, which never points to any element. This way, we can uniformly treat the elements of the list and say that the last element contains the pointer NIL. Notice that the notion of pointer need not be the same as the one from a programming language. Here it is simply an indication in one element where the the next one can be found in the list.

---

[2]Strictly speaking, we have the data structure "LIST of the list elements' type." However, this is clear from context and the list implementations do not depend on the type of the list elements. Thus, we will use just LIST, and similarly treat other data structures.

For a more mathematically inclined reader, we can say that a list is a sequence of zero or more elements $\langle x_1, \ldots, x_n \rangle$ of a given type whose index type is determined by a particular "successor" function. However, the real nature of the abstract data structure LIST is best grasped from its pictorial representation in Figure 1.

FIGURE 1. The abstract data structure LIST.

The number of elements n is said to be the size of the list. If $n \geq 1$, we say that x_1 is the first element and x_n is the last element. If $n = 0$, we have an empty list, one which has no elements.

One important property of the list is that its elements can be linearly ordered according to their position in the list. For $i = 1, 2, \ldots, n-1$, we say that x_i is the **predecessor** of x_{i+1}, and x_{i+1} is the **successor** of x_i. The first element, or **head**, has no predecessor, and the last element, or **tail**, has no successor.

To illustrate some common operations on lists, we represent each list element by an object with a field *key* and a pointer field *next*. The object may contain some other satellite data. Given a pointer x to an object in the list, we use the notation *field*(x) to refer to two concepts: to a field with the name *field* of the object pointed to by x, as well as to the content of the same field.[3] A particular usage will be clear, we hope, from the context. For example, *key*(x) and *next*(x) denote the *key* and *next* fields of the object pointed to by x, and also denote functions that return, respectively, the key and the pointer contained in these fields of the object. Moreover, we will often identify an object and a pointer that points to the object. Thus, instead of the clumsy phrase "a pointer x that points to an object", we will simply say "an object (or element) x," understanding that x is really a pointer to an object. Not only that, if the key data of an object x are irrelevant, we refer to them anonymously simply as x. Other data structures will be treated similarly.

A list has also a unique attribute object with a field *name* that contains the name of the list, and a pointer field *head* that points to the first element of the list. The attribute object is identified with the name of the list. Thus, a particular list L is determined by an attribute object whose *name* field contains L. In the same vain, the *head* field of the attribute object L will be identified by the notation *head*(L), which also gives the pointer to the head of the list L. Therefore, if *head*(L) = NIL, then the list L is empty.

3.1. Initializing a linked list. To initialize a linked list, we can use the following procedure LIST-MAKE(L) that just makes the list L empty.

---

[3] In the parlance of the C programming language, *field*(x) denotes both an l-value operator and a function applied on the pointer x. In a Pascal-like language, *field*(x) is expressed as $x \uparrow .field$.

> LIST-MAKE (L)
>   head(L) = NIL;
>   **return;**

Clearly, the running time of the LIST-MAKE procedure is $O(1)$.

3.2. Inserting into a linked list. Given (a pointer to) an object x and (a pointer to) an element p of the list L, the following procedure LIST-INSERT (L, x, p) inserts x in L after the element p.

> LIST-INSERT (L, x, p)
>   next(x) = next(p);
>   next(p) = x;
>   **return;**

Figure 2 shows how an element is inserted into a list.

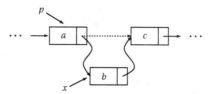

FIGURE 2. Inserting an element into a list.

The LIST-INSERT procedure cannot add an element to the beginning of a list. Here is the procedure that does just that.

> LIST-HEAD-INSERT (L, x)
>   next(x) = head(L);
>   head(L) = x;
>   **return;**

It is clear that the running time of the procedures LIST-INSERT and LIST-HEAD-INSERT on a list of n elements is $O(1)$.

3.3. Deleting from a linked list. Given (a pointer to) an element p of the list L, the procedure LIST-DELETE (L, p) removes the element in L which comes after the element p. The LIST-DELETE procedure also returns a pointer to the removed element so that the caller procedure can deallocate its recourses.

$$\text{LIST-DELETE}(L, p)$$
$$x = next(p);$$
$$next(p) = next(x);$$
$$\textbf{return } x; \quad [\![\, free(x) \,]\!]$$

Figure 3 shows how an element is removed from a list.

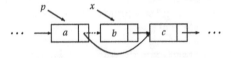

FIGURE 3. Removing an element from a list.

The LIST-DELETE procedure also cannot remove an element from the beginning of a list. Here is the procedure that does just that.

$$\text{LIST-HEAD-DELETE}(L)$$
$$x = head(L);$$
$$head(L) = next(x);$$
$$\textbf{return } x; \quad [\![\, free(x) \,]\!]$$

It is clear that the running time of the procedures LIST-DELETE and LIST-HEAD-DELETE on a list of n elements is $O(1)$.

3.4. Searching a linked list. The procedure LIST-SEARCH (L, k) finds the first element with the key k in the list L, returning a pointer to this element. If no element with the key k is found in the list, pointer NIL is returned.

$$\text{LIST-SEARCH}(L, k)$$
$$x = head(L);$$
$$\textbf{while } (x \neq \text{NIL}) \wedge (key(x) \neq k) \textbf{ do}$$
$$x = next(x);$$
$$\textbf{return } x;$$

To search a list of n elements, the procedure LIST-SEARCH (L, k) takes $O(n)$ time in the worst case, since it may have to search the entire list.

Notice that if we wish to insert or delete an element after some list element with a given key, we must first call the LIST-SEARCH procedure to find that element. Therefore, the running time for this insertion or deletion operation would be $O(n)$ in the worst case.

3.5. Doubly linked lists. The lists we have been considering so far are *singly linked*, because each element has one pointer field that indicates the next element. Such lists can be traversed in only one direction, as the LIST-SEARCH procedure demonstrates. In a number of applications however, we may wish to traverse a list both forwards and backwards efficiently. Or, given an element, we might wish to determine its predecessor and successor quickly. In such situations we may use a *doubly linked list* in which each element has a *key* field and two other pointer fields: *next* and *prev*. For every element x in the list, *next*(x) points to its successor, and *prev*(x) points to its predecessor. If *prev*(x) = NIL, the element x has no predecessor and is therefore the first element of the list. As before, the attribute object of a list L has the field *head*(L) that points to the head of the list. Figure 4 shows an example of a doubly linked list.

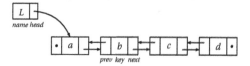

FIGURE 4. A doubly linked list.

Another advantage of doubly linked lists is that the basic operations can directly use as a parameter the element we want to work with, rather than the less natural pointer to the previous element. For example, the next procedure DLLIST-DELETE(L, x) removes the element x from the doubly linked list L. Other operations are left as an exercise.

DLLIST-DELETE(L, x)
    if *prev*(x) ≠ NIL **then**
        next(*prev*(x)) = *next*(x);
    else *head*(L) = *next*(x);
    if *next*(x) ≠ NIL **then**
        prev(*next*(x)) = *prev*(x);
    return x; ⟦ free(x) ⟧

Figure 5 shows how an element is deleted from a doubly linked list.

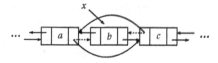

FIGURE 5. Deleting an element from a doubly linked list.

3.6. Implementing lists. In programming languages that support pointers natively, the task of implementing lists is straightforward. We can treat a list and its operations more or less in the same way as we did it in the previous section on a more abstract level.

In this section we will tackle the more challenging task of implementing lists in programming languages, such as Fortran, that do not have pointers. If we are working with such a language, we can represent lists by arrays and simulate pointers with integers that indicate positions in the arrays.[4]

3.7. Array representation of lists. We can represent a collection of objects that have the same fields by using one array for each field. For example, Figure 6 shows how to represent a doubly linked list from Figure 4 with three arrays. The array *key* holds the values of the keys of current elements in the list, and pointers are kept in the arrays *next* and *prev*. For a given array index x, the group $key(x)$, $next(x)$, and $prev(x)$ represent an element in the list. Under this interpretation, a pointer x is simply a common index into the *key*, *next*, and *prev* arrays. These integers that indicate positions in the arrays are called **cursors** (or **handles** in other context).

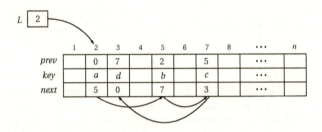

FIGURE 6. Multiple-array representation of a doubly linked list.

In Figure 6, the element a occupies the second common position, so $key(2) = a$. Its successor, the element b, appears in the common position 5, and so $key(5) = b$. Moreover, we have $next(2) = 5$ and $prev(5) = 2$. Other elements are similarly linked using cursors instead of pointers. The pointer NIL is usually represented by an integer (such as 0 or −1) that cannot possibly represent an actual index into the arrays. The variable L holds the index of the head of the list.

We can use similar idea to represent lists by single arrays. This approach actually mimics the memory management of the real computer. An object occupies a contiguous group of locations in the computer memory. A pointer to an object is simply the address of the first memory location of the object, and other memory locations within the object can be indexed by adding an offset to the pointer.

The same strategy can be used for implementing objects in programming languages that do not provide explicit pointer data types. For example, Figure 7 shows how a single array A can be used to represent the doubly linked list of Figures 4 and 6. An object occupies a contiguous subarray $A(j \ldots k)$. Each field of the object is determined by an offset in the range from 0 to $k - j$, and a pointer to the object is the index j. In Figure 7, the offsets corresponding to the *key*, *next*,

---

[4]The word "simulate" should be taken with a grain of salt, because in the computer pointers are really *implemented* as nonnegative integers (memory addresses) indicating positions in a huge array (the main memory).

and *prev* fields are 0, 1, and 2, respectively. Thus, given a pointer x, to read the value of *prev*(x), we add the offset 2 to the value x of the pointer, and so we need to read the content of the array element $A(x + 2)$.

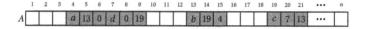

FIGURE 7. Single-array representation of a doubly linked list.

The single-array representation is flexible in that it permits object of different sizes to be stored in the same array. The problem of managing such a heterogeneous collection of objects is more difficult than the problem of managing a homogeneous collection, where all objects have the same fields. For our purposes, it will be sufficient to use the multiple-array representation of homogeneous objects.

3.8. Multiple-array C implementation of singly linked lists. Let us illustrate previous discussion by writing (almost) complete C code for managing singly linked multiple lists. We will use the multiple-array implementation of the lists. But now, in addition to the list operations, we have to take care of the problem of allocating and freeing unused list elements.

For simplicity, we assume that the elements of all lists only consist of the *key* and *next* fields of character and integer type, respectively. For all the lists under consideration we use two (global) arrays, the *key* array of character type and the *next* array of integer type. Size of these arrays is declared to be sufficiently large constant n. Notice that the type of the *key* array is arbitrary and depends on the particular application, whereas the *next* array must be of integer type.

Suppose that all lists at some moment use m ($m \leq n$) elements of the *key* array. Then remaining $n - m$ elements are free and can be used to represent list elements inserted into the lists in the future. We keep the free elements in another singly linked list, which we call *free list*. The free list uses only the *next* array, which holds the *next* pointers within the lists. The head of the free list is held in the global variable f. During the lifetime of the dynamic sets represented by the lists, the free list may be interlaced with the "real" nonempty lists.

We assume that positions of the heads of all lists are held in an appropriately declared global integer array L. Thus, a list l is determined by the value of $L[l]$. Figure 8 shows two lists, $L[1] = \langle a, b, c \rangle$ and $L[2] = \langle d, e \rangle$, sharing the *key* and *next* arrays of size $n = 10$. (Recall that array indices in C are enclosed in square brackets and start with 0, hence the NIL value is represented by -1.)

Notice that all elements of the *key* array that are not in either list are singly linked in the free list. This list is necessary so that we can obtain an empty element when we want to insert into some list, and so that we can have a place to put deleted elements for later use. Also notice that each element in the representation is either in some list or in the free list, but not in both.

The free list initially contains all n unallocated elements. New elements are allocated out of the free list, and removed elements are put back into the same

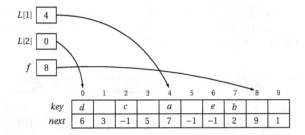

FIGURE 8. Multiple-array implementation of two singly linked lists.

list. Here is the C function init_free that initially links all the array elements into the free list.

```
void init_free ()
{
    int i;

    /* each element but last points to the next */
    for (i = 0; i < n-1; i++)
        next[i] = i+1;
    next[n-1] = -1;  /* the end of the free list */
    f = 0;           /* the head of the free list */
}
```

The C functions new_elem and free_elem perform the task of allocating new elements and freeing elements not needed anymore. When the free list is exhausted, the new_elem function signals an error. However, other functions omit checks for errors; the reader may include them as an exercise.

```
int new_elem ()
{
    int i;

    if (f = -1)
        return -1; /* out of memory */
    else {
        i = f;
        f = next[i];
        return i;
    }
}

void free_elem (i)
{
    next[i] = f;
    f = i;
}
```

Making a new list l only consists of initializing the array element $L[l]$ to -1. The function list_make is therefore trivial.

```
void list_make (int l)
{
    L[l] = -1;
}
```

To insert an element into a list, we first allocate an element by taking the first element of the free list. Then we modify the new element's *key* field, and finally place the element in the required position of the list by modifying the new element's *next* field.

We have two C functions for inserting a new element x into a list l. The function list_insert adds the new element after the list element indicated by position p, and the function list_head_insert does this at the beginning of the list l. Note that list_insert does not need to know which list it is actually working on, since the function forcefully links the new element to the p-element of whichever list it belongs to. We provide the corresponding parameter to the function only as a way of "safe" programming.

```
void list_insert (int l, char x, int p)
{
    int i;
    i = new_elem();
    key[i] = x;
    next[i] = next[p];
    next[p] = i;
}

void list_head_insert (int l, char x)
{
    int i;
    i = new_elem();
    key[i] = x;
    next[i] = next[L[l]];
    L[l] = i;
}
```

To delete an element from a list, we first unlink it from the list, and then free (deallocate) it by placing it at the beginning of the free list. We also have two C functions for deleting an element from a list l. The list_delete function deletes the element after the one indicated by p, and the list_head_delete function deletes the head element of the list l. Similarly to list_insert, the list_delete function does not need the input list.

```
void list_delete (int l, int p)
{
    int i;
    i = next[p];
    next[p] = next[i];
```

```
    free_elem(i);
}

void list_head_delete (int l)
{
    int i;
    i = L[l];
    L[l] = next[i];
    free_elem(i);
}
```

4. The abstract data structure STACK

A **stack** is a special kind of the list in which all insertions and deletions take place at one end, called the **top**. The intuitive model of a stack is a pile of poker chips on a table, a series of cut meat on a barbecue skewer, or a pile of plates used in cafeterias. In all these examples it is convenient to remove only the top object from the group and add a new object on the top. In other words, the order in which the objects are popped from the stack is the reverse of the order in which they were pushed onto the stack.

The insert operation on a stack is often called PUSH, and the delete operation is often called POP. On an abstract level, the attribute object of a stack contains the *name* field and another field *top* that indicates the top of the stack.

Every implementation of lists we mentioned works for stacks as well, since a stack with its operations is a special case of a list with its operations. In programming environments that provide pointer data type, the pointer-based implementation of lists will do for stacks unchanged. The LIST-HEAD-INSERT and LIST-HEAD-DELETE operations in the implementation are actually other names for the PUSH and POP operations, respectively. Moreover, the *top* field of a stack is equivalent to the *head* field of a list.

However, an array-based implementation of a stack can take advantage of the fact that insertions and deletions occur only at the top of stack. We can use a single array, anchor the bottom of the stack at the beginning of the array, and then let the stack grow toward the end of the array. This idea is illustrated in Figure 9.

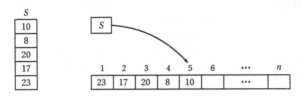

FIGURE 9. A stack S and its single-array representation.

As shown in Figure 9, we can represent a stack of at most n elements with an array A of size n. The stack is identified by a variable S that indicates the top of the stack, i.e., the position of the most recently inserted element. In addition, the stack occupies the subarray $A(1 \ldots S)$, where $A(1)$ is the element at the bottom of the stack and $A(S)$ is the element at the top.

When $S = 0$, the stack contains no elements and is **empty**. The stack can be tested for emptiness by a query operation STACK-EMPTY. If an empty stack is popped, we say that the stack **underflows**. When S equals n, the stack is **full**. The stack can be tested for fullness by a query operation STACK-FULL. If a full stack is pushed, we say that the stack **overflows**. Underflow and overflow conditions normally represent an error.

The stack operations for the array-based representation are easily implemented with a few lines of code.

```
STACK-MAKE(S)
  S = 0;
  return S;
```

```
STACK-EMPTY(S)
  if S = 0 then
    return TRUE;
  else
    return FALSE;
```

```
STACK-FULL(S)
  if S = n then
    return TRUE;
  else
    return FALSE;
```

```
PUSH(S, x)
  if STACK-FULL(S) then
    return "overflow";
  else
    S = S + 1;
    A(S) = x;
    return S;
```

```
POP(S)
  if STACK-EMPTY(S) then
    return "underflow";
  else
    x = A(S);
    S = S - 1;
    return S, x;
```

Each of the five stack operations takes $O(1)$ time.

5. The abstract data structure QUEUE

A **queue** is another special kind of the linked list in which elements are inserted at one end (the **rear** end) and deleted at the other end (the **front** end). The intuitive model of a queue is a line of people waiting in the box office, where people in the line are serviced (removed) from the front end, and new people take their place (get inserted) at the rear end. (No cutting into line occurs here!)

The insert operation on a queue is called ENQUEUE, and the delete operation is called DEQUEUE. The queue has the **head** (first) and the **tail** (last) element. When an element is enqueued, it takes place of the tail element of the queue, and the element dequeued is always the head element of the queue.

As for stacks, any list implementation is legal for queues. In environments with native pointer data type, the pointer-based implementation of a queue is easy and basically clones the list implementation. However, instead of using a carbon copy of the list implementation, we can make the ENQUEUE operation more efficient by keeping a pointer to the tail element. This way, each time we wish to enqueue an element, we don't have to run down the entire list from beginning to end.

For the array-oriented representations of a queue Q, one particularly interesting solution uses a single "circular" array A of size n, as suggested in Figure 10.

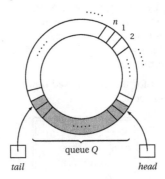

FIGURE 10. The circular-array representation of a queue.

We think of the array A as a circle, where the first position follows the last. The queue Q is found somewhere around the circle in consecutive locations, with the tail element somewhere clockwise from the head element. (Note that the meaning of "consecutive" must be taken in a circular sense.) The queue Q has a cursor *head* that indexes the first element, as well as a cursor *tail* that indexes the last element. Thus, the queue elements are in locations *head, head +* $1, \ldots, tail$, where we "wrap around" in the sense that location 1 immediately follows location n in a circular order.

To enqueue an element, we increment *tail* by one location clockwise and write the element in that position. To dequeue, we simply increment *head* by one location clockwise. Therefore, the whole queue moves in the clockwise direction as we enqueue and dequeue elements.

There is one subtlety that comes up in the representation of Figure 10 and any minor variation of that strategy (for instance, if *tail* points to the next location clockwise of the last element, rather than to that element itself). The problem is that there is no way to tell an empty queue from a full queue, short of maintaining an attribute that is true if and only if the queue is empty. (The interested reader should convince himself that this is indeed the case.) That's the reason why our representation will maintain a variable *length* that indicates the current number of queue elements in the array A. If we are not willing to maintain such an attribute, we must prevent the queue from ever filling the entire array A. In other words, we cannot let the queue grow bigger than $n - 1$, and wasting one array element to distinguish empty and full queues seems less natural than maintaining an explicit variable for that purpose.

Thus, besides the array A, our queue Q is identified by three additional variables *head, tail,* and *length*. Initially, we have *head* = 1, *tail* = n, and *length* = 0. If we use the modulus function for our circular viewpoint on the array A, the queue operations require only a few lines of code.

```
QUEUE-MAKE(Q)
   head = 1;
   tail = n;
   length = 0;
   return;
```

```
QUEUE-EMPTY(Q)
   if length = 0 then
      return TRUE;
   else
      return FALSE;
```

```
QUEUE-FULL(Q)
   if length = n then
      return TRUE;
   else
      return FALSE;
```

```
ENQUEUE(Q, x)
   if QUEUE-FULL(Q) then
      return "overflow";
   else
      tail = tail (mod n) + 1;
      A(tail) = x;
      length = length + 1;
      return;
```

```
DEQUEUE(Q)
   if QUEUE-EMPTY(Q) then
      return "underflow";
   else
      x = A(head);
      head = head (mod n) + 1;
      length = length - 1;
      return x;
```

Clearly, each of the five queue operations takes $O(1)$ time.

6. The abstract data structure SET

The set is the most fundamental notion of mathematics. All other concepts in mathematics are practically based on the set theory. We assume the reader is familiar with the mathematical notion of the set and the fact that, in mathematics, a set is actually not defined explicitly. Rather, a set is mathematically given by its properties. Also, we assume the working knowledge of elementary relationships among sets such as membership, subset, superset, and so on. Furthermore, we assume the reader's familiarity with basic operations on sets such as union, intersection and difference. In this section we consider data structures suitable for implementation of sets when the operations defined on sets are used from the mathematical perspective. In computer science, sets are used as the basis for many important algorithms and data structures. Some of these applications are discussed in more detail in Chapter 9.

As a data model, the set is a collection of *elements* or *objects*, collectively called *members*. Set elements are considered atomic in nature so that nothing can be a member of them in the set sense. In other words, set elements are assumed not to be sets, which is quite different from mathematics. The empty set that has no members is considered to be a set, and so cannot be a member of another set in our informal set theory.

All elements of a set are different, which means that no set can contain two copies of the same member. Set elements are usually values of primitive types, such as integers, characters, or strings. Nevertheless, it is sometimes convenient to use complex data types as the type of elements. Thus, set elements may be records or arrays, and in some cases they might be even more complicated data structures. In every case, however, all elements in any set are of the same type.

Another big and comforting difference from mathematics is that we assume the sets are finite. That is, for every set S, there is some integer n such that the set S has exactly n elements. This is denoted by $|S| = n$. Although infinite sets have very interesting mathematical properties, we won't consider them since they could not be stored in the computer's memory.

In this section we discuss two common ways for implementation of the set data model: the list and array representations.

6.1. List representation of sets. An easy way to represent a set is to use a linked list of the set elements. In our discussion we use the singly-linked lists, although we could equally well use the doubly-linked lists. For example, Figure 11 shows the set $S = \{x_1, x_2, \ldots, x_n\}$ represented as a linked list S.

FIGURE 11. A linked list representing the set $S = \{x_1, x_2, \ldots, x_n\}$.

Let us consider first how we can implement simple operations on a set such as insertion, deletion and membership test of an element. Clearly, for these set operations we can use ordinary list operations, respectively, LIST-INSERT, LIST-DELETE, and LIST-SEARCH from Chapter 3, slightly modified. Namely, if we wish to add an element x to, or delete it from, a set S, then the set operations SET-INSERT(S, x) and SET-DELETE(S, x) in full generality first need to check whether $x \in S$. Then we do nothing in the case of SET-INSERT if $x \in S$, or in the case of SET-DELETE if $x \notin S$. Thus, both insertion and deletion operation include membership test operation SET-MEMBER(S, x). And to test whether $x \in S$, we can obviously use unmodified LIST-SEARCH procedure. Since we observed in Section 3.4 that the running time of the LIST-SEARCH procedure is $O(n)$, each of these three set operations takes $O(n)$ time. The details of actual pseudocode for these set operations are left as an easy exercise.

We can make insertion operation run faster if we allow duplicates and simply insert a given element onto the front of the list representing a set.[5] However, if we do not check for the presence of an element in the list before inserting it, there may be several copies of an element on the list. Membership test operation is exactly the same as before, although we may have to search a longer list because of multiple occurrences of the set elements in the list. However, deletion of an element is different. Now we cannot stop the search for the given element when we first find it in the list, because there could be other copies of it further down the list. For a correct deletion operation therefore we must always search the entire list, rather than half of the list on average.

In summary, if we allow duplicates, then insertion operation takes constant time, but deletion and membership test operation run slower than before since

---

[5]Allowing duplicates in lists that represent sets may seem wrong, since sets in mathematical sense do not have duplicates. However, even if some element appears several times on a linked list, in the set operations it is treated logically as if it is only present once in the abstract set that the list data structure represents.

the lists are longer than when duplicates are not allowed. How much longer depends on how often we insert an element that is already present in the set. Anyway, this approach makes sense if insertions predominate. On the other hand, if deletions or membership tests predominate and we may insert duplicate elements, then we are probably better off checking for the presence of the given element before inserting it.

Finally, if we can be sure for some reason that we shall never insert an element already in the set, we can use both the fast insertion operation, where we blindly insert the given element, and the fast deletion operation, where we stop when we find the first occurrence of the element to be deleted.

The other basic set operations, such as union, intersection, and difference, can be easily implemented using the ordinary list operations as well. For example, to create a list C that represents the union of two sets A and B represented by two lists, we may first copy the list for A into the initially empty list C. For this step we use the list operation LIST-COPY(L_1, L_2) that makes the list L_2 an identical copy of the list L_1 (see Exercise 2). Then we examine each element of B to see whether it is in A. If not, we add the element to C; otherwise, we do nothing. This idea is sketched in the following algorithm SET-UNION(A, B, C) that implements the set operation $C = A \cup B$.

SET-UNION(A, B, C)
LIST-COPY(A, C);
for each $x \in B$ **do**
    if LIST-SEARCH(A, x) = NIL **then**
        LIST-HEAD-INSERT(C, x);
return C;

To estimate the running time of the SET-UNION algorithm, we notice that basically we must compare each element of one set with each element of the other. Suppose A has n elements and B has m elements. Then copying the list A to the initially empty list C in the SET-UNION algorithm clearly takes $O(n)$ time. Since the **for** loop is iterated m times, and each iteration takes $O(n)$ time, the loop takes $O(mn)$ time. The total time of the SET-UNION algorithm is therefore $O(n) + O(mn) = O(mn)$.

We leave similar algorithms for the set intersection and difference, each taking $O(mn)$ time, as an exercise for the reader.

6.2. Sorted-list representation of sets. If elements of a set S are drawn from a totally ordered universe, another possibility is to keep the set elements in the list representing S sorted in increasing order. In this case there is no reason to allow duplicates, since we must find anyway the proper place for an element when inserting it. Thus, to maintain the list sorted at all times, the SET-INSERT(S, x), SET-DELETE(S, x), and SET-MEMBER(S, x) operations basically have to scan the list from the beginning and go as far down the list as the position in which x should appear. In other words, if we encounter an element greater than x, then we know that x could not be later in the list. So, we insert x there in the case of the SET-INSERT operation, or discover that $x \notin S$ in the case of the SET-DELETE and SET-MEMBER operations. If $|S| = n$, this means that these operations take time proportional to $n/2$ on average; of course, their worst-case running time is still $O(n)$. For example, the following are details of the SET-INSERT procedure, while the other two procedures are left as an easy exercise.

```
SET-INSERT(S, x)
    p = head(S);
    if (p = NIL) ∨ (key(p) > key(x)) then
        LIST-HEAD-INSERT(S, x);
    else
        while (p ≠ NIL) ∧ (key(p) < key(x)) do
            q = p;
            p = next(q);
        if (p = NIL) ∨ (key(p) ≠ key(x)) then
            LIST-INSERT(S, x, q);
    return S;
```

The SET-INSERT procedure works as follows. The pointer p initially points to the beginning of the list representing the set S. We next insert x as the head of the list if $S = \emptyset$ or x is the smallest element, or skip all elements less than x and then insert x. The reader is advised to check the procedure's pointer handling to make sure this is correctly implemented.

We can perform set unions, intersections, and differences much faster when the lists representing the sets are sorted. For example, consider the computation of $C = A \cup B$, where the sets A and B are represented by sorted lists. The idea is to merge these sorted lists and construct a single sorted list containing all the elements (without duplicates) of the two given lists. One simple way to do this is to iteratively examine both lists "in parallel" from the front. This means that, in each step, we compare the heads of the two lists and

- if the current head elements are different, we copy the smaller onto the resulting list and advance the corresponding list; or
- if the current head elements are equal, we make only one copy of them onto the resulting list, but advance both lists.

This approach is easily implemented using common operations of the abstract data structure LIST. We should be careful though to not remove elements from the lists for A and B for the union, since we should not destroy A or B while uniting them. Instead, we must make copies of the two input lists and work with the copies to build the union list. Assuming that the attribute object of each list contains an additional pointer field *tail* pointing to the last element of the list (which is appropriately maintained by the list operations), actual details of the SET-UNION procedure are given by the following pseudocode.

```
SET-UNION(A, B, C)
  LIST-COPY(A, L₁); LIST-COPY(B, L₂);
  LIST-MAKE(C);⟦ head(C) = NIL and tail(C) = NIL ⟧
  while (head(L₁) ≠ NIL) ∧ (head(L₂) ≠ NIL) do
    if head(L₁) = head(L₂) then
      x = LIST-HEAD-DELETE(L₁);
      x = LIST-HEAD-DELETE(L₂);
    else if head(L₁) < head(L₂) then
      x = LIST-HEAD-DELETE(L₁);
    else
      x = LIST-HEAD-DELETE(L₂);
    LIST-INSERT(C, x, tail(C));
  while head(L₁) ≠ NIL do
    x = LIST-HEAD-DELETE(L₁);
    LIST-INSERT(C, x, tail(C));
  while head(L₂) ≠ NIL do
    x = LIST-HEAD-DELETE(L₂);
    LIST-INSERT(C, x, tail(C));
  return C;
```

If $|A| = n$ and $|B| = m$, the running time of the the the SET-UNION algorithm implemented in this way is $O(n + m)$. This follows from the fact that we scan the lists for A and B only once, and in each step we perform a constant amount of work advancing one of the lists (or both).

This idea for computing the union works for intersections and differences of sets as well. For example, in the case of intersection, we need to copy an element only if it appears on both lists. The required change is therefore that in each step we take an element into the intersection only if the current elements are the same. If they are different, the smaller cannot appear on both lists, and so we do not add anything to the intersection but merely advance the corresponding list. Similar reasoning also applies in the case of the set difference. Thus, when two sets are represented by sorted lists, the running time of taking their union, intersection, or difference is $O(n + m)$.

In fact, when the sets are represented by unsorted lists, usually it pays to sort the lists before performing these set operations. Namely, as we shall see, there are faster methods than those we discussed in Chapter 2 to sort a list of n elements taking only $O(n \log n)$ time. Thus, we can sort the two lists in $O(n \log n + m \log m)$ time, and then take the union of the sorted lists in $O(n + m)$ time. The total time for both steps is therefore

$$O(n + m) + O(n \log n + m \log m) = O(n \log n + m \log m).$$

The expression $O(n \log n + m \log m)$ can be greater than $O(nm)$, which is the time required for uniting two unsorted lists, but is less whenever n and m are, say, close in value, that is, whenever the sets are roughly of the same size.

6.3. Array representation of sets. When all sets of our interest are subsets of some reasonably small "universal set" U, we can use a representation of sets that is more efficient than the list representation discussed in the previous section. We first order the elements of the universal set U in some way so that each element of U is associated with a unique integer between 1 and n, where $|U| = n$. In other words, we can assume that $U = \{x_1, x_2, \ldots, x_n\}$. A set $S \subseteq U$ is then represented by the ***characteristic vector*** of S such that for each element $x_i \in U$, if $x_i \in S$, the i-th element of the vector is 1, and if $x_i \notin S$, then the i-th element of the vector is 0. Since the characteristic vector is an array of 0's and 1's, it is often referred to as the bit vector.

For example, let U be the set of colors listed in Table 1. The order of a color in the listing indicates the color's position in characteristic vectors of length 5. Thus, the set Green = {Blue, Yellow} is represented by the characteristic vector $G = 01100$; the set Pink = {Red, White} is represented by $P = 10001$; the set Brown = {Red, Blue, Yellow, Black} is represented by $B = 11110$; and the empty set $\emptyset$ is represented by $E = 00000$.

Order	Color
1	Red
2	Blue
3	Yellow
4	Black
5	White

TABLE 1. "Universal" set of colors.

The major advantage of this representation is that insertion, deletion, and membership test operations can be performed in constant time by directly addressing the appropriate position in a characteristic vector. The set union, intersection and difference operations can be performed in time proportional to the size of the universal set. For example, if $|U| = n$ and A, B, C are characteristic vectors of the sets A, B and $C = A \cup B$, the pseudocode for the algorithm SET-UNION(A, B, C) is a direct application of the fact that an element from U belongs to $A \cup B$ if and only if it belongs to A or B.

```
SET-UNION(A, B, C)
    for i = 1 to n do
        if (A(i) = 1) ∨ (B(i) = 1) then
            C(i) = 1;
        else
            C(i) = 0;
    return C;
```

Notice that the sizes of the sets involved in the union, intersection, and difference operations are not directly related to the size n of the universal set. However, if those sets have sizes that are reasonably large fractions of n, then the time for union, intersection, and difference is proportional to the sizes of the sets involved. That is better than the $O(n \log n)$ time for sorted lists, and much better than the $O(n^2)$ time for unsorted lists. On the other hand, in the case when the sets are much smaller than the universal set, the running time of these set operations is far greater than the sizes of the sets involved.

Exercises

1. Write the procedure LIST-LENGTH(L) that computes the length of a list L. Assume the pointer-based implementation of lists.

2. Write the procedure LIST-COPY(L_1, L_2) that creates the list L_2 as an identical copy of the list L_1 (leaving the list L_1 unchanged). Assume the pointer-based implementation of lists and analyze the running time of your procedure.

3. Write the procedure LIST-CONCAT(L_1, L_2) that concatenates two lists L_1 and L_2 into a third list L_3. Assume the pointer-based implementation of lists and analyze the running time of your procedure.

4. Write the procedure LIST-DELDUP(L) that removes all duplicates in a list L. Assume the pointer-based implementation of lists and analyze the running time of your procedure.

5. Write the procedure DLLIST-INSERT(L, x, p) that inserts element x after element p in a doubly linked list L. Assume the pointer-based implementation of lists.

6. A single array can be used to store two stacks, one growing up from the left end and the other growing down from the right end. Implement the stack operations using this single-array representation of two stacks. Analyze the running time of the stack operations.

7. Show how to implement a queue using two stacks. More precisely, write the procedures ENQUEUE, DEQUEUE, and QUEUE-EMPTY using only the procedures PUSH, POP, and STACK-EMPTY on two stacks that represent single queue. Analyze the running time of the queue operations.

8. Show how to implement a stack using two queues. More precisely, write the procedures PUSH, POP, and STACK-EMPTY using only the procedures ENQUEUE, DEQUEUE, and QUEUE-EMPTY on two queues that represent single stack. Analyze the running time of the stack operations.

9. **Virtual initialization of an array.** Given an array $A(1 \dots n)$ of size n, naive method to initialize the array elements to some default value d is:

INIT-ARRAY(A, n)
    for $i = 1$ to n **do**
        $A(i) = d;$

Clearly, this method requires $O(n)$ time and constant space (besides the space needed for the array itself). Two operations to access an element of the array—READ(A, i), which reads the value of the i-th element of A, and WRITE(A, i, x), which writes the value x into the i-th element of A—both take constant time:

READ(A, i)
   return $A(i)$;

WRITE(A, i, x)
   $A(i) = x$;
   return;

However, in some applications the initialization time $O(n)$ is unacceptable, and instead we need to "virtually" initialize the array in constant time. This is even more true in cases in which only a fraction of the (possibly large) array is ever used. Write the procedures INIT-ARRAY, READ, and WRITE that do what they are supposed to do, but so that each of them runs in constant time. (*Hint:* Use two additional arrays of size n, where one of them simulates a stack.)

10. Write the procedures SET-DELETE(S, x) and SET-MEMBER(S, x) if the list representing the set S is sorted.

11. Write the procedures SET-INTERSECTION(A, B, C) and SET-DIFFERENCE(A, B, C) to implement the operations of intersection and difference of two sets A and B. More precisely, they should compute $C = A \cap B$ and $C = A \setminus B$, respectively, if the sets are represented using

    (a) unsorted lists;
    (b) sorted lists;
    (c) characteristic vectors.

    In each case estimate the running time of the corresponding procedure.

12. Using abstract data structure operations from this chapter, implement in a more detailed pseudocode (or, better yet, in your favorite programming language) the SM algorithm from Section 1.7 in Chapter 2. (*Hint:* Make use of the data structures indicated at the conclusion of the running-time analysis of the algorithm.)

4

Tree Algorithms

A tree is a hierarchical structure that naturally arises in many real-life situations. Familiar examples of trees are the table of contents of a book, genealogies, and organizational charts. Trees of various types are also widely used in different areas of computer science. For example, almost all operating systems store files and directories in a tree or a treelike structure. Another application of trees, called the expression trees, can be found in compilers for representation of the syntactic structure of expressions in source programs. In addition to these direct applications, one of the most important uses of trees is in the implementation of other data structures. As we shall see, they can be used in the implementation of disjoint sets, priority queues, and so on.

1. Basic definitions

In general, trees are just a special case of graphs that we discuss in more detail in Chapter 8. However, in this chapter we won't consider these general trees, usually called *free trees*. We will focus rather on a restricted version of the trees called rooted trees. A ***rooted tree*** consists of a set of elements, called ***nodes***, and a set of pairs of distinct nodes connected by lines, called ***edges***. In figures we often depict a node by a letter or a number inside a circle around it. One of the nodes of a rooted tree is distinguished as the ***root***. We draw the root node on the top of a hierarchical structure induced by the "parenthood" relation on the nodes represented by the set of edges. Figure 1 shows an example of a rooted tree with 11 nodes, where the root node is n_1.

Notice that the tree of Figure 1 is *connected* in the sense that if we start at any node and follow the edges passing through different nodes, we can reach any other node of the tree. On the other hand, the tree is *acyclic* in the sense that starting at any node and following the edges to different nodes without repetition, we can never return to the starting node. In fact, these two properties characterize general, free trees—rooted trees have one additional property that

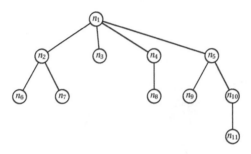

FIGURE 1. A rooted tree.

one node, the root, is special. Since throughout this chapter we consider only rooted trees, in the rest we will refer to them simply as trees.

To define trees more precisely, we first formalize the intuitive notion of a path in a tree. If $n_1, n_2, \ldots, n_k$ ($k \geq 1$) is a sequence of distinct nodes in a tree such that for $k > 1$ all pairs of nodes n_i and n_{i+1} with $i = 1, \ldots, k - 1$ are connected by an edge, then this sequence is called the **path** from node n_1 to node n_k. In other words, a path from node x to node y consists of all distinct nodes that we pass through following the sequence of edges from node x to node y. We can now define that a (rooted) tree is a set of nodes and a set of edges between them with the following properties:

(i) One node is distinguished as the root.
(ii) Every node c except the root is connected by an edge to exactly one other node p called the **parent** of c.
(iii) There is a unique path from the root to each node (and hence from each node to the root).

If p is the parent of c, we also say that c is a **child** of p. A node may have zero or more children, but every node other than the root has exactly one parent—the root is the only node with no parent. Moreover, the path from any node to the root is always unique. For example, n_1's children in the tree of Figure 1 are $n_2, n_3, n_4,$ and n_5. Because n_1 is the root, it has no parent, but all other nodes have parents (say, n_7's parent is n_2). Also, the unique path from the node, say, n_{10} to the root is the sequence of nodes n_{10}, n_5, n_1.

A tree T can be also defined recursively with an inductive definition that constructs larger trees out of smaller ones:

1. A single node by itself is a tree. This node is also the root of the tree.
2. If r is a new node and $T_1, T_2, \ldots, T_\ell$ are trees with roots $r_1, r_2, \ldots, r_\ell$, respectively, then the tree T is built by making the node r be the parent of the nodes $r_1, r_2, \ldots, r_\ell$ (see Figure 2). In this new tree T, node r is the root and $T_1, T_2, \ldots, T_\ell$ are subtrees of node r.

Every node x on the path from some node y to the root is called a (proper) **ancestor** of y, and y is called a **descendant** of x. Notice that any node is an ancestor and a descendant of itself. A **subtree** of a tree consists of a node and all its descendants, together with inherited edges. Children of the same node

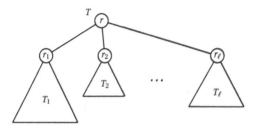

FIGURE 2. A tree T defined recursively.

(that is, nodes with a common parent) are called **siblings**, and a node without children is called a **leaf**.

The **length** of a path is one less than the number of nodes in the path. Thus, the length of the path from a node x to a different node y is the number of edges along the path. Moreover, the length of a path consisting of a single node is zero.

For example, in the tree of Figure 1, the length of the path from n_1 to n_{11} is three, and the length of the path consisting of only, say, n_4 is zero. Nodes n_2, n_3, n_4, and n_5 are all siblings. All leaves of the tree are n_3, n_6, n_7, n_8, n_9, and n_{11}. The subtree rooted at n_2 consists of the nodes n_2, n_6, and n_7, as well as inherited edges between them.

The **height of a tree** is the length of longest path from the root to a leaf. The **height of a node** is the height of the subtree rooted at that node. The **depth (level) of a node** is the length of the unique path from the root to that node. Thus, the depth (level) of the root is always 0, and the depth (level) of any node is one more than the depth (level) of its parent. The **depth of a tree** is the height of the tree. Nodes of a tree on level k are all nodes in the tree whose depth (level) is equal to k.

For example, depth (height) of the tree in Figure 1 is three. Node n_{10}'s depth is two, and its height is one. Nodes on level 2 are n_6, n_7, n_8, n_9, and n_{10}.

The children of each node in a rooted tree are usually ordered from left to right. Thus the two trees in Figure 3 are considered different, because the two children of the node n_1 appear in different order in the two trees. In this book we consider only such **ordered** rooted trees, which is understood implicitly when we refer to trees in the sequel.

FIGURE 3. Two distinct (ordered) trees.

1.1. Labeled trees. In computer science it is often useful to associate a label with each node of a tree, similarly to the spirit in which we have associated a data field with, say, a list element. That is, the label of a node is a key value that is "stored" at the node.

For example, Figure 4 illustrates a labeled tree in which the labels $A, B, C, D,$ E, F, G, H, I, J, K are given to the nodes of the tree of Figure 1. In the labeled tree, the node labels are shown inside the nodes, and the names of nodes are conventionally shown next to the nodes.

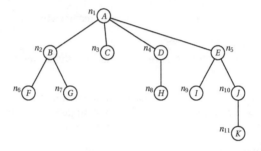

FIGURE 4. A labeled tree.

As another example, Figure 5 shows a labeled tree called the ***expression tree***. In an expression tree, the node labels are arithmetic operands or operators given inside the nodes.

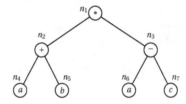

FIGURE 5. An expression tree.

The expression tree in Figure 5 represents the arithmetic expression $(a+b) *$ $(a-c)$. More precisely, a labeled tree is an expression tree if it satisfies two conditions:

1. Every leaf node is labeled by an arithmetic operand. For example, node n_4 represents the expression (consisting of the alone operand) a.
2. Every interior node is labeled by an arithmetic operator, such as $+$, $-$, or $*$. Suppose that an interior node x is labeled by a binary operator ω, and that its left child represents expression E_1 and its right child represents expression E_2. Then the node x represents expression $(E_1)\omega(E_2)$. We may remove the parenthesis if they are not necessary.

Thus, since in Figure 5 the node n_2 has operator $+$, and its children represent the expressions a and b, the node n_2 represents the expression $(a) + (b)$, or simply $a + b$. The node n_1 (that is, the entire expression tree) represents the expression $(a + b) * (a - c)$, because $*$ is the label of n_1, and its children n_2 and n_3 represent the expressions $a + b$ and $a - c$, respectively.

Notice that, strictly speaking, the label of a node is not the same as the name of the node, and theoretically we may even change the label of a node while the node's name remains the same. In practice however, nodes are frequently identified with their labels, not with their names. Moreover, in computer science the tree structure is used as a convenient way to organize a collection of data for easy retrieval. This means that a data item from the collection is represented by a tree node. Since each data item is identified by a unique value called the **key**, and on an abstract level the key is usually the only relevant information, we can think of the label of a node as being the key of an item that the node is representing. This is why in the rest of the book we follow the tradition and mostly refer to the tree nodes in terms of their keys, omitting any satellite data that a node may represent.

1.2. Node orderings of a tree. There are several ways to systematically order all nodes in a tree. The four most important orderings are called level-order, preorder, inorder, and postorder.

The **level-order list** (or **level-order traversal**) of nodes in a tree T consists of all the nodes on level 0 from left to right (i.e., just the root), followed by all the nodes on level 1 from left to right, then all the nodes on level 2 from left to right, and so on, up to all the nodes on the last level of T from left to right. The level-order traversal implements a more general technique known as *breadth-first search*.

The preorder, postorder, and inorder lists of nodes in a tree T are defined recursively using the inductive definition of the tree T:

1. If the tree T consists of a single node, then that node by itself makes the preorder, postorder, and inorder lists of nodes in T.
2. Otherwise, let T be a tree with root r and subtrees $T_1, T_2, \ldots, T_\ell$, as suggested in Figure 2.

   - The **preorder list** (or **preorder traversal**) of the nodes of T consists of the node r of T, followed by the nodes of T_1 in preorder, then the nodes of T_2 in preorder, and so on, up to the nodes of T_ℓ in preorder.
   - The **inorder list** (or **inorder traversal**) of the nodes of T consists of the nodes of T_1 in inorder, followed by the node r, followed by the nodes of $T_2, \ldots, T_\ell$, each group of nodes in inorder.
   - The **postorder list** (or **postorder traversal**) of the nodes of T consists of the nodes of T_1 in postorder, followed by the nodes of T_2 in postorder, and so on, up to the nodes of T_ℓ in postorder, and then all followed by the node r.

Figure 6 illustrates these four node orderings on a sample tree.

The level-order list of nodes of the tree in Figure 6 is obviously 1, 2, 3, 4, 5, 6, 7, 8, 9, 10.

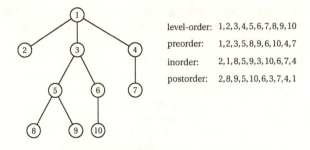

level-order: $1,2,3,4,5,6,7,8,9,10$

preorder: $1,2,3,5,8,9,6,10,4,7$

inorder: $2,1,8,5,9,3,10,6,7,4$

postorder: $2,8,9,5,10,6,3,7,4,1$

FIGURE 6. Four orderings of nodes in a tree.

For the preorder list, we first list 1, and then list in preorder each subtree rooted at 1's children 2, 3, and 4. The subtree rooted at 2 is a single node, so we simply list 2. Then we proceed to the second subtree of 1, i.e., the tree rooted at 3. We list 3, and then recursively list in preorder its subtrees rooted at 5 and 6. For the subtree rooted at 5, we list 5, and then recursively 8 and 9. For the subtree rooted at 6, we list 6, and then again recursively 10. Having done with the second subtree of the node 1, we proceed with its third subtree, i.e., the tree rooted at 4. We list 4, and then recursively list in preorder its subtrees. Since it has only one subtree, and that subtree is a single node, we finally list 7. Thus, the complete preorder list of the tree in Figure 6 is 1, 2, 3, 5, 8, 9, 6, 10, 4, 7.

Similarly, following the recursive definition, we find that the inorder list of the tree is 2, 1, 8, 5, 9, 3, 10, 6, 7, 4, and the postorder list is 2, 8, 9, 5, 10, 6, 3, 7, 4, 1. Notice that the order of the leaves in the three listings is always the same left-to-right ordering of the leaves. It is only the ordering of interior nodes and their relationship to the leaves that vary among the three listings.

As another example, consider the preorder, inorder, and postorder lists of nodes of an expression tree. These lists have certain syntactic meaning used primarily in lexical analysis of computer programs. The preorder list of an expression tree gives us what is known as the *prefix* notation of an expression, where the operator precedes its left and right operands. For example, the preorder list of nodes of the tree in Figure 5 is $* + ab - ac$.

Similarly, the postorder list of an expression tree gives us what is known as the *postfix* (or *Polish*) notation of an expression. For example, the postorder list of nodes of the tree in Figure 5 is $ab + ac - *$. Notice that no parenthesis are necessary in the prefix and postfix notation, as we can always deduce a correct subexpression by looking for the shortest prefix or suffix that is a legal prefix or postfix subexpression.

The inorder list of an expression tree gives us the *infix* notation we are used to, but with no parenthesis. For example, the inorder list of nodes of the tree in Figure 5 is $a + b * a - c$.

Let's now turn to more algorithmic implementation of all four types of node orderings of a tree. The procedure RT-LEVEL-ORDER(T) below implements the level-order traversal of nodes in a (rooted) tree T. We iteratively process nodes level by level, starting at the root and going top to bottom, left to right. In each step we maintain two queues Q_1 and Q_2, where Q_1 holds the nodes on the current level to be listed from left to right, and Q_2 holds their children on the next level. If $Q = \emptyset$ denotes the operation of making the queue Q empty and $Q_1 = Q_2$ the operation of copying Q_2 into Q_1, then the RT-LEVEL-ORDER procedure is given by the following pseudocode.

```
RT-LEVEL-ORDER(T)
    Q₁ = ∅;
    x = the root of T;
    ENQUEUE(Q₁, x);
    while QUEUE-EMPTY(Q₁) ≠ TRUE do
        Q₂ = ∅;
        repeat
            x = DEQUEUE(Q₁);
            write(x);
            for each child c of x, if any, in the left-to-right order do
                ENQUEUE(Q₂, c);
        until QUEUE-EMPTY(Q₁) = TRUE;
        Q₁ = Q₂;
    return;
```

Notice that we can implement the RT-LEVEL-ORDER procedure by using only one queue instead of two. This somewhat more efficient version is left as an exercise for the reader.

To implement the other node orderings of a tree, we closely follow their recursive definitions. The following are recursive procedures RT-PREORDER(x), RT-POSTORDER(x), and RT-INORDER(x) that take a node x as an argument and list nodes of the subtree rooted at x in the corresponding order.

```
RT-PREORDER(x)
    write(x);
    for each child c of x, if any, in the left-to-right order do
        RT-PREORDER(c);
    return;
```

```
RT-POSTORDER(x)
    for each child c of x, if any, in the left-to-right order do
        RT-POSTORDER(c);
    write(x);
    return;
```

```
RT-INORDER(x)
  if x is a leaf then
      write(x);
  else
      RT-INORDER(leftmost child of x);
      write(x);
      for each child c of x except for the leftmost, in the left-to-right order do
          RT-INORDER(c);
  return;
```

2. The abstract data structure TREE

Just as in the case of the basic data structures (lists, stacks, queues, sets), to treat trees as an abstract data structure, we need to give a representation of trees in programs and to implement a collection of useful operations on trees.

There is a great variety of operations that can be performed on trees. For example, besides the basic operations to add, remove, or find a child of a given node, we may have functions that return the parent, leftmost child, or right sibling of a given node. We might likewise want to get the node that is the root of a tree, as well as to traverse a tree in the preorder, inorder, postorder, or level-order. However, once we decide on a proper tree representation, these operations are easy to implement and so we won't go into further details about their actual implementation.

Consider then in more detail the possible ways to represent a tree using pointers and arrays. Perhaps the simplest representation of a tree, which naturally mirrors its logical structure, would be to have in each node a pointer to each child of the node. However, since we are to represent a general tree, i.e., the number of children per node may vary so greatly and is not bounded in advance, this approach might be infeasible because there would be too much wasted space. That's why we usually use two other tree representations that we discuss next.

2.1. The list-of-children representation of trees. One way to solve the problem of unknown number of children of a node in a tree is to build for each node a linked-list of its children. This approach is also important because it can be generalized to represent more complex structures such as graphs. Figure 7 suggests how the tree of Figure 6 may be represented in this way. There is a header array T, indexed by nodes, in which each element points to a linked list of nodes of a tree. The elements of the list to which $T(i)$ points to are the children of node n_i. Moreover, the left-to-right order of children is preserved by the order of list elements. For example, $T(5)$ points to the list $8 \rightarrow 9$ because nodes n_8 and n_9 are the children of node n_5.

One drawback with this representation is the inability to create large trees from smaller ones. The reason is that, while all trees can share one space for all linked lists of children (recall the multiple-lists representation from Section 3.8 in Chapter 3), each tree has its own header array for its nodes. Thus, to build

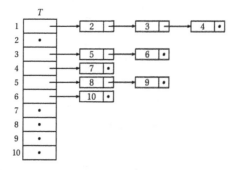

FIGURE 7. A list-of-children representation of the tree in Figure 6.

a new tree from a given node as the root and two given subtrees T_1 and T_2, we would have to copy T_1 and T_2 into a third tree and add a new node, the root, whose children are the roots of T_1 and T_2.

2.2. The leftmost-child, right-sibling representation of trees. Another solution to the problem of representing unbounded trees is given by the *leftmost-child, right-sibling representation* of each node. In this representation, each node keeps two pointers: one to its leftmost child (if it is not a leaf), and a second pointer to its right sibling (if it is not the rightmost sibling). This is illustrated in Figure 8 for the tree of Figure 4, where a downward arrow represents a pointer to the leftmost child, and a rightward arrow represents a pointer to the right sibling.

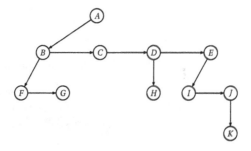

FIGURE 8. A pointer-based, leftmost-child, right-sibling representation of the tree in Figure 4.

The same tree of Figure 4 in an array-based, leftmost-child, right-sibling representation is shown in Figure 9.

All operations but parent-of type are straightforward to implement in the leftmost-child, right-sibling representation. If we do need to perform the parent-of operation efficiently, we can add another pointer to each node to indicate its parent directly in this representation.

	1	2	3	4	5	6	7	8	9	10	11	12	13	14	15
leftmost child	•	4	•	3	•		•			12	13	•	•		1
key	K	A	F	B	C		G			D	E	H	I		J
right sibling	•	•	7	5	10		•			11	•	•	15		•

FIGURE 9. An array-based, leftmost-child, right-sibling representation of the tree in Figure 4.

3. Binary trees

A **binary tree** is a tree in which no node can have more than two children. These children are called the **left child** and **right child**, because they are conventionally drawn extending, respectively, to the left and to the right of a node. In other words, a binary tree is a tree in which every node has either no children, a left child, a right child, or both a left and a right child.

Notice that the notion of the binary tree is somewhat different from the ordinary tree we defined in the previous section, because each child in a binary tree is designated as a left child or as a right child. This point is illustrated in Figure 10, with two different binary trees and one ordinary tree that looks very similar to both of the binary trees.

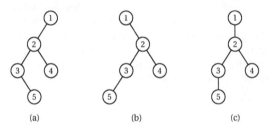

(a)   (b)   (c)

FIGURE 10. Two different binary trees and one ordinary tree.

Observe that the trees (a) and (b) of Figure 10 are not the same binary tree, nor is either in any sense comparable with the ordinary tree (c). For example, in the binary tree (a), the node 2 is the left child of 1, and 1 has no right child, while in the binary tree (b), 1 has no left child but has 2 as the right child. On the other hand, in the ordinary tree (c), there is no notion of left or right child—2 is just a child of 1, neither left nor right.

Despite of the small conceptual differences, the basic terminology of ordinary trees applies to the case of binary trees. All basic notions we defined for ordinary trees make sense for binary trees as well. Also, the standard node orderings of a binary tree are defined in the same spirit to those of an ordinary tree. For example, the inorder list of nodes of a binary tree T with root r, left subtree T_1, and right subtree T_2 is the inorder list of T_1 followed by r followed by the inorder list of T_2.

The representation of a binary tree is greatly simplified by the fact that we have an upper bound on the number of children of each node (which is, of course, two). For example, for an array-based representation of a binary tree with n nodes we can use two arrays L and R (or one array of records with two fields) of size n such that the elements $L(i)$ and $R(i)$ contain cursors to the left and right children of node n_i. A value of 0 in either element indicates the absence of a child.

A pointer-based representation of binary trees is even simpler: for each node we keep one pointer to its left child and one pointer to its right child (and one pointer to its parent if we wish). The absence of a child (or the parent) is indicated by the NIL pointer. Figure 11 illustrates these ideas.

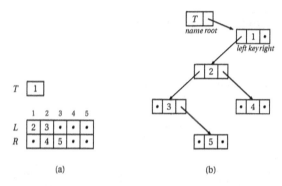

FIGURE 11. The array-based and pointer-based representations of the binary tree in Figure 10(a).

To make our discussion on efficiency of the binary tree operations more precise, we need to choose a concrete representation of binary trees. For the sake of concreteness therefore, we choose the pointer-based representation of binary trees. More precisely, a binary tree is organized as a linked data structure in which each node is represented by an object. The object contains a *key* field, which represents the key of a node, and pointer fields *left* and *right*, which point to objects representing the node's left and right children. If a child is missing, the appropriate field contains the value NIL. The object may contain some other satellite data, which is represented by appropriate fields.

The tree data structure has also a unique attribute object with a *name* field that contains the name of the tree, and a pointer field *root* that points to the root object. To simplify discussion, we often identify nodes with pointers that point to them. Thus, we say node x, meaning the object in the tree data structure pointed to by a pointer x. In addition, as in the case of list objects, we use the notation *field*(x) to refer to the field with name *field* of the object x, as well as to refer to the content of the same field. For example, given a (pointer to) node x, *left*(x) denotes the *left* field of x, and it also represents the pointer to the left child of the node (pointed to by) x. Likewise, given a tree name T, *root*(T) denotes the *root* field of the attribute object of the tree T. It also returns the pointer to the root object of the data structure representing the tree T. Thus, if *root*(T) = NIL, then the tree T is empty.

With this sort of tree representation we can, for example, write more precise procedures for the tree traversal techniques. In the special case of binary

trees, the preorder, inorder, and postorder traversal are greatly simplified, as illustrated in Figure 12. The numbers 1, 2, and 3 in the figure denote the order
in which we list the root and its left and right subtrees according to the corresponding type of traversal.

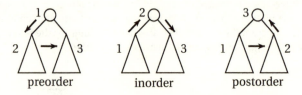

preorder   inorder   postorder

FIGURE 12. The binary tree traversals.

We present only the procedure for inorder traversal of binary trees and show
that its running time is proportional to the number of tree nodes. Other procedures are left as an easy exercise. The following procedure BT-INORDER(t) lists
the nodes of a binary subtree rooted at the node t in an inorder fashion. Clearly,
to list all nodes in the entire binary tree T, we just need to make the initial call
BT-INORDER($root(T)$).

```
BT-INORDER(t)
  if t ≠ NIL then
    BT-INORDER(left(t));
    write(key(t));
    BT-INORDER(right(t));
  return;
```

It is easy to show that inorder traversal of an n-node binary tree takes $\Theta(n)$
time. (In fact, this is true for any type of traversals that we defined.) This follows
from the fact that the BT-INORDER procedure is called exactly twice for each
node in the tree—once for its left child and once for its right child. Since the
time taken by the BT-INORDER procedure exclusive of its recursive calls is constant, the total time taken by inorder traversal is $2n \cdot \Theta(1) = \Theta(n)$. Arguing more
formally however, we can prove the following theorem.

THEOREM   *If x is the root of an n-node subtree, then the call* BT-INORDER(x)
takes $\Theta(n)$ *time.*

PROOF: Let $T(n)$ denote the time taken by the call BT-INORDER(x). If $n = 0$,
the call takes constant amount of time on an empty subtree (for the test whether
$x \neq$ NIL), and so $T(0) = c$ for some (small) positive constant c.

If $n > 0$, suppose that BT-INORDER procedure is called on a node x whose
left subtree has m nodes, and whose right subtree has consequently $n - m -$
1 nodes. Then $T(n) = T(m) + T(n - m - 1) + d$, where d is a positive constant
that accounts for the time taken by the call BT-INORDER(x) exclusive of the time
spent in recursive calls.

We claim that $T(n) = (c + d)n + c$ for $n \geq 0$, and argue by induction on n. For $n = 0$, the claim is clear since $T(0) = (c + d) \cdot 0 + c = c$. For $n > 0$, by induction hypothesis we can suppose that the claim is true for $T(m)$ and $T(n - m - 1)$ since $0 \leq m < n$ and $0 \leq n - m - 1 < n$. Then,

$$T(n) = T(m) + T(n - m - 1) + d$$
$$= \big((c + d) \cdot m + c\big) + \big((c + d) \cdot (n - m - 1) + c\big) + d$$
$$= (c + d) \cdot n + c - (c + d) + c + d$$
$$= (c + d)n + c. \blacksquare$$

One remark on our "programming style" is in order. We will often express simple algorithms, such as the BT-INORDER algorithm, in recursive form. Of course, any recursive algorithm can be implemented iteratively, at least by simulating recursion with a stack. In fact, in practice the iterative coding prevails over the recursive one, thus saving some of the overhead incurred by recursion. For example, the following is an iterative version of the same BT-INORDER algorithm.

```
BT-INORDER(t)
    S = ∅;   ⟦ initialize stack S ⟧
    repeat
        while t ≠ NIL do
            PUSH(S, t);
            t = left(t);
        t = POP(S);
        write(key(t));
        t = right(t);
    until (S = ∅) ∧ (t = NIL);
    return;
```

It should be rather easy to see how this works—the iterative version basically models recursion using the stack S. ($S = \emptyset$ denotes the operation of making the stack S initially empty.) However, we prefer recursion because the algorithms it produces, although less efficient (in an exact sense, not asymptotically), are almost always easier to understand and analyze.

4. Binary search trees

A **binary search tree** (or **BST** for short) is a binary tree in which each node x satisfies the following property: all nodes in the left subtree of x have keys less than the key of x, and all nodes in the right subtree of x have keys greater than the key of x. Here we make a simplified assumption that all the keys are mutually distinct. This property is called the **binary search tree property** (or **BST property**).

Assuming the pointer-based representation of a binary tree T, the BST property of the tree T can be expressed as follows. For every node x of T, if y is a node in the left subtree of x, then $key(y) < key(x)$; and if z is a node in the right subtree of x, then $key(x) > key(z)$.

Notice that in a binary search tree we implicitly assume that the keys of nodes are drawn from a totally ordered set. This is a set with "less than" relation defined for every pair of its members. Examples include integer and real numbers with usual less than order, as well as character strings with the lexicographic order.

Figure 13 illustrates two binary search trees that represent the same set of key values. As we shall see, the running time of most operations on a binary search tree is proportional to the height of the tree. In this sense, the tree on the right in Figure 13 is less efficient.

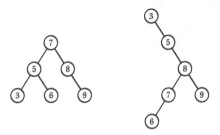

FIGURE 13. Two binary search trees representing the same set of keys.

4.1. BST operations. Typical operations on a binary search tree are finding, inserting, and deleting a node in the tree. Let us examine how these operations can be implemented.

4.1.1. *Searching a node.* First suppose that we want to search for a key k in a binary search tree T. If we compare k with the key of the root of T, we can take advantage of the BST property to locate k quickly, or determine that k is not present in T. If k is at the root of T, we are done. Otherwise, if k is less than the key of the root, by the BST property we know that k could be found only in the left subtree. And if k is greater than the key of the root, then it could be only found in the right subtree, again by the BST property.

Given a pointer t to some node in the binary search tree and given a key k, this informal description is implemented by the following recursive algorithm BST-SEARCH(t, k). It returns a pointer to a node with the key k, if one exists in the subtree whose root is t; otherwise, it returns NIL.

```
BST-SEARCH(t, k)
    if (t = NIL) ∨ (k = key(t)) then
        return t;
    else if k < key(t) then
        return BST-SEARCH(left(t), k);
    else
        return BST-SEARCH(right(t), k);
```

Of course, if we want to lookup up a key k in an entire binary search tree T, we should make the call BST-SEARCH$(root(T), k)$. Figure 14 shows an example of the BST-SEARCH algorithm in action.

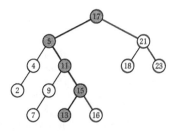

FIGURE 14. Searching for the node 13 in a sample tree.

As illustrated in Figure 14, the BST-SEARCH algorithm begins its search from the root and traces a path downward in the tree. For each node on the path, it compares the node's key with the searched value. If they are equal, the search terminates; otherwise, the search continues in the left or right subtree of the node depending on the test whether the node's key is greater or less than the searched value. The nodes encountered during the recursion form a path from the root down the tree to a leaf in the worst case, and so the running time of the BST-SEARCH algorithm is $O(h)$, where h is the height of the tree.

4.1.2. *Inserting a node.* The insertion and deletion operations modify a binary search tree, so we must be careful to make the changes in such a way that the BST property continuous to hold. Still, inserting a new node x into a binary search tree T is relatively straightforward.

To insert the new node x in T, we essentially try first to locate the x's key in T by applying the search procedure. If we find it, we are done, since we do not want to add a duplicate node violating our simplified assumption that T's keys are different. Otherwise, following the search procedure we eventually reach a node whose left or right child is empty indicating that x is not in T. It takes very little of convincing to see that this is always the right place to put the new node. So, we add x as a new leaf node to T and connect x as an appropriate child of the node that stopped the unsuccessful search.

Correctness of this procedure is probably easier to see in its recursive form (if you are used to recursive thinking). The following informal description incorporates the main idea. If T is an empty tree, it becomes a single node tree that solely consists of the new node x. If T is not empty, and its root has the key equal to the x's key, then x is already in the tree, hence we do nothing. On the other hand, if T is not empty and does not have the x's key at its root, we insert x into the left subtree of the root if its key is greater than the x's key, or insert x into the right subtree of the root if its key is less than the x's key.

The BST-INSERT algorithm below implements this sketch of the recursive insert operation. Given the pointer t to a node in a binary search tree and a pointer to the new node x, BST-INSERT(t, x) adds the node x to the subtree whose root is t. We assume that $key(x)$ has already been filled in and that $left(x) =$

right(*x*) = NIL. Again, we have to call BST-INSERT(*root*(*T*), *x*) if we want to insert the node *x* in the binary search tree *T*.

BST-INSERT(*t*, *x*)
  if *t* = NIL **then**
    t = *x*;
  else if *key*(*x*) < *key*(*t*) **then**
    left(*t*) = BST-INSERT(*left*(*t*), *x*);
  else if *key*(*x*) > *key*(*t*) **then**
    right(*t*) = BST-INSERT(*right*(*t*), *x*);
  else
    ⟦ Duplicate insertion is ignored ⟧;
  return *t*;

One technical difficulty in the recursive BST-INSERT algorithm is that when the new node replaces currently empty node, we need to back up and update one of the pointer fields in the parent of the new node. That's why we use the usual trick of passing pointer information up the tree by returning a pointer to the resulting subtree after insertion. In other words, the procedure returns a pointer to the root of the subtree with the newly added node. However, this also introduces one of the reasons for relative inefficiency of the recursive implementation because only the last node needs the pointer update. At all other nodes the procedure is returning the same pointer that is passed in, hence no real change is taking place.

Figure 15 shows how the BST-INSERT algorithm works on an example binary search tree. It begins from the root and follows a path down the tree just like in the case of searching. The path eventually ends with a node whose left or right child should host the new node. Thus, just like the search operation, the BST-INSERT algorithm takes $O(h)$ time on a tree of height h.

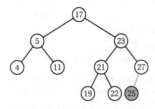

FIGURE 15. Inserting the node 25 into a sample tree.

4.1.3. *Deleting a node.* Deleting a node *x* from a binary search tree is a little more complicated than finding and inserting a node. In the first place, we may have to locate the node *x* in the tree, and if there is no such node, we are done, since we have nothing to remove. If *x* is a leaf, we can simply remove the leaf. However, if *x* is an interior node, we cannot remove that node, because we would disconnect the tree. Thus, in this case, we have to restructure the tree so that the BST property is maintained and yet *x* is removed.

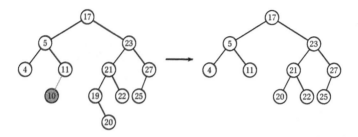

FIGURE 16. Removing the node 10 from a sample tree.

Figure 16 shows an example of removing a leaf node from a sample tree.

In general, if x is an interior node to be removed, we further consider two cases. First, if x has only one child, we can replace x by that child. That is, we can make that child be a new child of the parent of x, in place of the old child x. This general transformation is illustrated in Figure 17.

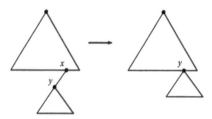

FIGURE 17. Removing the node x with one child from a binary search tree.

Figure 18 shows a concrete example of removing a one-child node from a sample tree.

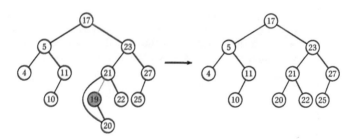

FIGURE 18. Removing the node 19 from a sample tree.

Second, if x has two children, one strategy is to find the smallest node y in the right subtree of x. (Another possibility is to find the largest node in the left subtree.) Then we replace x by y and remove y from the right subtree, as suggested in Figure 19.

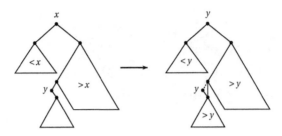

FIGURE 19. Removing the node x with two children from a binary search tree.

Observe that the BST property continuous to hold, because x is greater than any node in the left subtree, and so is y or any node from the right subtree. Thus, as far as the left subtree is concerned, y is a suitable node to replace x. But y is also a suitable node as far as the right subtree is concerned, because y is the smallest node in the right subtree.

Figure 20 illustrates the operation of removing a node with two children from a sample tree.

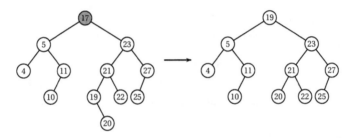

FIGURE 20. Removing the node 17 from a sample tree.

Before presenting pseudocode for the delete operation, we give first a utility procedure BST-DELETEMIN(t) that removes the smallest node from a nonempty subtree rooted at t. It also returns pointer to the deleted smallest node in a global variable d for the calling procedure to make use of it.

```
BST-DELETEMIN(t)
    if left(t) = NIL then
        d = t;
        return right(t);
    else
        left(t) = BST-DELETEMIN(left(t));
        return t;
```

In the BST-DELETEMIN procedure, we first locate the smallest node by following left children from t until we find a node y whose left child is NIL. Figure 21

suggests why such a node is the smallest one. Namely, following the path from
t to y, which is represented by the left edge of the subtree rooted at t, we know
that all ancestors of y on the path, as well as all the nodes in their right subtrees,
are larger than y. So are all the nodes in the y's right subtree, which exhaust all
the nodes different from y in the subtree rooted at t.

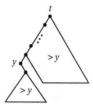

FIGURE 21. Finding the smallest node in a binary search tree.

Once we find the smallest node in the subtree rooted at t, it is straightforward to delete it because the smallest node has no left subtree, and so we always
have an easy case of deletion.

Now we present pseudocode for the algorithm BST-DELETE(t, k) that implements the procedure for deleting a node with the key k from a binary search
tree given by the pointer t to its root. We assume that we have available a global
variable d in which we store pointer to the deleted node so that the calling procedure has access to the node to recycle it. The variable d may be initialized
to NIL before calling BST-DELETE$(root(T), x)$ if we need to check after the call
whether the key k is really found in T.

```
BST-DELETE(t, k)
    if t ≠ NIL then [ search for k ]
    if k < key(t) then
        left(t) = BST-DELETE(left(t), k);
    else if k > key(t) then
        right(t) = BST-DELETE(right(t), k);
    else          [ k = key(t), remove t ]
        d = t;
    if left(t) = NIL then
        return right(t);
    else if right(t) = NIL then
        return left(t);
    else      [ neither child is NIL ]
        right(t) = BST-DELETEMIN(right(t));
        key(t) = key(d);
        [ copy also other satellite data ]
    return t;
```

The BST-DELETE algorithm begins again from the root of a binary search
tree and follows a path down the tree to first find the node to be deleted. It then

removes the node by applying appropriate strategy depending on the number of its children. The largest amount of work is required if the node has two children, in which case the algorithm follows a path further down the tree to find the smallest node in the right subtree of the node to be deleted. In all cases therefore, the path traversed is no longer than the height of the tree. Thus, like the search and insert operations, the BST-DELETE algorithm takes $O(h)$ time on a tree of height h.

4.2. Efficiency of the BST operations. We have argued that all three basic operations on a binary search tree run in $O(h)$ time, where h the height of the tree. But the height of a binary search tree varies, as nodes are inserted and deleted. Thus, to make our analysis more precise, we need to estimate the height of a typical binary search tree with n nodes.

In the worst case, all the nodes will be arranged in a single path. This would result, for example, when n keys are inserted in increasing order into an initially empty tree. Other degenerate cases include single paths that take turns left or right at some interior nodes. Figure 22 illustrates these worst-case scenarios.

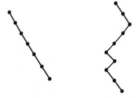

FIGURE 22. The "worst" binary search trees.

The height of all these n-node degenerate binary search trees is clearly $n-1$. We can thus conclude that the BST operations take $O(n)$ time in the worst case.

In the best case, a binary search tree represents a *full* binary tree. In such a tree, all leaves appear on the last level and all interior nodes have exactly two children. Figure 23 shows a full binary tree with 15 nodes.

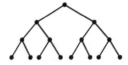

FIGURE 23. A full binary tree.

In a full binary tree of height h, an easy induction on the height shows that the number of nodes is $n = 2^{h+1} - 1$. Thus, in such a tree we have $n + 1 = 2^{h+1}$ or, equivalently, $h + 1 = \log(n + 1)$, i.e., $h = \log(n + 1) - 1$. In other words, the running time of the BST operations is $O(\log n)$ in the best case.

In practice, neither of these two cases is common, so the question arises where is the typical case between these two extremes. Is it closer to the linear or logarithmic time in the number of nodes? Fortunately, it can be shown that the BST operations typically take logarithmic time.

More precisely, one can show that the average height of a randomly built binary search tree with n nodes is $O(\log n)$. This means that we build a binary search tree from an empty tree by only inserting n nodes in a random order. That is, given a set of n (distinct) numbers, if we take a random permutation of these numbers and construct a binary search tree by inserting them one at a time, we end up with a binary search tree whose height is $O(\log n)$. Since the proof of this claim requires relatively advanced probabilistic argument, we present it in the next separate section that can be skipped on the first reading.

4.3. The expected height of binary search trees. Since the efficiency of the BST operations is in a direct proportion with the height of binary search trees, in this section we discuss their expected height in more detail. In general, the height clearly depends on the sequence of insertions and deletions used to build a BST. However, the structure of a BST can vary wildly with an arbitrary sequence of insertions and deletions, and so not much is known for the general case. Somewhat surprisingly, the situation is better if we allow only insertions.

Our goal therefore is to bound the expected height of a BST built by a random sequence of insertions only. The trees obtained in this way are called **random binary search trees**. More precisely, given a set $S = \{x_1, x_2, \ldots, x_n\}$ of $n \geq 1$ (distinct) keys, we take a random permutation of these keys, and then construct a random BST by inserting all the keys sequentially one at a time in the order of the permutation. Of course, to calculate the expected height of such a tree, we make a natural assumption on the distribution of the random permutations so that all $n!$ permutations are equally likely to be chosen.

An equivalent recursive definition of the random BST for n keys from S is the following. If $n = 1$, that is, S is a singleton set, then the tree consists of a single node whose key is the only key from S. Otherwise, if $n > 1$, we first choose a random key $x \in S$ (each key in S with the same probability $1/n$), create the root node with the key x, and then construct left and right subtrees of the root as random binary search trees for the sets $S_l = \{y \in S : y < x\}$ and $S_g = \{y \in S : y > x\}$, respectively. If S_l or S_g is empty, then the respective subtree does not exist (i.e., it has no node at all). Figure 24 illustrates this recursive construction of a random BST.

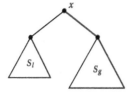

FIGURE 24. A random binary search tree for n keys from S.

Figure 25 displays all binary search trees for three keys, each with its probability to occur as a random BST.

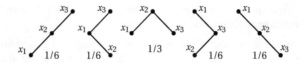

FIGURE 25. Random binary search trees for $x_1 < x_2 < x_3$.

Since the height of a tree is the largest depth of a leaf node in the tree, what we are after in this setting is the expected maximum depth of leaves in a random BST for n keys. Before we get to an estimate of the expected height, let's first introduce some notions. First, define the ***rank*** of a key $x \in S$ to be its position in the sorted list of the keys from S in increasing order. In other words, the key of rank k is the k-th smallest key in S. More formally, the rank of $x \in S$, denoted $r(x)$, is the number of elements in S less than x plus 1, i.e.,

$$r(x) = |\{y \in S : y < x\}| + 1.$$

Next, for $n = 1, 2, \ldots$ and $i = 1, 2, \ldots, n$, let $Y_{i,n}$ be the random variable for the depth of the key of rank i in a random BST for n keys. Also, let $I_{i,n}$ be the indicator random variable that indicates whether the key of rank i in a random BST for n keys is a leaf, i.e.,

$$I_{i,n} = \begin{cases} 1, & \text{the key of rank } i \text{ is a leaf in a random BST for } n \text{ keys} \\ 0, & \text{otherwise}. \end{cases}$$

Then

$$X_n = \max_{1 \le i \le n} I_{i,n} \cdot Y_{i,n}$$

is the random variable for the height of a random BST for n keys.

Now, to calculate $\mathbb{E}(X_n)$, the idea is to use the recursive definition of a random BST and get an easy recurrence equation for $\mathbb{E}(X_n)$ in terms of the expected heights of the smaller subtrees of the root. However, it's not clear how to do this directly, since the height of the whole tree is one larger than the greater height of the heights of the subtrees.

That's why we consider another random variable Z_n that represents the sum of the exponential depths of the leaves of a random BST for n keys. This way we will be able to work with a random variable involving a sum, rather than with a random variable involving a maximum. Anyway,

$$Z_n = \sum_{i=1}^{n} I_{i,n} \cdot 2^{Y_{i,n}},$$

and, for convenience, we define $Z_0 = 0$. Clearly, $\mathbb{E}(Z_1) = 1$ and $\mathbb{E}(Z_2) = 2$.

Let's see if $\mathbb{E}(Z_n)$ is easier to get. Given that the root of a random BST for $n > 1$ keys has rank k, we know that the left subtree of the root contains the $k - 1$ keys smaller than the root, and the right subtree contains the $n - k$ keys larger than the root. The depth of each leaf in the subtrees is exactly one less than the depth of the node viewed as a leaf in the whole tree. Moreover, the structure of both random subtrees is independent of the particular keys and depends only on their number. Namely, having chosen the root of rank k, the left subtree is randomly built on the $k - 1$ keys whose ranks are less than k. This subtree is just like any other randomly built BST on $k-1$ keys. Similarly, the right subtree is randomly built on the $n-k$ keys whose ranks are greater than k. Again, its structure is just like the structure of any other randomly built BST on $n-k$ keys. Hence, for a fixed k, $(1 \le k \le n)$, and taking an empty sum to be 0, we can write

$$Z_n \mid \text{the root has rank } k = \sum_{i=1}^{k-1} I_{i,n} \cdot 2^{Y_{i,n}} + \sum_{i=k+1}^{n} I_{i,n} \cdot 2^{Y_{i,n}}$$

$$= \sum_{i=1}^{k-1} I_{i,k-1} \cdot 2^{Y_{i,k-1}+1} + \sum_{i=1}^{n-k} I_{i,n-k} \cdot 2^{Y_{i,n-k}+1}$$

$$= 2 \sum_{i=1}^{k-1} I_{i,k-1} \cdot 2^{Y_{i,k-1}} + 2 \sum_{i=1}^{n-k} I_{i,n-k} \cdot 2^{Y_{i,n-k}}$$

$$= 2(Z_{k-1} + Z_{n-k}).$$

Since any key from the n keys has the same probability $1/n$ to be the root of a random BST, for a fixed k between 1 and n we have that $\mathbb{P}\{\text{the root has rank } k\} = 1/n$. If we additionally make use of an important property of the expectation called the *linearity of expectation*, which states that expected value of a (finite) sum of random variables is the sum of expected values of these variables, we can write

$$\mathbb{E}(Z_n) = \sum_{k=1}^{n} \mathbb{E}(Z_n \mid \text{the root has rank } k) \cdot \mathbb{P}\{\text{the root has rank } k\}$$

$$= \sum_{k=1}^{n} 2\big(\mathbb{E}(Z_{k-1}) + \mathbb{E}(Z_{n-k})\big) \cdot \frac{1}{n}$$

$$= \frac{2}{n} \sum_{k=1}^{n} \big(\mathbb{E}(Z_{k-1}) + \mathbb{E}(Z_{n-k})\big)$$

$$= \frac{4}{n} \sum_{i=0}^{n-1} \mathbb{E}(Z_i).$$

Putting $z_n = \mathbb{E}(Z_n)$, we get $z_0 = 0$, $z_1 = 1$, and, for $n \ge 2$, the recurrence equation

(2) $$z_n = \frac{4}{n} \sum_{i=0}^{n-1} z_i.$$

How can we find the solution to this recurrence equation? For $n \ge 2$, let's first multiply the equation (2) by n to get

(3) $$n z_n = 4 \sum_{i=0}^{n-1} z_i.$$

Substituting $n-1$ for n in (3) we have, for $n \ge 3$,

(4) $$(n-1) z_{n-1} = 4 \sum_{i=0}^{n-2} z_i.$$

Next, subtracting (4) from (3) yields

$$n z_n - (n-1) z_{n-1} = 4 \sum_{i=0}^{n-1} z_i - 4 \sum_{i=0}^{n-2} z_i$$

$$= 4 z_{n-1}.$$

Consequently, for $n \geq 3$, $nz_n = (n-1)z_{n-1} + 4z_{n-1} = (n+3)z_{n-1}$, or

$$z_n = \frac{n+3}{n}z_{n-1},$$

that is,

(5) $$z_n = \frac{(n+3)(n+2)(n+1)}{(n+2)(n+1)n}z_{n-1}.$$

Now, repeatedly substituting the general formula (5) for $z_{n-1}, z_{n-2}, \ldots, z_3$ in the equation for z_n in (5), we get

$$
\begin{aligned}
z_n &= \frac{(n+3)(n+2)(n+1)}{(n+2)(n+1)n} \cdot \frac{(n+2)(n+1)n}{(n+1)n(n-1)}z_{n-2} \\
&= \frac{(n+3)(n+2)(n+1)}{(n+1)n(n-1)}z_{n-2} \\
&= \frac{(n+3)(n+2)(n+1)}{(n+1)n(n-1)} \cdot \frac{(n+1)n(n-1)}{n(n-1)(n-2)}z_{n-3} \\
&= \frac{(n+3)(n+2)(n+1)}{n(n-1)(n-2)}z_{n-3} \\
&\;\;\vdots \\
&= \frac{(n+3)(n+2)(n+1)}{5 \cdot 4 \cdot 3}z_2 \\
&= \frac{(n+3)(n+2)(n+1)}{5 \cdot 4 \cdot 3} \cdot 2 \\
&= \frac{(n+3)(n+2)(n+1)}{30}.
\end{aligned}
$$

Therefore, $\mathbb{E}(Z_n) = \Theta(n^3)$, that is, we have found the expected sum of the exponential depths of the leaves of a random BST. We are not done yet though, since we seek an upper bound on the expected height. To get that, first observe

$$2^{\max_{1 \leq i \leq n} I_{i,n} \cdot Y_{i,n}} \leq \sum_{i=1}^{n} I_{i,n} \cdot 2^{Y_{i,n}},$$

that is, the power of the maximum depth among all leaves is not greater than the sum of the powers of the depths of all leaves. In other words, $2^{X_n} \leq Z_n$, hence

(6) $$\mathbb{E}(2^{X_n}) \leq \mathbb{E}(Z_n).$$

Second, we use Jensen's inequality stating that for any convex function $f(x)$ we have $\mathbb{E}(f(X)) \geq f(\mathbb{E}(X))$, provided the expectations exist. (A real function $f(x)$ is **convex** if for all a and b and every λ such that $0 \leq \lambda \leq 1$ it holds $f(\lambda a + (1-\lambda)b) \leq \lambda f(a) + (1-\lambda)f(b)$.) Then, since $f(x) = 2^x$ is a convex function, we get

(7) $$\mathbb{E}(2^{X_n}) \geq 2^{\mathbb{E}(X_n)}.$$

Consequently, taking logarithm of (7) and combining it with (6), we get

(8) $$\mathbb{E}(X_n) \leq \log \mathbb{E}(2^{X_n}) \leq \log \mathbb{E}(Z_n).$$

Since we have already shown $\mathbb{E}(Z_n) = \Theta(n^3)$, from (8) it follows finally that $\mathbb{E}(X_n) = O(\log n)$.

5. Binary heaps

Binary heaps are binary trees in which the nodes satisfy a special property, just like in the binary search trees. Unlike binary search trees however, the corresponding binary trees that make binary heaps must also have a special shape and be as full (balanced) as possible. More precisely, we call a binary tree **complete** if it has nodes on all levels except possibly the lowest. On the lowest level, where some leaves might be missing, we require that all leaves that are present are to the left of all missing leaves. Figure 26 shows the general shape and a concrete example of a complete binary tree.

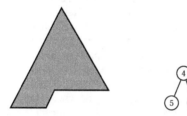

FIGURE 26. A complete binary tree.

A **binary heap** (or simply a **heap**) is a complete binary tree in which the key of each node is less than the keys of the children of the node. In other words, if each node includes the *key* field that stores the key of the node, then $key(x) < key(y)$ and $key(x) < key(z)$, where y and z are children of x. This property is called the **heap property**.[1]

Notice that the heap property implies that the key of every node x in a heap is less than the keys of all nodes in the subtree rooted at the node x. Also, the smallest node (i.e., the node having the minimal key) must be at the root of a heap, but smaller nodes need not appear at lower levels than larger nodes. For example, in the heap in Figure 26 nodes 4, 5, and 6 appear at higher level than node 7. Another observation is that nodes on every path from some node down the tree appear in sorted increasing order.

In fact, heaps that satisfy the heap property as defined previously are called min-heaps. We could appropriately define max-heaps in which the key of each node is greater than the keys of its children. There is nothing special about favoring one form to another. They are both used in practice and all considerations with obvious modifications apply equally well in both cases.

5.1. Heap operations. Typical operations on a heap include deleting the minimum node (and returning its key), inserting a new node into the heap, and initially making a heap out of a complete binary tree whose nodes appear in arbitrary order. Let's examine how these operations can be implemented.

---

[1]Note that the heap property makes sense for an ordinary (non-binary) tree as well. That is, we can say that the condition is met if the key of each node is less than the keys of *all* children of the node.

5.1.1. *Deleting the minimum node.* Returning the key of the smallest node is straightforward, because we observed that it must be at the root. However, if we simply remove the root, we no longer have a tree. Thus, we must somehow find a new key for the root, keeping the tree almost full and preserving the heap property. This is achieved in two phases. First, to maintain the tree as full as possible, we take the key of the rightmost leaf on the lowest level, remove the leaf, and temporary put its key at the root. The resulting tree remains almost full, but the heap property of the new root may be violated. So in the second phase, to preserve the heap property, we move the new key of the root down the tree to its proper place. This means that we push it as far down as it will go, by exchanging it with the key of one of the current node's children having smaller key. More precisely, suppose the current node is a violator of the heap property. This means that the current node is greater than one or both of its children. We can exchange it with one of its children, but we must be careful which one. If we exchange the violator with the smaller of its children, then we are sure not to introduce a violation of the heap property between two former children, one of which has now become the parent of the other. Doing this in turn for the initial key, it arrives eventually either at a leaf or at a node whose both children have larger keys. Figure 27 illustrates this process of deleting the minimum node from the heap of Figure 26.

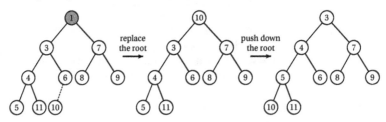

FIGURE 27. Deleting the minimum node from a heap.

Observe that the delete operation takes $O(\log n)$ time if it is applied on an n-node heap. This follows from the fact that the process of forcing the temporary root key down the tree is repeated at most h times, where h is the height of the tree. Since $h \le \log(n + 1)$ in an n-node complete binary tree, and each step in the process takes a constant time per node, the total time is $O(\log n)$.

5.1.2. *Inserting a node.* Let us now consider how we can insert a new node into a binary heap, keeping in mind that the resulting heap should be as full as possible and that the heap property should be invariant. One strategy is to place the new node on the lowest level as far left as possible, starting a new level if the current lowest level is all filled. This keeps the tree balanced, but again the heap property of the new node's parent may be violated. To restore it, we push the new node up the tree, by exchanging it with its parent if the parent has smaller key. We repeat this process until the new node arrives either at the root or has larger key than its parent. Figure 28 shows how node 2 is inserted into the heap of Figure 26.

The time taken by the insert operation is proportional to the length of the path that the new node travels from its initial place on the lowest level to its

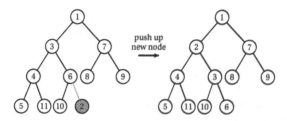

FIGURE 28. Inserting a new node into a heap.

proper place. Clearly, the distance is bounded by the height of the heap, and so the time is $O(\log n)$ for an n-node heap.

5.1.3. *Making a heap.* Suppose that the node keys in a complete binary tree appear in an arbitrary order, i.e., the tree does not satisfy the heap property. Our task is to "heapify" the tree so that it becomes a heap. An obvious method to establish the heap property is to imagine that, starting with an empty heap, we are inserting into it the nodes one by one. (Here we can go by taking the nodes from, say, the level-order list of nodes of the given tree.) However, this approach is not very efficient since for an n-node tree it takes $O(n \log n)$ time. This is so because we execute n insertions that take $O(\log n)$ time each. There is a more clever solution that takes only $O(n)$ time, which we discuss in the next section in which we consider a concrete implementation of a heap.

5.2. Array implementation of a heap. Since a heap is an almost full binary tree, it can be efficiently represented in an unusual way as an array. The idea is borrowed from the level-order list of nodes of a tree. That is, in this representation we are filling the array with heap nodes level by level, starting from the root and going from left to right within a level. Specifically, the first element of the array holds the root, the second element holds its left child, the third element holds its right child, the fourth element holds the left child of the left child of the root, and so on. For example, Figure 29 shows how a heap with 10 nodes is represented by an array A.

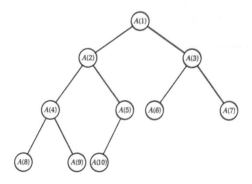

FIGURE 29. The array implementation of a 10-node heap.

Thus, for $i \geq 1$, the left child of node in $A(i)$, if it exists, is at $A(2i)$, and its right child, if it exists, is at $A(2i + 1)$. An equivalent interpretation is that the parent of node $A(i)$ is $A(\lfloor i/2 \rfloor)$ for $i > 1$.

To simplify discussion, we assume that the keys of the heap nodes are integers. We also ignore any additional data the nodes might contain so that the heap can be represented merely as an integer array. In practice however, if each node includes satellite data, we can represent a heap by an array of records of appropriate type. Better yet, since all heap operations rely on a lot of data swapping, we would represent a heap by an array of pointers to static records, so that we need to move only these pointers. Therefore, in the discussion that follows we suppose that our heap is represented by an integer array A exactly as given in Figure 29. The actual size of the heap is kept in a global variable n, in order to avoid the parameter list clutter of the discussed procedures. Moreover, we take it for granted that the array size is always big enough to accommodate an n-node heap, and so we omit any explicit bound checking for overflow errors.

We saw that the insert operation is based on the process of moving the new element up the tree toward its right place. We say that the new node *bubbles up* to its right position in the tree. The following procedure HEAP-BUBBLE-UP(A, i) recursively bubbles up the node $A(i)$ in the heap represented by the array A.

HEAP-BUBBLE-UP(A, i)
  $j = \lfloor i/2 \rfloor$;   〚 j is the parent of i 〛
  if $(j > 0) \wedge (A(i) < A(j))$ **then**
    SWAP$(A(i), A(j))$;
    HEAP-BUBBLE-UP(A, j);
  return;

Using this algorithm, the following procedure HEAP-INSERT(A, x) implements the insert operation that adds the new node x to an n-node heap represented by the array A.

HEAP-INSERT(A, x)
  $n = n + 1$; $A(n) = x$;
  HEAP-BUBBLE-UP(A, n);
  return;

The delete operation moves the temporary root node down the tree toward its right place. We say that the temporary node has been *sifted down* to its right position in the tree. The following is recursive procedure HEAP-SIFT-DOWN(A, i) that sifts down the node $A(i)$ in an n-node heap represented by the array A.

```
HEAP-SIFT-DOWN(A, i)
    j = 2i;      [[ j is the left child of i ]]
    if (j < n) ∧ (A(j + 1) < A(j)) then
        j = j + 1;
    [[ j is the smaller child of i, move the child up? ]]
    if (j ≤ n) ∧ (A(i) > A(j)) then
        SWAP(A(i), A(j));
        HEAP-SIFT-DOWN(A, j);
    return;
```

It is now easy to write the procedure HEAP-DELETEMIN(A) that in an n-node min-heap represented by the array A deletes the root node and returns its key.

```
HEAP-DELETEMIN(A)
    x = A(1);
    A(1) = A(n);  n = n - 1;
    HEAP-SIFT-DOWN(A, 1);
    return x;
```

It remains to see how we can create an initial heap from an array whose elements appear in arbitrary order. We mentioned an obvious, albeit inefficient solution, which starts with an empty heap and inserts into it the array elements one by one. We leave it as an exercise for the reader.

A better strategy is to successively make bigger heaps starting from the nodes on the last level. Those are all leaves and each represent a heap by itself. Then we make heaps from the subtrees whose roots are the nodes on the level immediately above the bottommost level. For this, in fact, we need to sift only the roots down. The subtrees rooted at the nodes on the next upper level are then made into heaps by sifting down their roots. This is enough since their left and right subtrees are already heaps. By repeating these steps level by level in the bottom-up direction until we get to the global root, we eventually build entire heap. Figure 30 illustrates the process of making a heap in this fashion.

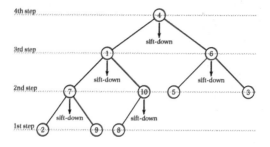

FIGURE 30. Making an initial heap.

This idea is implemented by the following, remarkably simple and efficient procedure HEAP-MAKE(A) that makes a heap out of the n elements of the array A appearing in an arbitrary order.

> HEAP-MAKE(A)
> **for** $i = \lfloor n/2 \rfloor$ **downto** 1 **do**
> HEAP-SIFT-DOWN(A, i);
> **return;**

The running time of the HEAP-MAKE algorithm is $O(n)$. To prove this a somewhat surprising result, we will calculate a tight bound on the number of times the HEAP-SIFT-DOWN algorithm is called. Since the work done by the HEAP-SIFT-DOWN algorithm on a single node, exclusive of recursive calls, is constant, in this way we will estimate the time taken by the HEAP-MAKE algorithm.

Without loss of generality, we consider the worst case when the heap is given by a full binary tree of height h for some $h \geq 0$. Equivalently, the size of the array A that represents the heap equals $n = 2^{h+1} - 1$. Furthermore, although the HEAP-MAKE algorithm does not try to sift down the leaves of the tree, we will assume that it is not optimized to skip this redundant processing. In other words, this is as if the **for** loop started from n instead of $\lfloor n/2 \rfloor$.

Then, the HEAP-SIFT-DOWN algorithm is first called once on 2^h leaves. Next, on 2^{h-1} nodes on the level immediately above the last level, HEAP-SIFT-DOWN algorithm is called at most twice, since in this phase it sifts down the roots of subtrees whose height is one. On the next level above, which contains 2^{h-2} nodes, for each of these nodes the HEAP-SIFT-DOWN algorithm is called at most three times, since now it sifts down the roots of subtrees whose height is two. Continuing in this way, when we eventually get to the global root of the entire tree, we see that the HEAP-SIFT-DOWN algorithm is called at most $h + 1$ times, since finally it sifts down the root of the entire tree whose height is h. Figure 31 illustrates this argument and shows the node sections of the array A with corresponding number of times the HEAP-SIFT-DOWN algorithm is called on subtrees rooted at these nodes.

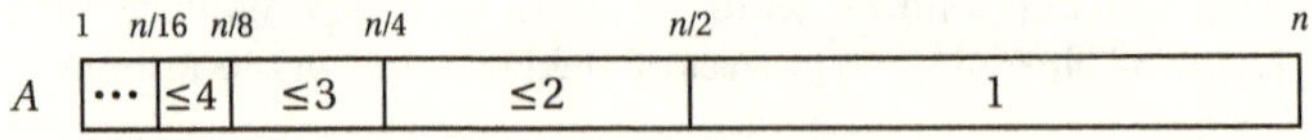

FIGURE 31. Counting the number of HEAP-SIFT-DOWN calls.

Therefore, if $S(h)$ denotes the total number of the HEAP-SIFT-DOWN calls for a full binary tree of height h, then

$$S(h) \leq 2^h \cdot 1 + 2^{h-1} \cdot 2 + 2^{h-2} \cdot 3 + \cdots + 2^1 \cdot h + 2^0 \cdot (h+1)$$
$$= \sum_{i=1}^{h+1} 2^{h+1-i} \cdot i$$
$$= 2^{h+1} \sum_{i=1}^{h+1} \frac{i}{2^i}.$$

To bound the last sum, we bound its infinite counterpart $\sum_{i=1}^{\infty}(i/2^i)$:

$$\sum_{i=1}^{\infty}\frac{i}{2^i} = \frac{1}{2} + \frac{2}{4} + \frac{3}{8} + \frac{4}{16} + \cdots$$

$$= \frac{1}{2} + \left(\frac{1}{4} + \frac{1}{4}\right) + \left(\frac{1}{8} + \frac{1}{8} + \frac{1}{8}\right) + \left(\frac{1}{16} + \frac{1}{16} + \frac{1}{16} + \frac{1}{16}\right) + \cdots .$$

Next, to bound this infinite sum, we first rearrange it in a "triangle sum", then sum the rows, and finally sum the last column following the equality signs:

$$\frac{1}{2} + \frac{1}{4} + \frac{1}{8} + \frac{1}{16} + \frac{1}{32} + \cdots = 1$$

$$\frac{1}{4} + \frac{1}{8} + \frac{1}{16} + \frac{1}{32} + \cdots = \frac{1}{2}$$

$$\frac{1}{8} + \frac{1}{16} + \frac{1}{32} + \cdots = \frac{1}{4}$$

$$\frac{1}{16} + \frac{1}{32} + \cdots = \frac{1}{8}$$

$$\vdots$$

Thus,

$$\sum_{i=1}^{\infty}\frac{i}{2^i} = \sum_{i=0}^{\infty}\frac{1}{2^i} = 2,$$

and so

$$\sum_{i=1}^{h+1}\frac{i}{2^i} \le \sum_{i=1}^{\infty}\frac{i}{2^i} = 2.$$

Consequently, the running time of the (slower) HEAP-MAKE algorithm is

$$T(n) = S(h) \cdot O(1)$$
$$\le 2^{h+1} \cdot 2 \cdot O(1)$$
$$= 2(n+1) \cdot O(1)$$
$$= O(n),$$

as desired.

Exercises

1. Answer the following questions about the tree in Figure 32.

   (a) Which nodes are leaves?
   (b) Which node is the root?
   (c) Which node is the parent of node C?
   (d) Which nodes are children of node C?
   (e) Which nodes are ancestors of node E?
   (f) Which nodes are descendants of node E?
   (g) What is the depth of node C?
   (h) What is the height of node C?

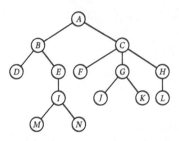

FIGURE 32

2. List the nodes of the tree in Figure 32 in level-order, preorder, postorder, and inorder.

3. Draw tree representation for the expression $\big((a+b)+c*(d+e)+f\big)*(g+h)$.

4. Prove the following claims about trees:

   (a) Every tree with n nodes has $n-1$ edges.
   (b) The ***degree*** of a node in a tree is the number of children it has. In any binary tree, the number of leaves is one more than the number of nodes of degree two.
   (c) A binary tree is a ***proper*** binary tree if every node has either zero or two children. If such a tree has n non-leaf nodes, then it has $n+1$ leaves. Conclude that any proper binary tree has an odd number of nodes.
   (d) Recall that a full binary tree has leaves only on the last level, and every non-leaf node has exactly two children. If the height of a full binary tree is h, then the number of nodes is $2^{h+1}-1$.

5. Give the RT-LEVEL-ORDER algorithm from Section 1.2 so that it uses only one queue instead of two.

6. If a general tree is represented by the leftmost-child, right-sibling, and parent pointers, give nonrecursive preorder, postorder, and inorder traversal algorithms.

7. Assuming the pointer-based representation of binary trees, give preorder and postorder traversal algorithms for binary trees.

8. For each of the following key sequences determine the binary search tree obtained when the keys are inserted into an initially empty tree, one by one in the given order.

   (a) $1,2,3,4,5,6,7$.
   (b) $4,2,1,3,6,5,7$.
   (c) $1,6,7,2,4,3,5$.

9. For each of the binary search trees obtained in the previous exercise, deter-mine the binary search tree obtained when the root is removed.

10. Suppose we are given a sorted sequence of n keys $x_1 \le x_2 \le \cdots \le x_n$, to be inserted into a binary search tree. For simplicity, assume n is a power of 2.

   (a) What is the minimum height of a binary tree that contains n nodes?
   (b) Devise an algorithm to insert the given keys into a binary search tree so that the height of the resulting tree is minimized.
   (c) What is the running time of your algorithm?

11. Suppose that each node in a binary search tree is augmented with a pointer to its parent. Write an efficient algorithm BT-SUCCESSOR(x) that returns the node containing the successor key of the node x's key in the sorted order of all keys. What is the running time of your algorithm? (*Hint:* The successor of x is the minimum node of the x's right subtree, if the subtree exists; other-wise, it is the lowest (first) ancestor of x whose left child is also an ancestor of x.)

12. Under the same assumptions as in the previous exercise, write an efficient al-gorithm BT-PREDECESSOR(x) that returns the node containing the predeces-sor key of the node x's key in the sorted order of all keys. What is the running time of your algorithm?

13. Give an efficient algorithm to select the k-th smallest key in a binary search tree. For example, given a tree with n nodes, $k = 1$ selects the smallest key, $k = n$ selects the largest key, and $k = \lceil n/2 \rceil$ selects the median key.

14. Let $H_n = \sum_{i=1}^{n} \frac{1}{i}$ be the n-th *harmonic number*. It is well-known that $H_n = \ln n - \gamma + o(1)$, where $\gamma = 0.5772\ldots$ is the Euler's constant.

   (a) Show that the expected depth of the smallest key in a random BST is $H_n - 1$.
   (b) Show that the expected depth of the k-th smallest key in a random BST is $H_{n-k+1} + H_k - 2$.
   (c) Conclude that the expected depth of any key in a random BST is $O(\log n)$.

15. Prove that the function $f(x) = 2^x$ is convex.

16. For each of the following key sequences determine the heap obtained when the keys are inserted into an initially empty heap, one by one in the given or-der.

   (a) $0, 1, 2, 3, 4, 5, 6, 7, 8, 9.$
   (b) $3, 1, 4, 1, 5, 9, 2, 6, 5, 4.$
   (c) $2, 7, 1, 8, 2, 8, 1, 8, 2, 8.$

17. For each of the heaps obtained in the previous exercise, determine the heap obtained after deleting the minimum key.

18. Assuming an array implementation of a heap, devise an algorithm to find the *largest* key in a min-heap. What is the running time of your algorithm? (*Hint:* First show that the largest key must be in one of the leaves.)

19. To sort a sequence of n numbers $x_1, x_2, \ldots, x_n$ in increasing order, apply the following algorithm that works in two phases.

> **1. Insertion phase:** Insert the numbers $x_1, x_2, \ldots, x_n$, in this order, in a binary search tree. (The tree is initially empty.)
>
> **2. Listing phase:**   List the nodes of the final tree.

   (a) In what order should the nodes be listed in the listing phase: level-order, preorder, postorder, or inorder?
   (b) What is the *average* time complexity of the insertion phase?
   (c) What is the time complexity of the listing phase?
   (d) What is the total *average* time complexity of the algorithm?
   (e) If the input sequence is already sorted initially, what is the total running time of the algorithm?
   (f) If we use a binary heap instead if a binary search tree, what is the time complexity of the insertion phase?
   (g) How should the listing phase be modified in the case of a binary heap?
   (h) What is the time complexity of this modified listing phase?

20. Given an array A of n numbers, describe the **heapsort** algorithm that sorts A in increasing order using the heap data structure. What is the running time of the heapsort?

5

Divide-and-Conquer Algorithms

The *divide-and-conquer* method (probably one of the oldest military strategies!) applied to the design of algorithms may be informally described as follows: given a problem to be solved, break it into several simpler subproblems, independently solve each of the subproblems, and then combine their solutions into a solution to the original problem. If the subproblems are similar to the original, then we may be able to use the same procedure to solve the subproblems recursively. Of course, as for any recursive algorithm, after a finite number of subdivisions we must encounter a trivial subproblem that can be solved outright. Otherwise, a recursive algorithm will continue subdividing the problem forever, without finding a solution.

Put it another way, the divide-and-conquer approach involves three steps at each level of the recursion:

1. **Divide** the problem into a number of independent subproblems;
2. **Conquer** the subproblems by solving them recursively;
3. **Combine** the solutions to the subproblems into a solution to the original problem.

In this chapter we show how to design and analyze divide-and-conquer algorithms through a number of applications, starting with a simple game and ending with the problem of multiplying large integers. We also consider some techniques for solving recurrence equations that describe the running time of the divide-and-conquer algorithms.

1. Towers of Hanoi

The Towers of Hanoi problem is a simple puzzle in which initially n disks of mutually different sizes are placed on one of three poles in decreasing order from the bottom. The disks are all to be moved one at a time to another pole using the third one as a spare. In addition, only the top disk of one pole may be moved to the top of other pole, and no disk may be placed on top of a smaller one. Figure 1 shows the initial configuration of the game for $n = 4$ if we imagine that the poles are arranged in the corners of a triangle.

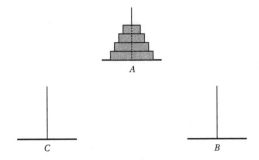

FIGURE 1. Triangle arrangement of the poles.

More precisely, the Towers of Hanoi game consists of three poles A, B, C and n disks of different size (radius) $d_1 < d_2 < \cdots < d_n$. Initially, all the disks are stacked on the pole A in decreasing order, i.e., the largest disk d_n is at the bottom and the smallest disk d_1 is on the top. The goal of the game is to move all the disks to the pole C subject to the following two rules:

1. Only the top disk may be removed from one pole and placed onto the top of another pole;
2. No larger disk is above a smaller one at any time.

Note that these rules imply that only one disk can be moved at a time, and all disks at all times must rest on the poles.

The game can be solved by the following simple algorithm. On odd-numbered moves, move the smallest disk to the next pole clockwise. On even-numbered moves make the only legal move not involving the smallest disk. This algorithm is concise and correct, but it is hard to understand why it works and even harder to come up with it.

Consider instead the following divide-and-conquer approach. The problem of moving n disks from A to C using B as a spare can be reduced to two similar subproblems for $n - 1$ disks. The first one is moving the $n - 1$ smallest disks from A to B using C as a spare, and the second one is moving the $n - 1$ smallest disks from B to C using A as a spare. This reduction makes the **divide** part of the divide-and-conquer method. In the **conquer** part, solutions to the two subproblems of moving the $n-1$ smallest disks are obtained recursively. Notice that in this case we don't need to do much work in the divide step, because the first subproblem is determined simply by a proper choice of the arguments of the recursive procedure, and the second subproblem is automatically constructed by solving the first one. In general however, constructing subproblems of an original problem may require more complicated processing. The **combine** step for the Towers of Hanoi problem is now easy. We use solution to the first subproblem to move the $n-1$ smallest disks from pole A to pole B using pole C as a spare, and expose the largest disk on pole A. Then we move that largest disk from A to C. Finally, we use solution to the second subproblem to move the $n-1$ smallest disks from B to C using A as a spare. Clearly, this way we will have solved the original problem for n disks. Figure 2 illustrates the results of each of the main steps in the solution, starting from the initial configuration for $n = 5$ disks.

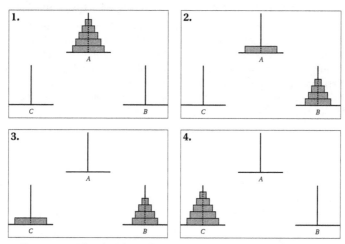

FIGURE 2. The devide-and-conquer solution to the Towers of Hanoi game.

Notice how this recursive algorithm is conceptually simple to understand and prove correct, as well as, we would like to think, to invent it in the first place. It is probably the ease of discovery of divide-and-conquer algorithms that makes this technique so important, although in many cases these algorithms are also more efficient than others.

If we assume that an operation $move(x, y)$ accomplishes moving the top disk of pole x onto the top of pole y, then the following straightforward recursive algorithm TOWERS-OF-HANOI(n, a, b, c) moves n disks from a to c using b as a spare.

```
TOWERS-OF-HANOI (n, a, b, c)
    if n = 1 then
        move(a, c);
    else
        TOWERS-OF-HANOI (n − 1, a, c, b);
        move(a, c);
        TOWERS-OF-HANOI (n − 1, b, a, c);
    return;
```

1.1. Analysis of the TOWERS-OF-HANOI algorithm. We showed in Chapter 2 how to analyze the running time of algorithms that do not call themselves recursively. We also saw in the example of Euclid's algorithm that the analysis of a recursive algorithm may be rather peculiar. Fortunately, the running time of most recursive algorithms can be obtained in a standard way using **recurrence equations** or **recurrences**. They describe the overall running time of a recursive algorithm on an input of size n in terms of the algorithm's running time on inputs of size less than n. We can then apply rich mathematical apparatus to solve the recurrences and derive bounds on the performance of a recursive algorithm.

Consider the TOWERS-OF-HANOI algorithm and let $T(n)$ be its worst-case running time for n disks. (The reader should by now recognize that the input size for this problem is the number of disks.) It's not hard to convince ourselves that, in any reasonable implementation, the *move* operation is a basic operation. Furthermore, for simplicity, we may assume it takes one unit of time. Since for $n = 1$ the algorithm executes only the *move* operation, and for $n > 1$ it executes itself twice with $n - 1$ disks plus one *move* operation, we have

$$T(n) = \begin{cases} 1, & \text{if } n = 1 \\ 2T(n-1)+1, & \text{if } n > 1. \end{cases}$$

Solving a recurrence equation like this means obtaining a closed-form formula for $T(n)$ not involving any $T(m)$ for $m < n$. One way to do this is the iterative substitution method, in which we repeatedly substitute the general expression given by $T(n)$ for each $T(n-1), T(n-2), \ldots, T(2)$ on the right-hand side of the recurrence for $n > 1$. In the end, we replace the initial condition $T(1)$ and simplify the resulting formula.

Applying this method in the case $n > 1$ of the TOWERS-OF-HANOI algorithm, we obtain the following solution for $T(n)$:

$$T(n) = 2 \cdot T(n-1) + 1$$
$$= 2 \cdot (2T(n-2)+1) + 1 = 2^2 \cdot T(n-2) + 2 + 1$$
$$= 2^2 \cdot (2T(n-3)+1) + 2 + 1 = 2^3 \cdot T(n-3) + 2^2 + 2 + 1$$
$$\vdots$$
$$= 2^{n-1} \cdot T(1) + 2^{n-2} + \cdots + 2^2 + 2 + 1$$
$$= 2^{n-1} \cdot 1 + \sum_{i=0}^{n-2} 2^i = 2^{n-1} + \frac{2^{n-1}-1}{2-1} = 2 \cdot 2^{n-1} - 1$$
$$= 2^n - 1.$$

Notice that the expression for $T(n)$ represents actually the number of moves in the solution to the Towers of Hanoi game with n disks.

2. Merge sort

Although we have defined the sorting problem in terms of an array, the merge sort algorithm is best described in terms of a list of numbers. There is nothing essential that prevents us from using the merge sort on an array. Simply it is easer to describe and comprehend it on a more abstract level of lists. Otherwise, we would have to take care of an undue level of detail with manipulating the array indices. Moreover, we could even take another approach: for a given array to be sorted, create a list of its elements, and copy the sorted list of elements back into the array, if needed. These two operations are easy to implement and take time proportional to the size of the array, which does not affect asymptotic efficiency of the merge sort algorithm.

The **merge sort** algorithm for sorting a list of n elements closely follows the divide-and-conquer paradigm. Informally, the algorithm first divides the n-element list of numbers to be sorted into two sublists of roughly $n/2$ elements each. This is the first **divide** phase of the divide-and-conquer paradigm. The algorithm then recursively sorts each of the half-sized lists separately. This is the second **conquer** phase. Finally, it merges the two $n/2$-element sorted sublists to complete the sorting of the original list of n elements. This is the third **combine** phase. We now explain these steps (in the reverse order) in more detail.

To merge two sorted lists in the third phase means to produce a single sorted list containing all the elements of the two given lists (and no other elements). For example, given two sorted lists $\langle 1,2,3,7,8 \rangle$ and $\langle 1,2,6,9 \rangle$, the single sorted list obtained by merging these lists is $\langle 1,1,2,2,3,6,7,8,9 \rangle$. Notice that the operation of merging two lists assumes that they are already sorted.

One simple way to merge two lists is to examine them sequentially from the beginning to the end. In each step then, we compare the heads of the two lists, choose the smaller element (ties can be broken arbitrarily) as the next element for the merged list, and remove the chosen element from its list. The following is pseudocode that implements this method of merging two sorted lists L_1 and L_2 into a single sorted list L. It assumes that the attribute object of a list contains an additional pointer field *tail* pointing to the last element of the list. (Of course, all the basic list operations should appropriately maintain the *tail* pointer.)

```
LIST-MERGE(L₁, L₂, L)
    head(L) = NIL; tail(L) = NIL;
    while (head(L₁) ≠ NIL) ∧ (head(L₂) ≠ NIL) do
        if head(L₁) ≤ head(L₂) then
            x = LIST-HEAD-DELETE(L₁);
        else
            x = LIST-HEAD-DELETE(L₂);
        LIST-INSERT(L, x, tail(L));
    while head(L₁) ≠ NIL do
        x = LIST-HEAD-DELETE(L₁);
        LIST-INSERT(L, x, tail(L));
    while head(L₂) ≠ NIL do
        x = LIST-HEAD-DELETE(L₂);
        LIST-INSERT(L, x, tail(L));
    return L;
```

To break a list L into two half-sized lists L_1 and L_2, we simply repeatedly remove the head of the given list and alternatively put it on the first or the second list.

LIST-BREAK(L, L_1, L_2)
  $head(L_1)$ = NIL; $head(L_2)$ = NIL;
  $i = 1$;
  while $head(L) \neq$ NIL **do**
    x = LIST-HEAD-DELETE(L);
    if $i = 1$ **then**
      LIST-HEAD-INSERT(L_1, x);
    else
      LIST-HEAD-INSERT(L_2, x);
    $i = -i$;
  return L_1, L_2;

By making use of the LIST-BREAK and LIST-MERGE algorithms as subroutines and applying the divide-and-conquer approach, the algorithm MERGE-SORT(L) that sorts the input (nonempty) list L is easy to implement.

MERGE-SORT(L)
  if $head(L) \neq tail(L)$ **then** ⟦ L is not a one-element list ⟧
    LIST-BREAK(L, L_1, L_2);
    MERGE-SORT(L_1);
    MERGE-SORT(L_2);
    LIST-MERGE(L_1, L_2, L);
  return L;

Let us use the MERGE-SORT algorithm to see how it works on a list of single-digit numbers ⟨8, 6, 9, 2, 2, 1, 1, 7, 3⟩. For succinctness, to represent lists in the example we will omit the brackets and comma between digits. Thus, the input list is $L = 869221173$. First, the list is split into two lists L_1 and L_2. These lists consist of every other element of the original list, starting from the first and second element, respectively. That is, $L_1 = 89213$ and $L_2 = 6217$. Then, these lists are recursively merge sorted, resulting in lists $L_1 = 12389$ and $L_2 = 1267$. Finally, the LIST-MERGE procedure produces the sorted list $L = 112236789$.

Of course, the sorting of the two smaller lists does not occur by magic, but instead by the systematic application of the recursive algorithm. The MERGE-SORT algorithm first splits the input list, and then recursively continues to split the lists until each list is a one-element list. Figure 3(a) shows this recursive splitting of the lists from the example. Then the split lists are merged in pairs, going up the tree, until the entire list is sorted. This process is shown in Figure 3(b). However, it is worth noting that the splits and merges occur in a mixed order, i.e., not all splits are followed by all merges. For instance, the sublist $L_1 = 89213$ is completely split and merged before anything is done with the sublist $L_2 = 6217$.

2.1. Analysis of the MERGE-SORT algorithm. Consider the MERGE-SORT algorithm and again let $T(n)$ be its worst-case running time for an input list of size n. If $n = 1$, the algorithm takes a constant number of steps. In the case $n > 1$, the algorithm executes the LIST-BREAK and LIST-MERGE procedures and

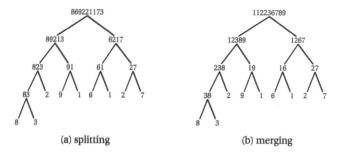

(a) splitting   (b) merging

FIGURE 3. Recursive splitting and merging.

calls itself recursively two times. Since the running time of the LIST-BREAK and LIST-MERGE procedures is clearly $\Theta(n)$, we can write the recurrence equation for $T(n)$ as

$$T(n) = \begin{cases} c_1, & \text{if } n = 1 \\ 2T(n/2) + c_2 n, & \text{if } n > 1. \end{cases}$$

The constant c_1 represents the constant running time of the MERGE-SORT algorithm on a list of size $n = 1$. In the case $n > 1$, the constant c_2 represents the hidden constant in the running time $\Theta(n)$ of the split and the merge part, plus the time taken by the test at the beginning of the algorithm. Observe that the above equation applies only if n is a power of 2. However, it is easy to extend the argument for every n, so we will generally omit this issue to avoid undue notational clutter.

To solve the merge-sort recurrence we use another method called guess-and-verify. This time we simply guess a solution and then try to verify, usually by induction, that our guess was correct. Moreover, since we are interested only in the asymptotic running time of an algorithm, we need not even look for the exact solution. It suffices to provide a function that just upper bounds the running time.

For example, for the running time of the MERGE-SORT algorithm, we make an educated guess that $T(n) \leq an\log n + b$ for some constants a and b to be determined later. From $T(n) = 1$ in the recurrence equation for $n = 1$, it follows that we can pick $b = c_1$. If for the induction step we assume that $T(m) \leq am\log m + b$ holds for all m such that $1 \leq m < n$, we need to prove that $T(n) \leq an\log n + b$. Indeed, taking into account the recurrence expression for $n > 1$ and the inductive hypothesis, we can write

$$T(n) = 2T(n/2) + c_2 n \leq 2 \cdot \left(a\frac{n}{2}\log\frac{n}{2} + b \right) + c_2 n$$
$$= an\log n - an + 2b + c_2 n.$$

Thus, if we choose $a = c_2 + b = c_2 + c_1$, we obtain

$$T(n) \leq an\log n - (c_2 + b)n + 2b + c_2 n$$
$$= an\log n - c_2 n - bn + 2b + c_2 n$$
$$= an\log n - b(n - 2)$$

$$\leq an \log n + b,$$

as required. In other words, the running time of the MERGE-SORT algorithm is $O(n \log n)$, which is asymptotically much faster than the running time of the simple sorting algorithms discussed in Chapter 2.

3. Quick sort

The quick sort algorithm also closely follows the divide-and-conquer paradigm. Unlike sorting by merging though, most of the work in the nonrecursive part of the algorithm is spent in constructing the subproblems rather than combining their solutions.

The **quick sort** algorithm divides an array to be sorted into two subarrays by choosing one array element about which the original array is partitioned. The element chosen is called the **pivot**, and we would like that the pivot breaks the array into two roughly equally sized parts. This criterion translates to the best case for the pivot being the median element of the array, because it is greater than about half the elements and less than about the other half.

Specifically, let A be the given array of n elements to be sorted. If p is the value of the chosen pivot element, then the pivot is moved to its right place in the sorted array, all elements less than or equal to the pivot are moved to its left, and all elements greater than the pivot are moved to its right. This formally means that the array elements are permuted so that the following three conditions are satisfied:

- $A(k) = p$ for some k between 1 and n;
- $A(i) \leq p$ for each i such that $1 \leq i \leq k$; and
- $A(i) > p$ for each i such that $k + 1 \leq i \leq n$.

Figure 4 summarizes this permutation after the divide phase of the quick sort.

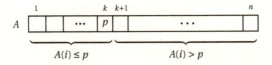

$$A(i) \leq p \qquad\qquad A(i) > p$$

FIGURE 4. Partitioning of an array by the quick sort.

The two sections of the original array on either side of the pivot are then sorted by recursive application of the algorithm. The final result is a completely sorted array, because all elements in the first section are less than all elements in the second section. Thus, no further combine step is necessary.

As we can see, picking a good pivot element is the key step in the algorithm. This is so because if the sizes of the two sections on either side of the pivot are equal, then sorting each of them takes about the same time, which balanced with the partition time gives overall the most efficient time for the quick sort. Taking the median value of the array is seemingly the best choice for the pivot, but finding the median takes more time than it is worth. For this reason, one usually makes some simple choice for the pivot in the hope that the chosen element is near the median. What is remarkable is that this proves to be a prudent approach, because in practice the quick sort algorithm runs very fast on average. This simple choice for the pivot element includes picking the first, last, middle, or even a random element of a (sub)array. In this section we consider partitioning of a subarray when we choose its first or last element as the pivot. Choice of the middle element is left as an exercise for the reader (see Exercise 2), and in Chapter 11 we discuss selection of a random pivot element.

Let's first consider partitioning of a subarray when we choose its first element as the pivot. Suppose we want to partition the subarray $A(l \ldots r)$, where $1 \leq l \leq r \leq n$, and let the pivot p be its its first element, i.e., $p = A(l)$. To rearrange the elements of the subarray around the pivot p, we scan the subarray just once, but starting from both ends. If i is the index of elements examined going from the left end, and j is the index of elements examined going from the right end, then i and j are initialized to $l + 1$ and r, respectively. The index i is then incremented until we encounter $A(i) > p$, and index j is decremented until we encounter $A(j) \leq p$. Figure 5 illustrates this process.

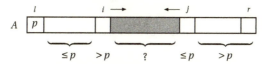

FIGURE 5. Pivoting with the first element of a subarray.

Thus, from the left end of the subarray we skip all elements less than or equal to p, and from the right end we skip all elements greater than p. Then, $A(i)$ and $A(j)$ are swapped to put them on their correct side with respect to the pivot. This process continues as long as $i < j$. When the indices i and j cross (that is, $i \geq j$), it signals that we have scanned the entire subarray. What is left then is to swap $A(l)$ and $A(j)$ to put the pivot element in its correct position. The following is pseudocode that implements this procedure for partitioning the subarray $A(l \ldots r)$ about its first element as the pivot. It also returns the index of the pivot in the rearranged subarray.

```
PARTITION (A, l, r)
  [ Assumes A(r + 1) = +∞ ]
  p = A(l);
  i = l + 1; j = r;
  while i < j do
    while A(i) ≤ p do i = i + 1;
    while A(j) > p do j = j − 1;
    swap(A(i), A(j));
  swap(A(l), A(j));
  return j;
```

In the PARTITION algorithm we have a technical problem that we solved by ignoring it. Namely, when we scan the subarray for the first time, starting from

its left end and looking for an element greater than the pivot p, we may get past
the right end of the subarray. This happens if the pivot is the largest element
of the subarray, or if all subarray elements are identical. To avoid introducing
additional test that prevents us from falling off the end of the subarray and thus
breaking simplicity of the algorithm, we have simply assumed that there is always a sentinel element $A(r + 1) = +\infty$ at the end of the subarray. Actual details
are left for the reader as an easy exercise.

Consider now partitioning a subarray when we choose its last element as the
pivot. This time as the pivoting proceeds, we scan the subarray only in one direction (from left to right) and partition the subarray into four (possibly empty)
regions as shown in Figure 6.

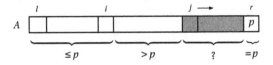

FIGURE 6. Pivoting with the last element of a subarray.

Each region satisfies certain property, indicated underneath the corresponding region in Figure 6. Having this in mind, we can write the following PARTITION
algorithm to partition a subarray around its last element as the pivot. Correctness of the algorithm is easy to see because all the properties are maintained as
the loop invariant in the algorithm.

```
PARTITION(A, l, r)
  p = A(r);
  i = l - 1;
  for j = l to r - 1 do
    if A(j) ≤ p then
      i = i + 1;
      swap(A(i), A(j));
  swap(A(i + 1), A(r));
  return i + 1;
```

We can now use either version of the PARTITION procedure to implement
the algorithm QUICK-SORT(A, l, r) that sorts the subarray $A(l \ldots r)$, where $1 \le l \le
r \le n$. Explicit bounds of the subarray as arguments are necessary in this case,
because of the recursive nature of the algorithm. However, it is clear that we can
sort the entire array A of size n by making the initial call QUICK-SORT$(A, 1, n)$ in
the program.

QUICK-SORT(A, l, r)
if $l < r$ **then**
$\quad k = $ PARTITION(A, l, r);
$\quad$ QUICK-SORT$(A, l, k - 1)$;
$\quad$ QUICK-SORT$(A, k + 1, r)$;
return A;

3.1. Analysis of the QUICK-SORT algorithm. The running time of the QUICK-SORT algorithm depends on whether the subarray partitioning is well balanced, which in turn depends on which subarray element is chosen as the pivot. If the subarrays are evenly divided at each level of the recursion tree, the algorithm runs asymptotically as fast as merge sort in $O(n \log n)$ time. On the other hand, if the partitioning is highly unbalanced, it can degrade to run asymptotically as slow as insertion sort in $O(n^2)$ time. Fortunately, in average case, "good" splits prevail over "bad" splits, and the algorithm runs in $O(n \log n)$ time.

Since the running time of the quick sort is not a clear-cut situation, let us try to develop an intuition behind its performance. First of all, it is clear that both versions of the PARTITION algorithm take time proportional to the size of the subarray they are called upon. Namely, in both cases, we scan the subarray just once. In the first case when we choose the first element as the pivot, we are scanning the subarray from both ends, but we stop when the corresponding indices cross over. In the second case when we choose the last element as the pivot, we are scanning the subarray simply from the left to right end. Thus, the running time of the algorithm PARTITION(A, l, r) is $\Theta(r - l)$. In particular, the initial partition ($l = 1, r = n$)) of the input array takes $\Theta(n)$ time.

The worst-case performance of quick sort occurs when we somehow manage to select the largest subarray element as the pivot in the partitioning procedure. This happens in the first version of the PARTITION algorithm when the input array is initially sorted in the decreasing order, and in the second version when the input array is already sorted in the increasing order. Then, at each recursive call, we would divide the subarray into two smaller subarrays: one with a single element (the pivot element) and the other with everything else. Therefore, the recurrence equation that describes the running time of the QUICK-SORT algorithm $T(n)$ in the worst case is

$$T(n) = T(n-1) + T(1) + \Theta(n)$$
$$= T(n-1) + \Theta(n),$$

since obviously the running time $T(1)$ of the algorithm on a one-element array is constant. It is straightforward to show using, for instance, the iterative substitution method that the solution to this recurrence is $O(n^2)$. Therefore, the worst-case running time of quick sort is no better than that of any simple sorting scheme.

In the best-case partitioning scenario, when at every recursive level of the algorithm we split the subarray into two, roughly half-sized subarrays, the recurrence for the running time is

$$T(n) = 2T(n/2) + \Theta(n),$$

which is identical to the merge sort recurrence. Therefore, the best-case running time of quick sort is $O(n \log n)$.

The average-case running time of quick sort is much closer to the best case than to the worst case. Before we formally prove this fact, let's first give an informal explanation of why this is the case. Suppose that the partitioning algorithm always produces, say, a 90-10 percent split, which seems quite unbalanced. Then we have the recurrence equation

$$T(n) = T(9n/10) + T(n/10) + \Theta(n)$$

for the running time of quick sort. But the solution to this recurrence is still $T(n) = O(n \log n)$, which may be verified by induction, although with a larger constant hidden in the asymptotic notation. Thus, with apparently highly unbalanced proportional split at every recursion level, quick sort takes asymptotically $O(n \log n)$ time—the same time as if the split were exactly in the middle. In fact, any split of constant proportionality, even 99-1 percent split, at every recursion level yields $O(n \log n)$ running time.

While it is unlikely that for a random input array the partitioning always produces proportional split at every level, we expect that some of the splits will be well balanced and that some will be fairly unbalanced. In other words, in the average case, good and bad splits will alternate at recursion levels. However, we expect many more good splits than bad splits, as our informal analysis has shown that seemingly bad splits are actually good. Thus, even if a really bad split occurs before good splits, the corresponding subarrays will bi quickly sorted. Namely, the added cost is dominated by additional partitioning, which is absorbed into the total cost, not by additional recursive calls.

Now we turn to a rigorous proof of the fact that, on average, the running time of the QUICK-SORT algorithm is $O(n \log n)$. Actually, the idea of the proof is very similar to the proof that the average height of a binary search tree is $O(\log n)$.

At the very beginning, let's clarify our assumptions about the input array and the algorithm. First, without loss of generality, we assume that the elements of the input array are all distinct. Therefore, the elements of the input array are exactly the elements of a set $S = \{x_1, x_2, \ldots, x_n\}$ in some order. There are $n!$ possible orders in which these elements might appear, and we assume that all the permutations are equally likely. Specifically, the first element in a random permutation can be of any rank,[1] and so for each $i = 1, 2, \ldots, n$ the probability that the first element in a random permutation is of rank i is $1/n$.

Second, we assume that the QUICK-SORT algorithm partitions the input array around the first element. It is clear that the average performance of quick sort is not affected by the choice of the position of the pivot element, so equally well it can be the last or middle element, or any element in a fixed (relative) position for that matter.

Third, to make our mathematics less cumbersome, we measure performance of QUICK-SORT by counting the number of pairwise comparisons of array elements the algorithm makes. It is easy to see that pairwise comparisons are the

---

[1] Recall that the rank of an element is its position in the sorted array.

major operations performed by QUICK-SORT, and so its running time is proportional to the total number of comparisons.

Let $\overline{T}(n)$ be the average number of pairwise comparisons of array elements that are made by QUICK-SORT, where the average is taken over all $n!$ of the possible input permutations of the set S. An equivalent way of looking at $\overline{T}(n)$ is to consider the number of comparisons of QUICK-SORT as a random variable $T(n)$ defined on the random permutations of S, where each of the $n!$ permutations occurs with equal probability $1/n!$. Then $\overline{T}(n)$ is the expected value of the random variable $T(n)$, that is, $\overline{T}(n) = \mathbb{E}(T(n))$.

Consider the total number of pairwise comparisons that QUICK-SORT$(A, 1, n)$ makes on a random permutation of S. This includes the comparisons done in the PARTITION procedure, as well as the comparisons that are done in the two recursive calls that follow selection of the pivot and rearranging the array about it. Since (the first version of) the PARTITION procedure takes the first array element and compares it to each of the remaining elements of the array, it is clear that the number of comparisons involved in this step is exactly $n - 1$. After the partitioning step, the number of comparisons in the two recursive calls depends on how the pivot has split the input array. Figure 7 illustrates the case when the pivot occupies the i-th position in the array after execution of the PARTITION procedure.

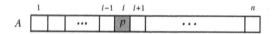

FIGURE 7. Position of the pivot p after splitting the input array.

The two recursive calls following the PARTITION call independently process the subarrays $A(1 \ldots i - 1)$ and $A(i + 1 \ldots n)$, and the pivot p in its final position is not involved in any further comparison whatsoever. Therefore, the number of pairwise comparisons that are involved in the two recursive calls are $T(i - 1)$ and $T(n - i)$, respectively. All of this implies that if the rank of the pivot is i, $(1 \le i \le n)$, then

$$T(n) \mid \text{the pivot has rank } i = n - 1 + T(i - 1) + T(n - i),$$

and hence by the linearity of expectation

$$\mathbb{E}(T(n) \mid \text{the pivot has rank } i) = n - 1 + \mathbb{E}(T(i - 1)) + \mathbb{E}(T(n - i))$$
$$= n - 1 + \overline{T}(i - 1) + \overline{T}(n - i),$$

where $\overline{T}(0) = 0$. Since any element in S is equally likely to be chosen as the pivot in a random permutation of S, for each value of $i = 1, 2, \ldots, n$ we have that the pivot has rank i with probability $1/n$. Thus,

$$\overline{T}(n) = \mathbb{E}(T(n)) = \sum_{i=1}^{n} \mathbb{E}(\overline{T}(n) \mid \text{the pivot has rank } i) \cdot \mathbb{P}\{\text{the pivot has rank } i\}$$
$$= \sum_{i=1}^{n} \left(n - 1 + \overline{T}(i - 1) + \overline{T}(n - i) \right) \cdot \frac{1}{n}$$

$$= n - 1 + \frac{1}{n} \sum_{i=1}^{n} \left(\overline{T}(i-1) + \overline{T}(n-i) \right)$$

$$= n - 1 + \frac{2}{n} \sum_{i=0}^{n-1} \overline{T}(i).$$

It follows that the average number of comparisons $\overline{T}(n)$ satisfies the recurrence equation

(9)
$$\overline{T}(n) = \begin{cases} 0, & n = 0 \\ n - 1 + \dfrac{2}{n} \sum_{i=0}^{n-1} \overline{T}(i), & n \geq 1. \end{cases}$$

To solve (9), we first simplify it by getting rid of the summation sign. For $n \geq 1$, multiplying both sides of (9) by n gives

$$n\overline{T}(n) = n(n-1) + 2 \sum_{i=0}^{n-1} \overline{T}(i).$$

If we subtract the right-hand side of this recurrence with $n-1$ on the left from the one corresponding to n, the summation sign disappears:

$$n\overline{T}(n) - (n-1)\overline{T}(n-1) = n(n-1) + 2 \sum_{i=0}^{n-1} \overline{T}(i) - (n-1)(n-2) - 2 \sum_{i=0}^{n-2} \overline{T}(i)$$

$$= 2(n-1) + 2\overline{T}(n-1).$$

Rearranging the terms we get

$$n\overline{T}(n) = 2(n-1) + 2\overline{T}(n-1) + (n-1)\overline{T}(n-1)$$

$$= (n+1)\overline{T}(n-1) + 2(n-1),$$

and dividing both sides by $n(n+1)$ yields

$$\frac{\overline{T}(n)}{n+1} = \frac{\overline{T}(n-1)}{n} + \frac{2(n-1)}{n(n+1)}.$$

The last recurrence equation gives this sequence of equations:

$$\frac{\overline{T}(n)}{n+1} = \frac{\overline{T}(n-1)}{n} + \frac{2(n-1)}{n(n+1)}$$

$$\frac{\overline{T}(n-1)}{n} = \frac{\overline{T}(n-2)}{n-1} + \frac{2(n-2)}{(n-1)n}$$

$$\vdots$$

$$\frac{\overline{T}(2)}{3} = \frac{\overline{T}(1)}{2} + \frac{2 \cdot 1}{2 \cdot 3}$$

$$\frac{\overline{T}(1)}{2} = \frac{\overline{T}(0)}{1} + \frac{2 \cdot 0}{1 \cdot 2}.$$

Summing up the left-hand and right-hand sides of these equations gives

$$\frac{\overline{T}(n)}{n+1} = 2 \sum_{i=1}^{n} \frac{i-1}{i(i+1)}$$

$$= 2 \sum_{i=1}^{n} \left(\frac{2}{i+1} - \frac{1}{i} \right)$$

$$= 2 \sum_{i=1}^{n} \frac{1}{i} + \frac{4}{n+1} - 4$$

$$= 2 H_n + \frac{4}{n+1} - 4,$$

where $H_n = \sum_{i=1}^{n} 1/i$ is the n-th harmonic number. Hence,

$$\overline{T}(n) = 2(n+1) H_n + 4 - 4(n+1)$$
$$= 2n H_n + 2 H_n + 4 - 4n - 4$$
$$= 2n H_n - 2(2n - H_n)$$
$$\leq 2n H_n.$$

Finally, since $H_n = \ln n - \gamma + o(1)$, where $\gamma \approx 0.58$ is the Euler's constant, that is, $H_n = \Theta(\log n)$, the last inequality implies $\overline{T}(n) = O(n \log n)$. Therefore, QUICK-SORT is really quick on average, even though it takes quadratic time in the worst case.

4. Divide-and-conquer recurrences

In this section we pause with algorithms and explore some general techniques for solving the kinds of recurrence equations that arise in the analysis of divide-and-conquer algorithms. Typically, we let a function $T(n)$ denote the maximum running time of such an algorithm on an input of size n, and characterize $T(n)$ using an equation that relates $T(n)$ to values of the function T for problem sizes smaller than n. The goal is to find a formula (or at least an upper bound) for $T(n)$ expressed only in terms of the input size n.

For example, in the case of the MERGE-SORT algorithm we have obtained the recurrence

(10)
$$T(n) = \begin{cases} c_1, & \text{if } n = 1 \\ 2T(n/2) + c_2 n, & \text{if } n > 1 \end{cases}$$

for some constants c_1 and c_2, taking the simplifying assumption that n is a power of 2. In fact, throughout this section, we take the simplifying assumption that n is an appropriate power so that we can avoid using floor and ceiling functions. Every asymptotic statement we make about recurrence equations will still be true for every n, but justifying this fact formally involves long and boring proofs.

We showed by an easy induction that $T(n) = O(n \log n)$ for the merge-sort recurrence, though in general we might get a recurrence equation that is more challenging to solve than this one. Thus, it is useful to consider the divide-and-conquer recurrences in their own right and develop some general insight into their solutions.

4.1. The iterative substitution method. One way to solve a divide-and-conquer recurrence equation is to use the *iterative substitution* method, which we already met in the case of the TOWERS-OF-HANOI algorithm. In using this method, we assume that the input size n is fairly large and then we substitute the general form of the recurrence for each occurrence of the function T on the right-hand side of the expression for $T(n)$. For example, taking for simplicity in

(10) that $c_1 = c_2 = c$ (e.g., taking larger of the two constants) and performing one such substitution in the recurrence, yields the equation

$$T(n) = 2\left(2T\left(n/2^2\right) + c \cdot n/2\right) + cn$$
$$= 2^2 T\left(n/2^2\right) + 2cn.$$

Plugging the general equation for $T\left(n/2^2\right)$ in again yields the equation

$$T(n) = 2^2 \left(2T\left(n/2^3\right) + c \cdot n/2^2\right) + 2cn$$
$$= 2^3 T\left(n/2^3\right) + 3cn.$$

The hope in applying the iterative substitution method is that, at some point, we will see a pattern that can be converted into a general closed-form solution (with single T appearing only on the left-hand side). In the case of the merge-sort recurrence, the general form is clearly

$$T(n) = 2^i T\left(n/2^i\right) + icn.$$

Notice that the general form of this equation shifts to the base case, $T(1) = c$, when $n = 2^i$, that is, when $i = \log n$, which implies

$$T(n) = cn + cn\log n,$$

that is, $T(n) = O(n\log n)$.

In a general application of the iterative substitution technique, we hope that we can determine a general pattern for $T(n)$ and that we can also figure out when the general form of $T(n)$ shifts to the base case.

From a mathematical point of view, there is one point in the use of the iterative substitution technique that involves a bit of a logical jump. This jump occurs at the point where we try to characterize the general pattern emerging from a sequence of substitutions. Often, as was the case with the MERGE-SORT recurrence equation, this jump is quite reasonable. Other times however, it may not be so obvious what a general form for the equation should look like. In these cases, the jump may be more dangerous. To be completely safe in making such a jump, we must fully justify the general form of the solution, possibly through induction. Combined with such a justification, the iterative substitution method is completely correct and often useful way of characterizing recurrence equations.

4.2. The recursion tree method. The iterative substitution method involves plugging in the recursive part of an equation for $T(n)$, and then often "chugging" through a considerable amount of algebra to get this equation into a form where we can infer a general pattern. However, it is very easy to get lost in all the symbolic manipulation, so the ***recursion tree*** method offers a more visual way to see what is going on at each recursion level. Using this technique, any recurrence is represented by a tree, where each substitution step of the recurrence takes us one level deeper in the tree.

Let's introduce the idea of a recursion tree via the recurrence for the MERGE-SORT algorithm (which we simplified by taking all constants to be 1 and assuming n is a power of 2):

(11)
$$T(n) = \begin{cases} 1, & \text{if } n = 1 \\ 2T(n/2) + n, & \text{if } n > 1. \end{cases}$$

Recall the algorithmic interpretation of the recurrence: to merge-sort an array of size n we must merge-sort 2 subarrays of size $n/2$ and do n units of additional work to perform the associated nonrecursive merge. We now draw the recursion tree for the merge-sort recurrence starting with the root node that represents the original problem and labeling it with the work done in the nonrecursive part to solve it, which is n. The fact that we break the problem into two subproblems is denoted by drawing two children of the root node. We also label the children with the work done in the nonrecursive part to solve each of them, which is $n/2$. At the next level, we have four nodes as each of the two subproblems on the previous level is further split into two subproblems. Each of these four children has label $n/4$, which is the nonrecursive work done to solve each of them. This continues to the bottom level of the tree, representing the nodes for problems of size 1 and labeled with the work that comes from the base case. Figure 8 shows the resulting recursion tree.

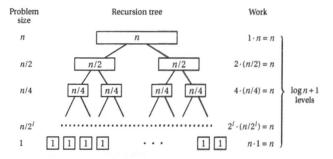

FIGURE 8. The recursion tree for the merge-sort recurrence.

Notice that on the right of the Figure 8 we summed up the work done at each level of the recursion tree. At level i (the top level is level 0), there are 2^i subproblems of size $n/2^i$, each taking $n/2^i$ additional time in the nonrecursive part of its solution, for a total of $2^i \cdot (n/2^i) = n$. Since the tree has $\log n + 1$ levels $(0, 1, 2, \ldots, \log n)$, and each level contributes the same amount n of additional time, the total running time is $T(n) = n(\log n + 1) = n \log n + n$. This is exactly what we got by the iterative substitution method as the solution to the merge-sort recurrence.

The important thing is that we now know exactly how many levels there are in the recursion tree, and how much work is done at each level. Once we know this, we can sum the total amount of work done over all the levels, giving us the solution to recurrence (11). The total work done throughout the tree is the solution to our recurrence, because the tree simply models the process of iterating the recurrence. Thus, the solution to recurrence (11) is $T(n) = n \log n + n$.

Note that the recursion tree method is just the iterative substitution method in disguise, although it helps visually in analyzing summations. Let's enforce this with one more recurrence (assuming as usual n is a power of 2):

(12)
$$T(n) = \begin{cases} c, & \text{if } n = 1 \\ 3T(n/2) + n^2, & \text{if } n > 1, \end{cases}$$

where $c > 0$ is a constant.

We can algorithmically interpret recurrence (12) (ignoring the base case) as "to solve a problem of size n we must solve 3 subproblems of size $n/2$ and do n^2 units of additional work." When drawing the recursion tree, we again label each node with the amount of additional (nonrecursive) work for the corresponding subproblem size. At level 0 (the top level), the additional work is n^2. At level 1, we have three nodes whose additional work is $(n/2)^2$ each, for a total of $3(n/2)^2 = (3/4)n^2$. At level 2, we have nine nodes whose additional work is $(n/4)^2$ each, for a total of $9(n/4)^2 = (3/4)^2 n^2$. This is easy to extrapolate for general level i, where we have 3^i nodes whose additional work is $\left(n/2^i\right)^2$ each, for a total of $3^i\left(n/2^i\right)^2 = (3/4)^i n^2$. Figure 9 shows the resulting recursion tree for recurrence (12).

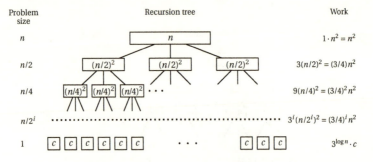

FIGURE 9. The recursion tree for recurrence (12).

The bottom level is different from the other levels. In the other levels, the work is described by the recursive part of the recurrence, which in this case is $T(n) = 3T(n/2) + n^2$. At the bottom level, the work comes from the base case, which in this case is $T(1) = c$.

Again, to get the total work throughout the recursion tree, i.e, the solution to recurrence (12), we sum up the work over all levels of the tree. Observing that the number of levels is $\log n + 1$ $(0, 1, 2, \ldots, \log n)$ and the number of nodes at the bottom level is $3^{\log n}$ leads to the following solution. In the summation below we have used the log identity $a^{\log_b n} = n^{\log_b a}$, as well as the formula for the geometric series

$$\sum_{i=0}^{m} x^i = \frac{x^{m+1} - 1}{x - 1}, \quad \text{if } x \neq 1.$$

$$T(n) = n^2 + (3/4)n^2 + (3/4)^2 n^2 + \cdots + (3/4)^{\log n - 1} n^2 + c3^{\log n}$$

$$= n^2 \sum_{i=0}^{\log n - 1} (3/4)^i + c3^{\log n}$$

$$= n^2 \cdot \frac{(3/4)^{\log n} - 1}{3/4 - 1} + c3^{\log n}$$

$$= n^2 \cdot \frac{1 - n^{\log 3}/n^2}{1/4} + cn^{\log 3}$$

$$= 4n^2 - 4n^{\log 3} + cn^{\log 3}$$
$$= 4n^2 + (c-4)n^{\log 3}.$$

Thus, the exact solution to recurrence (12) is $T(n) = 4n^2 + (c-4)n^{\log 3}$. If we wanted just an asymptotic expression, since $\log_2 3 \approx 1.58$, we see that $T(n) = \Theta(n^2)$.

4.3. The guess-and-verify method. Another technique for solving divide-and-conquer recurrences is the **guess-and-verify** method, which we used when we analyzed the MERGE-SORT algorithm in Section 2.1. This technique involves first making an educated guess as to what a closed-form solution to a given recurrence might look like, and then justifying that guess, usually by induction. Of course, by using this approach we may have to try several solutions until we get one that is a good upper bound for a given recurrence equation. In other words, if the verification of our current guess fails, probably we need to use a faster-growing function. But even if we succeed, if our current guess is justified too easily, then it is possible that we need to use a slower-growing function.

For example, consider the following recurrence equation (where a and b are some positive constants):

(13)
$$T(n) = \begin{cases} a, & \text{if } n = 1 \\ 2T(n/2) + bn\log n, & \text{if } n > 1. \end{cases}$$

This looks very similar to the merge-sort recurrence, so we might make the following as our first guess: $T(n) = O(n\log n)$. Then, we want to show by induction that the solution to the recurrence (13) satisfies $T(n) \le cn\log n$ for some positive constant c and every sufficiently large n.

We can certainly choose c and n_0 large enough to make this true for the base case $n = n_0$. Namely, if we let $c = a + b$ and $n_0 = 2$, then

$$T(2) = 2T(1) + b \cdot 2 \cdot \log 2 = 2a + 2b = 2(a+b) = 2c = c \cdot 2 \cdot \log 2.$$

For inductive step when $n > n_0$, if we assume by the inductive hypothesis that our first guess is true for the arguments smaller than n, then for the argument n we can write

$$T(n) = 2T(n/2) + bn\log n$$
$$\le 2\left(c\frac{n}{2}\log\frac{n}{2}\right) + bn\log n$$
$$= cn(\log n - \log 2) + bn\log n$$
$$= cn\log n - cn + bn\log n.$$

Thus, to make the induction work, we would need to have

$$cn\log n - cn + bn\log n \le cn\log n,$$

that is, $bn\log n \le cn$, or equivalently $\log n \le c/b$. But there is no way we can make this last inequality true for every $n \ge 2$. Thus, our first guess was not successful.

We must therefore try a faster-growing function for $T(n)$ as our second guess: $T(n) = O(n^2)$. In this case, to show $T(n) \le cn^2$ for some positive constant c and every sufficiently large n, we can choose $c = 2(a+b)$ and $n_0 = 1$ to easily see that the base case holds. For inductive step when $n > n_0$, if we assume by the inductive hypothesis that our second guess is true for the arguments smaller than n, then for the argument n we have

$$T(n) = 2T(n/2) + bn\log n$$

$$\leq 2 \cdot c\frac{n^2}{4} + bn\log n$$

$$\leq c\frac{n^2}{2} + \frac{c}{2} \cdot n^2$$

$$= cn^2.$$

Thus in this case the induction works. However, since this solution seems too easy to prove, let us finally try a not-so-fast growing (that is, the "right") function: $T(n) = O(n\log^2 n)$. Now, to show $T(n) \leq cn\log^2 n$ for some positive constant c and every sufficiently large n, in this case we can choose $c = a + b$ and $n_0 = 2$. Again it is easy to see that the base case holds. For inductive step when $n > n_0$ we have

$$T(n) = 2T(n/2) + bn\log n$$

$$\leq 2\left(c\frac{n}{2}\log^2\frac{n}{2}\right) + bn\log n$$

$$= cn(\log^2 n - 2\log n + 1) + bn\log n$$

$$\leq cn\log^2 n - 2cn\log n + cn + cn\log n$$

$$= cn\log^2 n - cn\log n + cn$$

$$= cn\log^2 n - cn(\log n - 1)$$

$$\leq cn\log^2 n.$$

Therefore, it follows that $T(n)$ is indeed $O(n\log^2 n)$.

4.4. The master method. Previous techniques for solving recurrence equations are *ad hoc* methods and require mathematical sophistication in order to be used effectively. There is, nevertheless, one quite general method for solving divide-and-conquer recurrence equations that does not require explicit use of induction. This **master method** is a "cookbook" method for determining the asymptotic solutions of a wide variety of recurrence equations of the form

$$T(n) = aT(n/b) + f(n),$$

where $a \geq 1$ and $b > 1$ are constants, and $f(n)$ is an asymptotically positive function.

Such a recurrence describes the running time of a divide-and-conquer algorithm that divides a problem of size n into a subproblems of size at most n/b each, solves each subproblem recursively, and then combines the subproblem solutions into a solution to the entire problem. The total cost of dividing the problem into subproblems and combining the subproblem solutions is described by the function $f(n)$. Thus, it is indeed a general form of the divide-and-conquer recurrence equations.

The master method requires memorization of three cases, but then the solution to many recurrences can be determined quite easily, often by simply writing down the answer. Each case is distinguished by comparing $f(n)$ to the special function $n^{\log_b a}$ in the following theorem.

THEOREM (MASTER THEOREM) *Let $T(n)$ be defined on the nonnegative integers by the recurrence*

$$T(n) = \begin{cases} d, & n \le n_0 \\ aT(n/b) + f(n), & n > n_0 \end{cases}$$

where $n_0 \ge 1$ is an integer constant, $a \ge 1$, $b > 1$, and $d > 0$ are real constants, and $f(n)$ is a function that is positive for $n \ge n_0$.

Case 1: *If $f(n) = O(n^{\log_b a}/n^\epsilon)$ for some constant $\epsilon > 0$, then $T(n) = \Theta(n^{\log_b a})$.*
Case 2: *If $f(n) = \Theta(n^{\log_b a})$, then $T(n) = \Theta(n^{\log_b a} \log n)$.*
Case 3: *If $f(n) = \Omega(n^{\log_b a} \cdot n^\epsilon)$ for some constant $\epsilon > 0$, and if $af(n/b) \le \delta f(n)$ for some constant $\delta < 1$ and all sufficiently large n, then $T(n) = \Theta(f(n))$.*

We will not prove the master theorem as stated here in its full generality. The proof is not hard, but it's rather involved and technical. Instead, we will give first intuition behind the master theorem, and then show a simplified version of it that covers most of the recurrences arising in the analysis of algorithms.

Intuitively, the master theorem says that the solution to the recurrence is determined by larger of the two functions $f(n)$ and $n^{\log_b a}$. If, as in case 1, the function $n^{\log_b a}$ is larger, then the solution is $T(n) = \Theta(n^{\log_b a})$. If, as in case 3, the function $f(n)$ is larger, then the solution is $T(n) = \Theta(f(n))$. If, as in case 2, the two functions have the same growth rate, we multiply either by a logarithmic factor so that the solution is $T(n) = \Theta(n^{\log_b a} \log n) = \Theta(f(n) \log n)$.

There are also some technical details that must be understood beyond this "big picture." In case 1, not only must $f(n)$ be smaller than $n^{\log_b a}$, it must be *polynomially* smaller. That is, $f(n)$ must be asymptotically smaller than $n^{\log_b a}$ by a factor of n^ϵ for some (small) constant $\epsilon > 0$. In case 3, not only must $f(n)$ be larger than $n^{\log_b a}$, it must be polynomially larger and, in addition, satisfy the regularity condition that $af(n/b) \le \delta f(n)$.

It is important to realize that the three cases do not cover all the possibilities for $f(n)$. There is a gap between cases 1 and 2 when $f(n)$ is smaller than $n^{\log_b a}$ but not polynomially smaller. Similarly, there is a gap between cases 2 and 3 when $f(n)$ is larger than $n^{\log_b a}$ but not polynomially larger. If the function $f(n)$ falls into one of these gaps, or if the regularity condition in case 3 is not satisfied, the master theorem cannot be used to solve the general divide-and-conquer recurrence.

We can intuitively see how each of the three cases is derived if we apply the iterative substitution method to the general divide-and-conquer recurrence equation. Assuming n is a power of b and $n_0 = 1$, we get

$$\begin{aligned} T(n) &= aT(n/b) + f(n) \\ &= a(aT(n/b^2) + f(n/b)) + f(n) \\ &= a^2 T(n/b^2) + af(n/b) + f(n) \\ &= a^3 T(n/b^3) + a^2 f(n/b^2) + af(n/b) + f(n) \end{aligned}$$

$$\vdots$$

$$= a^{\log_b n} T(1) + \sum_{i=0}^{\log_b n - 1} a^i f(n/b^i)$$

$$= n^{\log_b a} \cdot d + \sum_{i=0}^{\log_b n - 1} a^i f(n/b^i),$$

where the last substitution is based on the identity $a^{\log_b n} = n^{\log_b a}$. Indeed, this is where the special function comes from. If we now consider the last two terms, we can see that case 1 comes from the situation when $f(n)$ is small and the first term above dominates. Case 2 denotes the situation when each product of the terms in the above summation is proportional to $n^{\log_b a}$, so the characterization of $T(n)$ is $n^{\log_b a}$ times a logarithmic factor. Finally, case 3 denotes the situation when the first term is smaller than the second, and the summation above is a sum of geometrically decreasing terms that start with $f(n)$. Hence, $T(n)$ is itself proportional to $f(n)$.

To make this discussion more concrete, we give a simplified version of the master theorem in which basically $f(n)$ is a polynomial function of n.

THEOREM (SIMPLIFIED MASTER THEOREM)   *Let $T(n)$ be defined on the non-negative integers by the recurrence*

$$T(n) = \begin{cases} d, & n \le n_0 \\ aT(n/b) + cn^k, & n > n_0 \end{cases}$$

where $n_0 \ge 1$ is an integer constant, and $a \ge 1$, $b > 1$, $c > 0$, $d > 0$, and $k \ge 0$ are real constants.

Case 1: *If $a > b^k$, then $T(n) = \Theta(n^{\log_b a})$.*
Case 2: *If $a = b^k$, then $T(n) = \Theta(n^k \log n)$.*
Case 3: *If $a < b^k$, then $T(n) = \Theta(n^k)$.*

PROOF: (Sketch) We make the simplified assumption that n is a power of b (and $n_0 = 1$), so that floors and ceilings do not blur the general picture. We note though that it is reasonable to do so, and the proof with messy details extends for general n. Now, from the discussion above and taking $f(n) = cn^k$, it follows that

$$T(n) = dn^{\log_b a} + \sum_{i=0}^{\log_b n - 1} a^i \cdot c\left(n/b^i\right)^k$$

$$= dn^{\log_b a} + cn^k \cdot \sum_{i=0}^{\log_b n - 1} \left(\frac{a}{b^k}\right)^i.$$

Since the summation on the right-hand side is a geometric series, the claim now follows by considering the cases on the value of the ratio a/b^k. First, if $a = b^k$, i.e., $k = \log_b a$, then

$$T(n) = dn^k + cn^k \log_b n = \Theta(n^k \log n).$$

Second, if $a \neq b^k$, then

$$T(n) = dn^{\log_b a} + cn^k \cdot \frac{(a/b^k)^{\log_b n} - 1}{a/b^k - 1}.$$

Thus, if $a < b^k$, hence $k > \log_b a$ and $0 < a/b^k < 1$, then

$$T(n) = dn^{\log_b a} + cn^k \cdot \Theta(1) = \Theta(n^k),$$

since the second term is dominant and $\sum_{l=0}^{m} x^l = \Theta(1)$ if $0 < x < 1$.

On the other hand, if $x > 1$ then $\sum_{l=0}^{m} x^l = \Theta(x^m)$, and so if $a > b^k$, hence $a/b^k > 1$, then

$$T(n) = dn^{\log_b a} + cn^k \cdot \Theta\left(\left(a/b^k\right)^{\log_b n}\right)$$

$$= dn^{\log_b a} + cn^k \cdot \Theta\left(\frac{n^{\log_b a}}{n^k}\right)$$

$$= \Theta\left(n^{\log_b a}\right). \blacksquare$$

To solve a recurrence by the master method, we simply determine which case (if any) of the master theorem applies and write down the answer. For example, consider the recurrence

$$T(n) = 4T(n/2) + n.$$

If we use the simplified master theorem in this simple case, since $a = 4$, $b = 2$, and $k = 1$ implies $a > b^k$, we can immediately conclude that $T(n) = \Theta(n^{\log_2 4}) = \Theta(n^2)$. Of course, we get the same result by using the most general master theorem, although we have to work a little bit harder to recognize the correct case. Namely, we have $a = 4$, $b = 2$, $f(n) = n$, and thus $n^{\log_b a} = n^{\log_2 4} = n^2$. Since $f(n) = O(n^2/n^\epsilon)$ for $\epsilon = 1$, we can apply case 1 of the master theorem to get $T(n) = \Theta(n^2)$.

As an another example, consider the recurrence

$$T(n) = 2T(\sqrt{n}) + \log n = 2T(n^{1/2}) + \log n.$$

Unfortunately, this equation is not in a form that allows us to use the master method. However, we can put it into such a form by introducing new variable with the substitution $m = \log n$. Then $n = 2^m$ and we can write

$$T(2^m) = 2T(2^{m/2}) + m.$$

Letting $S(m) = T(2^m)$ we get the recurrence for the function $S(m)$:

$$S(m) = 2S(m/2) + m.$$

Now, this recurrence equation allows us to use the master method, which implies that $S(m) = \Theta(m \log m)$. Substituting back for $T(n)$ gives us that $T(n) = \Theta(\log n \log \log n)$.

Finally, the master method does not apply for the recurrence

$$T(n) = 2T(n/2) + n \log n,$$

although it has the proper form: $a = 2$, $b = 2$, $f(n) = n \log n$, and $n^{\log_b a} = n^{\log_2 2} = n$. It might seem that case 3 should apply, because $f(n) = n \log n$ is asymptotically larger than $n^{\log_b a} = n$. However, it is not polynomially larger, since the ratio $f(n)/n^{\log_b a} = (n \log n)/n = \log n$ is asymptotically less than n^ϵ for any positive constant ϵ. Consequently, the recurrence falls into the gap between case 2 and case 3, hence its solution must be sought by some other method. It turns out that by using the same variable substitution as in the previous example and applying the iterative substitution method, we can show in this case that $T(n) = \Theta(n \log^2 n)$.

5. Multiplication of large integers

As a practical application of the master theorem, consider the problem of multiplying two n-bit integers. Suppose that the number of bits n is large (say, $n > 100$), so that the integers cannot be handled directly by the arithmetic unit of standard computers. In other words, we cannot assume that the common arithmetic operations on large integers are basic operations that take a constant time for execution. Obviously, the time for their execution depends on the length of the representation of large integers. That is, it's a function of the number of bits used to represent a large integer in the base-2 number system. Multiplication of large integers has applications, for instance, to cryptography, where large integers are used in encryption schemes.

Given two large integers A and B represented in binary with n bits each, we can easily compute $A + B$ and $A - B$ using the common algorithm taught in elementary school. It is clear that the algorithm takes $O(n)$ time if we count single-digit additions and subtractions as basic operations. However, if we recall that the elementary school algorithm for multiplication involves computing n partial products of length n, we see that its running time is $O(n^2)$ if we count single-digit multiplications and additions as basic operations. Using the divide-and conquer technique though, we can design a subquadratic-time algorithm for multiplying two n-bit integers.

For simplicity, suppose n is a power of 2. If this is not the case, we can pad A and B to the left with zeros. Applying the divide-and-conquer approach, we divide the bit representations of A and B in two halves of $n/2$ bits each. One half represents the higher-order bits and the other represents the lower-order bits. If we denote the higher-order halves of A and B by I and K, respectively, and similarly the lower-order halves by J and L, we can write

$$A = I2^{n/2} + J$$

and

$$B = K2^{n/2} + L.$$

This halving of A and B is illustrated in Figure 10, where the numbers are given in decimal instead of binary.

The product of A and B can now be rewritten as

$$A \cdot B = \left(I2^{n/2} + J\right) \cdot \left(K2^{n/2} + L\right),$$

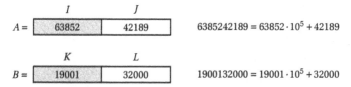

FIGURE 10. Dividing n-digit integers into $n/2$-digit halves.

or

(14) $$A \cdot B = IK \cdot 2^n + (IL + JK) \cdot 2^{n/2} + JL.$$

This formula suggests an easy divide-and-conquer algorithm for computing the product $A \cdot B$ as follows. Divide the n-bit numbers A and B in half, recursively compute the products involving resulting halves in (14) (IK, IL, JK, JL), and then combine the subproducts according to (14). We can terminate the recursion when we get down to the multiplication of two 1-bit numbers, which is straightforward.

Observe that multiplying a binary number by a power of 2, say 2^m, is trivial—it can be simply done by shifting the number left by m positions (or padding the number to the right with m zeros). Thus, multiplying an n-bit integer by 2^m takes $O(n + m)$ time if we count single-digit shifts as basic operations.

Now, to recursively compute $A \cdot B$ using the expansion (14) for the product, we have to perform four multiplications of $n/2$-bit integers (IK, IL, JK, JL), three additions of integers with at most $2n$ bits (corresponding to the three + signs), and two shifts of integers with roughly n bits (multiplications by 2^n and $2^{n/2}$). As each of these additions and shifts take $O(n)$ time, the running time of the divide-and-conquer algorithm $T(n)$ can be characterized by the recurrence

$$T(n) = \begin{cases} c, & n = 1 \\ 4T(n/2) + cn, & n > 1, \end{cases}$$

for some constant c. To solve this recurrence equation, we can apply the simplified master theorem with $a = 4$, $b = 2$, and $k = 1$. Since we are then in the first case of the theorem, we see that $T(n) = \Theta(n^{\log_2 4}) = \Theta(n^2)$. Unfortunately, this is no better than the elementary school algorithm.

However, the master method gives us some insight into how we can get an asymptotic improvement. If we can somehow decrease the number a of subproblems, i.e., the number of recursive calls, then we will reduce the complexity of the special function $n^{\log_b a}$ used in the running time. It may be a surprise that we can do so, but consider the following product of two $n/2$-bit integers:

(15) $$(I - J) \cdot (L - K) = IL - JL - IK + JK.$$

This is admittedly a strange product to consider, but comparing (14) we see that the product (15) in the expanded form contains the sum of two subproducts we need to compute ($IL + JK$), as well as two subproducts (JL and IK) we need to compute recursively anyway. Therefore, the product of two n-bit integers A and

B can be written as

(16) $$A \cdot B = IK \cdot 2^n + \left((I - J) \cdot (L - K) + JL + IK \right) \cdot 2^{n/2} + JL,$$

and hence computed using only three multiplications of $n/2$-bit integers (IK, JL, and $(I - J) \cdot (L - K)$), six additions or subtractions, and two shifts. Since all but the multiplications take $O(n)$ time, the running time $T(n)$ of a recursive algorithm for multiplication of two n-bit integers using formula (16) is given by the recurrence

$$T(n) = \begin{cases} c, & n = 1 \\ 3T(n/2) + cn, & n > 1, \end{cases}$$

whose solution by the simplified master theorem is $T(n) = O\left(n^{\log_2 3}\right) = O\left(n^{1.585}\right)$.

Although the divide-and-conquer algorithm for multiplication of two n-bit integers using (16) is now an easy exercise, for the sake of completeness and the fact that it requires multiplication of negative and positive $n/2$-bit integers, we give the complete algorithm MULTIPLY(A, B, n). The numbers A and B it takes as arguments are signed n-bit integers, and n is a power of 2. The algorithm uses operations $sign(X)$ and $abs(X)$ that return the sign and absolute value, respectively, of the bit representation of the number X. In addition, the algorithm uses operations $copy\text{-}left(X, m)$ and $copy\text{-}right(X, m)$ that copy left or right m bits of X, as well as $shift\text{-}left(X, m)$ that shifts X left for m positions. Finally, the expression of the form $s * X$ in the algorithm makes the number X to be of the sign s.

```
MULTIPLY(A, B, n)
  s = sign(A) ∧ sign(B);
  A = abs(A); B = abs(B);
  if n = 1 then
      return s * (A ∧ B);
  else
      I = copy-left(A, n/2);
      J = copy-right(A, n/2);
      K = copy-left(B, n/2);
      L = copy-right(B, n/2);
      X = MULTIPLY(I, K, n/2);
      Y = MULTIPLY(I − J, L − K, n/2);
      Z = MULTIPLY(J, L, n/2);
      P = s * (shift-left(X, n) + shift-left(X + Y + Z, n/2) + Z);
      return P;
```

We have designed a divide-and-conquer algorithm for integer multiplication that is asymptotically faster than the elementary school method, taking $O\left(n^{1.585}\right)$ time against $O\left(n^2\right)$. We can actually do even better than this, achieving a running time $O(n \log n \log \log n)$, by using a more complex divide-and-conquer algorithm called the *fast Fourier transform*. However, in our everyday life we still use the straightforward quadratic-time method for two reasons. First, it is much simpler to learn. And second, the constants hidden in the asymptotic notation are such that for small numbers (up to about 100 digits) the elementary school method is superior.

Exercises

1. The elements of a game are n counters divided into several piles. Assume n is a power of 2, i.e, $n = 2^k$ for some $k \in \mathbb{N}$. One move of the game consists of the following: choose two piles A and B having a and b counters, respectively, and then if, say, $a \geq b$ move b counters from the pile A onto the pile B. The goal of the game is to perform a sequence of the moves such that all counters end up in one pile. Give an algorithm that solves this game and determine the running time (that is, the number of moves) of your algorithm as a function of the number of counters n.

2. Write pseudocode for the PARTITION procedure of quick sort that selects the middle element of a subarray as the pivot. Argue that the running time of quick sort is not affected by this choice of the pivot element.

3. Describe a divide-and-conquer algorithm that determines the largest and second-largest element of an array A of size n. How many comparisons of array elements does your algorithm perform?

4. **Celebrity problem.** Suppose there are $n \geq 1$ persons in a room. A *celebrity* is someone who knows no other person but himself/herself, and everyone else knows him/her. If the relationships of persons knowing each other are represented in an appropriate way, the celebrity problem asks to give an efficient algorithm to determine the celebrity, if one exists, or to report that there is no such person. More precisely, denote by $x_1, x_2, \ldots, x_n$ the n persons and represent their relationships by an $n \times n$ matrix A such that

$$A(i, j) = \begin{cases} 1, & \text{if } x_i \text{ knows } x_j \\ 0, & \text{otherwise} \end{cases}$$

for $i, j = 1, 2, \ldots, n$. Devise a divide-and-conquer algorithm that returns the index of the celebrity person among n persons. If there is no celebrity, the algorithm should return 0. What is the running time of your algorithm? (*Hint:* Observe that if x knows y, then x is not a celebrity; otherwise, y is not a celebrity.)

5. **Fibonacci numbers.** The Fibonacci sequence of numbers is defined by the following recurrence equation:

$$F_0 = 0, \; F_1 = 1,$$
$$F_n = F_{n-1} + F_{n-2}, \; n \geq 2.$$

Thus, the n-th ($n \geq 2$) Fibonacci number is equal to the sum of previous two Fibonacci numbers. The initial part of the sequence is

$$0, 1, 1, 2, 3, 5, 8, 13, 21, 34, 55, \ldots.$$

(a) Write a divide-and-conquer algorithm that, given n, computes the n-th Fibonacci number by directly using the definition of the Fibonacci sequence. Estimate the running time of your algorithm.

(b) Prove that Fibonacci numbers satisfy the following equations for every integer $k \geq 1$:

$$F_{2k-1} = (F_k)^2 + (F_{k-1})^2$$
$$F_{2k} = (F_k)^2 + 2F_k F_{k-1}.$$

(c) Write another divide-and-conquer algorithm that computes the n-th Fibonacci number using the previous equations. What is the running time of this algorithm?

(d) Use the matrix

$$M = \begin{pmatrix} 0 & 1 \\ 1 & 1 \end{pmatrix}$$

and its n-th power M^n to write yet another divide-and-conquer algorithm that computes the n-th Fibonacci number. What is the running time of this algorithm?

6. Assuming n is a power of 3, solve the recurrence

$$T(n) = \begin{cases} 1, & n < 3 \\ 4T(n/3) + n^2, & n \geq 3 \end{cases}$$

using

(a) the iterative substitution method;
(b) the recursion tree method;
(c) the (simplified) master theorem.

7. **"Russian peasant" multiplication.** Consider the following algorithm that computes the product of two nonnegative integers x and y:

```
RUSSE(x, y)
  p = 0;
  while x > 0 do
    if x is odd then p = p + y;
    x = ⌊x/2⌋;
    y = 2·y;
  return p;
```

(a) Argue that the RUSSE algorithm correctly computes the product of x and y.

(b) What is the running time of the algorithm if the input numbers are small, and what if the numbers are large?

(c) Give a divide-and-conquer algorithm that uses the same idea for multiplication of x and y.

(d) What is the running time of the divide-and-conquer algorithm if the input numbers are small, and what if the numbers are large?

8. Give an algorithm that performs division of large integers. That is, given large positive integers a and b, the algorithm computes quotient q and remainder r such that $a = qb + r$, where $0 \le r < b$. What is the running time of your algorithm?

9. If x is a large n-bit integer, give an algorithm that computes $\lfloor\sqrt{x}\rfloor$ by trying successive approximations to $\lfloor\sqrt{x}\rfloor$. What is the running time of your algorithm?

10. If x is a large n-bit integer, give a divide-and-conquer algorithm that computes $\lfloor\sqrt{x}\rfloor$. What is the running time of your algorithm?

11. Consider the following algorithm that computes the number of nodes in a binary tree rooted at node x:

```
BT-NodeCount(x)
   if x = NIL then
      return 0;
   else
      return 1 + BT-NodeCount(left(x)) + BT-NodeCount(right(x));
```

(a) If a binary tree S has n nodes, give the recurrence equation for the running time $T(n)$ of the call BT-NodeCount(root(S)).
(b) Prove by induction that $T(n) = O(n)$.

12. In an unsorted array A of n *different* numbers, a pair of distinct elements $A(i)$ and $A(j)$ is called an ***inversion*** if $i < j$ and $A(i) > A(j)$. For example, in the array $[5,3,10,2,8,9,20]$ there are 6 inversions: $(5,3)$, $(5,2)$, $(3,2)$, $(10,2)$, $(10,8)$, $(10,9)$.

(a) Write a divide-and-conquer algorithm that computes the total number of inversions in the array A.
(b) Give and solve the recurrence equation for the running time of your algorithm.

13. Give and analyze a divide-and-conquer algorithm that solves the maximum contiguous subsequence sum problem from Section 1.6 in Chapter 2. (*Hint:* Divide the input array into two halves and observe that the maximum-sum contiguous subsequence can either reside entirely in the first half, or reside entirely in the second half, or begin in the first half but end in the second half. In the last case find it with a right-to-left scan and a left-to-right scan from the middle element.)

14. Given a set of n teams in some sport, a ***round-robin tournament*** is a collection of games in which each team plays each other team exactly once. Design

a divide-and-conquer algorithm for constructing a round-robin tournament assuming n is a power of 2.

15. Given two complex numbers $a + bi$ and $c + di$, show how their product $(a + bi)(c + di) = (ac - bd) + (ad + bc)i$ can be computed using only three multiplications.

16. Given two linear polynomials $ax + b$ and $cx + d$, their product

$$(ax + b)(cx + d) = acx^2 + (ad + bc)x + bd$$

can be computed with four multiplications of their coefficients: $ac, ad, bc,$ and bd.

   (a) Using this observation, describe a divide-and-conquer algorithm that computes the product of two n-degree polynomials

$$A_n(x) = a_n x^n + a_{n-1} x^{n-1} + \cdots + a_1 x + a_0$$

     and

$$B_n(x) = b_n x^n + b_{n-1} x^{n-1} + \cdots + b_1 x + b_0.$$

     Assuming that the polynomials $A_n(x)$ and $B_n(x)$ are given by the arrays of their coefficients, the result of the algorithm should be another array representing coefficients of the polynomial $A_n(x) \cdot B_n(x)$.

   (b) Give and solve the recurrence equation for the running time of the algorithm in (a).

   (c) Show how the product of two linear polynomials $ax + b$ and $cx + d$ can be computed with only *three* multiplications of the coefficients.

   (d) Based on the solution to (c), describe another divide-and-conquer algorithm that computes the product of two n-degree polynomials.

   (e) Give and solve the recurrence equation for the running time of the second algorithm.

6

Greedy Algorithms

The *greedy method* is typically used to solve optimization problems, that is, problems that involve searching through a set of configurations to find one that minimizes or maximizes an objective function defined on these configurations. A configuration that minimizes or maximizes the objective function is called an *optimal solution* for the problem. In simple words, an optimization problem is one in which you want to find not just *a* solution, but the *best* solution.

The main idea of the greedy method, as the name implies, is to make a series of greedy choices in order to construct an optimal solution (or close to optimal solution) for a given problem. Thus, to solve a given optimization problem, we proceed step by step making a sequence of choices. The sequence starts from some well-understood starting configuration, and then it is iteratively built by choosing the best configuration from all of those that are currently possible in each step.

It is clear why this approach is called "greedy": a solution (hopefully optimal) is built from partial solutions step by step, and at every step is chosen one of the best possible candidates. Moreover, in this single-minded "take the best you can get right now" strategy, the choice is made without worrying about the future: if any of the best candidates is promising, it is included in the solution for good. A promising candidate is one that extends the current partial solution to a feasible solution. A *feasible solution* is a partial solution that is possible to complete to at least one global solution. Also, a greedy algorithm never changes its mind: once a candidate is thrown away, it is never reconsidered.

A greedy algorithm always makes the choice that looks best at the moment, but we should emphasize that the greedy approach does not always lead to an optimal solution. Just as in life, a greedy strategy may produce a good result for a while, yet the overall result may be poor. But there is a class of optimization problems for which "the greed is good." These problems are characterized by two properties:

1. *Greedy-choice property.* This is the property that a global optimal configuration can be reached by a series of locally optimal choices (that is, choices that are the best from among the possibilities available at a time).
2. *Optimal-substructure property.* This property assures that an optimal solution to the problem contains an optimal solution to subproblems.

In this chapter we convey the idea behind the greedy approach using two simple problems. Other more important examples can be found in Chapter 10, where we study problems related to weighted graphs.

1. Change-making problem

Consider the problem of making change: we have available an unlimited number of coins whose values are 1, 5, 10, and 25 units, and we need to return an amount $n \geq 0$ in change. Moreover, the number of coins returned should be minimal.

Almost without thinking, we would first select the largest coin whose value is not greater than n, then select the largest coin whose value is not greater than the difference of n and the value of the first coin, and so on. For example, if $n =$ 68, we select first the largest coin whose value is not greater than 68 (a 25-unit coin), and subtract its value from 68 getting 43. We then select the largest coin whose value is not greater than 43 (another 25-unit coin), and subtract its value from 43 getting 18. This procedure is repeated until we get to the zero amount that needs to be converted in change. In this way, we convert the amount of 68 units into two 25-unit coins, one 10-unit coin, one 5-unit coin, and three 1-unit coins.

This algorithm uses the greedy strategy by selecting the largest promising coin at any particular stage. A promising candidate in this case is a coin whose value does not exceed the current amount for which we are to make change.

In pseudocode, the following is the greedy algorithm MAKE-CHANGE(n) that returns change for an amount of $n \geq 0$ units. The number of 1-unit coins returned is g_1, the number of 5-unit coins returned is g_2, the number of 10-unit coins returned is g_3, and the number of 25-unit returned coins is g_4.

```
MAKE-CHANGE(n)
    g₁ = 0; g₂ = 0; g₃ = 0; g₄ = 0;
    while n ≥ 25 do g₄ = g₄ + 1; n = n − 25;
    while n ≥ 10 do g₃ = g₃ + 1; n = n − 10;
    while n ≥ 5 do g₂ = g₂ + 1; n = n − 5;
    while n ≥ 1 do g₁ = g₁ + 1; n = n − 1;
    return g₁, g₂, g₃, g₄;
```

We have written the MAKE-CHANGE algorithm so that it closely mimics the "everyday's life algorithm." However, observe that the **while** loops actually compute remainders when a correspondong amount is divided by the coin values 25, 10, 5, and 1, respectively. Thus, we can write the sequence of the **while** loops as the following sequence of assignments:

$$g_4 = \lfloor n/25 \rfloor;$$
$$g_3 = \lfloor (n - 25g_4)/10 \rfloor;$$
$$g_2 = \lfloor (n - 25g_4 - 10g_3)/5 \rfloor;$$
$$g_1 = n - 25g_4 - 10g_3 - 5g_2;$$

In other words, the MAKE-CHANGE algorithm is a constant-time algorithm.

1.1. Optimality of the MAKE-CHANGE algorithm. We noted that a greedy solution is not necessarily an optimal solution. Moreover, frequently it is not outright clear whether a greedy solution is indeed an optimal one. In many cases therefore we need a formal mathematical argument of this fact, or a counterexample that shows suboptimality of the greedy approach.

Even in the simple case of the MAKE-CHANGE algorithm, it is not obvious that it does produce an optimal solution. For example, the greedy algorithm for making change produces an overall optimal solution only because of the special values of the coins. If the coins had values, say, 1, 5, and 12 units, and we were to make change of the amount of 15 units, then the greedy solution would first select one 12-unit coin and then three 1-unit coins, for a total of four coins. However, three 5-unit coins is a better solution.

Let's now turn to a rigorous proof of optimality of the greedy algorithm using formal mathematical statements and tools. Suppose that we have an unlimited number of coins with values from the set $\{1, 5, 10, 25\}$. The change-making problem can be rephrased as an optimization problem as follows. Given a nonnegative integer n, minimize the sum $n_1 + n_2 + n_3 + n_4$ subject to $n = 1 \cdot n_1 + 5 \cdot n_2 + 10 \cdot n_3 + 25 \cdot n_4$ and each n_i is a nonnegative integer.

Given a nonnegative integer n, any 4-tuple (n_1, n_2, n_3, n_4) of nonnegative integers such that $n = 1 \cdot n_1 + 5 \cdot n_2 + 10 \cdot n_3 + 25 \cdot n_4$ is called a *solution* for n. A solution that furthermore minimizes the sum $n_1 + n_2 + n_3 + n_4$ is called a *minimal (optimal) solution*. For any nonnegative integer n, it is clear that there are only finitely many solutions, hence a minimal solution always exists.

We first observe that the 4-tuple (g_1, g_2, g_3, g_4) returned by the (revised) algorithm MAKE-CHANGE(n) is a solution for n. We need to show that (g_1, g_2, g_3, g_4) is a minimal solution. To do this formally, we first prove two easy lemmas. The first lemma basically says that it is always better to replace five 1-unit coins by one 5-unit coin, two 5-unit coins by one 10-unit coin, and three 10-unit coins by one 25-unit coin and one 5-unit coin.

LEMMA 1   *If (a, b, c, d) is a minimal solution for n, then $a \le 4$, $b \le 1$, and $c \le 2$.*

PROOF: For each inequality involving a, b, and c in the claim, we argue that assuming to the contrary would lead us to a contradiction with minimality of the solution (a, b, c, d).

Well, if $a \ge 5$, then we can write $a = 5k + a'$ for some $k \ge 1$ and $0 \le a' < 5$. This implies $a = 5k + a' > k + a'$ since $k \ge 1$. But then the 4-tuple $(a', k + b, c, d)$ is a solution for n, because all of its components are obviously nonnegative integers, and

$$25d + 10c + 5(k+b) + 1a' = 25d + 10c + 5k + 5b + a'$$
$$= 25d + 10c + 5b + (5k + a')$$
$$= 25d + 10c + 5b + a$$
$$= n.$$

Moreover,

$$a' + (k+b) + c + d = (a' + k) + b + c + d < a + b + c + d.$$

Thus, $(a', k+b, c, d)$ is a strictly better solution than (a, b, c, d), which is a contradiction.

Similarly, if $b \geq 2$, then we can write $b = 2k + b'$ for some $k \geq 1$ and $0 \leq b' < 2$. By the same token, $b > k + b'$ since $k \geq 1$, and the solution $(a, b', k+c, d)$ for n contradicts the minimality of (a, b, c, d).

Finally, if $c \geq 3$, then we can write $c = 3k + c'$ for some $k \geq 1$ and $0 \leq c' < 3$. Again, we have that $c > 2k + c'$ since $k \geq 1$. The solution for n that works now is $(a, k+b, c', k+d)$. For,

$$25(k+d) + 10c' + 5(k+b) + 1a = 25k + 25d + 10c' + 5k + 5b + a$$
$$= 25d + 30k + 10c' + 5b + a$$
$$= 25d + 10(3k + c') + 5b + a$$
$$= 25d + 10c + 5b + a$$
$$= n.$$

Also,

$$a + (k+b) + c' + (k+d) = a + b + (2k + c') + d < a + b + c + d,$$

again a contradiction. ∎

The second lemma says that if (a, b, c, d) is a minimal solution for n, then d is the quotient and $10c + 5b + a$ is the remainder of n divided by 25.

LEMMA 2   *If (a, b, c, d) is a minimal solution for n, then $10c + 5b + a < 25$.*

PROOF: We again argue by contradiction. If $10c + 5b + a \geq 25$, then by Lemma 1 it must be the case that b and c attain their upper bounds, i.e., $b = 1$ and $c = 2$. But then the solution $(a, 1, 2, d)$ could not be optimal, because the solution $(a, 0, 0, d+1)$ would be strictly better. ∎

The last fact we need is a well-known theorem from Number Theory, which we take without proof:

THEOREM   *Given two integers n and $m > 0$, there exist unique integers q and r such that $n = q \cdot m + r$ and $0 \leq r < m$. In fact, $q = \lfloor n/m \rfloor$ and $r = n - qm$.*

Now we can prove that the greedy algorithm produces an optimal solution for the change-making problem. Let (a, b, c, d) be a minimal solution for n. Then $n = 25d + 10c + 5b + a$ and, by Lemma 2, $10c + 5b + a < 25$. Taking $m = 25$ in the previous Theorem, we can immediately conclude that $d = \lfloor n/25 \rfloor = g_4$. Consider next the integer $n - 25g_4$. Because $n - 25g_4 = n - 25d = 10c + 5b + a$, and $5b + a < 10$ by Lemma 1, it follows again from the previous Theorem that $c = \lfloor (n - 25g_4)/10 \rfloor = g_3$. Similarly, repeating the same argument for the number $n - 25g_4 - 10g_3$, we see that $b = g_2$ and $a = g_1$.

This concludes the proof of optimality of the greedy solution. Actually, it shows more than that: the greedy solution is a *unique* optimal solution.

The last remark we make is that our argument does not work in the case when we don't have an unlimited number of coins of each value. This is so because those allegedly minimal solutions that lead to a contradiction in our lemmas may happen to contain coins that are not available at all. Nevertheless, the same ideas apply, and by more careful analysis and formula juggling one can prove that the greedy solution remains optimal.

2. Classroom-scheduling problem

Suppose we have one classroom and a set $C = \{1, 2, \ldots, n\}$ of n classes that can use that classroom. Each class i is given by its start time s_i and its finish time f_i, where $s_i < f_i$. Class i must start at time s_i and it is guaranteed to be finished by time f_i. Two classes i and j are nonconflicting if their meeting intervals do not overlap, that is, if $f_i \leq s_j$ or $f_j \leq s_i$. Two classes can be scheduled to meet in the classroom only if they are nonconflicting. The ***classroom-scheduling problem*** we consider here is to schedule as many classes as possible in the classroom in a nonconflicting way. This is another example of an optimization problem, because we need to maximize a target function representing the number of nonconflicting classes.

Figure 1 illustrates seven classes, given by the pairs of start and finish times, of which we can schedule at most four classes $\{1, 2, 6, 7\}$ in the classroom.

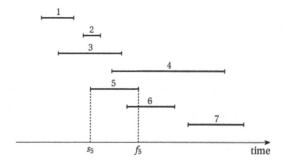

FIGURE 1. Seven classes given by their start and finish times.

The main idea for a solution to the classroom-scheduling problem is to schedule classes one by one and leave as much time as possible after each scheduled class with the hope that there is a higher chance of putting more classes within the longest remaining time. This idea can be implemented using the greedy approach. In this case, the greedy solution is represented by a set of (partially) scheduled classes G. Initially, we let the set G be empty. After that, in each step, we choose a promising class among those unscheduled, say p, whose finish time is the earliest. A promising class is one of those that are nonconflicting with any class in G. We then add p to G and repeat the process as long as we have a promising class to choose from unscheduled classes. The following is a pseudocode for the greedy approach.

CLASS-SCHEDULE(C)
   $G = \emptyset$; ⟦ greedy solution ⟧
   while there is a promising class in $C \setminus G$ **do**
      Chose a promising class p from $C \setminus G$ with the earliest finish time;
      $G = G \cup \{p\}$;
   return G;

Thus, the greedy choice in each step of this algorithm is the selection of a promising class with the earliest finish time. The intuition is that this will leave the longest free time for the classroom in each step, and this in turn will produce the largest number of classes that can be scheduled in the classroom.

To analyze the running time of the CLASS-SCHEDULE algorithm, we give its implementation in more detail. First, we assume that the set C of n classes is represented by an array whose elements are pairs of start and finish times of classes. Thus, $C(i)$ contains the pair (s_i, f_i), where s_i and f_i are, respectively, start and finish time of the class i. Second, to enable easy selection of a nonconflicting class with the earliest finish time, we sort the array C upfront in ascending order of finish time. Without loss of generality and for the sake of avoiding double subscripts, we can assume that as the result of the initial sorting we have obtained the order of classes so that $f_1 \le f_2 \le \cdots \le f_n$. Then, we repeatedly check if the start time of the next class is after the finish time of the last class added to the solution set G of scheduled classes. If so, the class is not in conflict with any class in G and has the earliest finish time, hence we include the class in the solution set G. Otherwise, the actual class is in conflict with at least one class from G, and so it cannot be part of the solution. The following is a pseudocode for this implementation.

CLASS-SCHEDULE(C)
   Sort C in ascending order of finish time;
   ⟦ Without loss of generality, assume $f_1 \le f_2 \le \cdots \le f_n$ ⟧
   $G = \{1\}$; ⟦ the earliest finish time class belongs to the greedy solution ⟧
   $j = 1$;
   for $i = 2$ **to** n **do**
     if $s_i > f_j$ **then**
       $G = G \cup \{i\}$;
       $j = i$;
   return G;

Figure 2 shows how this algorithm works for an example array of classes sorted in ascending order of finish time.

i	j	s_i	f_j	G
				$\{1\}$
2	1	4	3	$\{1,2\}$
3	2	2	6	$\{1,2\}$
4	2	7	6	$\{1,2,4\}$
5	4	5	10	$\{1,2,4\}$
6	4	9	10	$\{1,2,4\}$
7	4	13	10	$\{1,2,4,7\}$
8	7	11	15	$\{1,2,4,7\}$
9	7	10	15	$\{1,2,4,7\}$

C

i	1	2	3	4	5	6	7	8	9
s_i	1	4	2	7	5	9	13	11	10
f_i	3	6	8	10	12	14	15	16	18

FIGURE 2. Progress of the CLASS-SCHEDULE algorithm on the array C of classes sorted in ascending order of finish time.

Since an efficient sorting of the array C takes $O(n \log n)$ time and the **for** loop clearly takes $O(n)$ time, the total running time of the CLASS-SCHEDULE algorithm is $O(n \log n) + O(n) = O(n \log n)$.

2.1. Optimality of the CLASS-SCHEDULE algorithm. Recall that a greedy algorithm does not necessarily give the best solution. However, the CLASS-SCHEDULE algorithm does produce an optimal solution. This is certainly not obvious in this case, so we must show that the number of classes scheduled by this algorithm is the maximum possible. To prove this, we adopt the strategy of showing that the greedy solution is as good as any better solution. Specifically, we will argue that any better solution can be transformed to the greedy solution while keeping the same number of classes.

To this end, we first show that any solution set of classes $B_0 = \{ b_i \mid 1 \le i \le |B_0| \}$ better than the greedy solution $G = \{ g_i \mid 1 \le i \le |G| \}$, i.e., one with $|B_0| \ge |G|$, can be transformed into another better solution B such that $G \subseteq B$ and $|B_0| = |B|$.

Well, assume the classes in both G and B_0 are listed in ascending order of finish time. Suppose that G and B_0 agree on the first $m - 1$ classes and disagree on the m-th class. This means that the greedy algorithm chooses class $g_m = k$ in the corresponding step as the m-th element of G, while the better solution B_0 has the class $b_m = u \ne k$ as the m-th element. Observe that $m \le |G|$; otherwise, there is nothing to prove. Denote further $b_{m-1} = g_{m-1} = j$ and $b_{m+1} = v$. Then $f_j < s_k$ since $g_{m-1} \in G$. Also, $f_k \le f_u$ by greediness, and $f_u < s_v$ since $b_{m+1} \in B_0$. This implies $f_k < s_v$. Figure 3 illustrates this argument.

So, if B_1 is the set obtained by substituting the class k for the class u in B_0, i.e., $B_1 = (B_0 \setminus \{u\}) \cup \{k\}$, then B_1 is an alternative better solution than G since $|B_1| = |B_0| \ge |G|$. Moreover, B_1 and G disagree for one element less than B_0 and G.

Repeating now the similar argument for the sets B_1 and G, we obtain another better solution B_2 such that $|B_2| = |B_1|$, and B_2 and G disagree for two elements less than B_0 and G. Continuing in this way, we can incrementally transform B_0

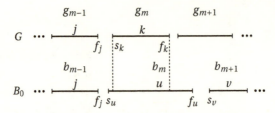

FIGURE 3. Transforming a better solution into the greedy solution.

until we get a better solution B that agrees with G on every class in G, that is, B simply extends G. More precisely, we can construct a sequence of transformations $B_0 \rightarrow B_1 \rightarrow \cdots \rightarrow B_\ell \rightarrow B$, where each B_i is a better solution than G and $|B_0| = |B_1| = \cdots = |B_\ell| = |B|$. Moreover, B is a better solution than G such that $G \subseteq B$.

But this is not possible unless $G = B$, since otherwise it would contradict the way the CLASS-SCHEDULE algorithm works. Namely, if $G \subset B$, then any $p \in B \setminus G$ would have been a promising class that the CLASS-SCHEDULE algorithm would have included in G.

Therefore, for any solution B_0 better than G, there is a solution B such that $|B_0| = |B| = |G|$. But this demonstrates that there can be no *strictly* better solution than the greedy solution, i.e., the greedy solution is optimal.

Exercises

1. The old Britain's coin system, which was in use before 1971 when decimal currency was introduced, had coin denominations (in units of a penny) $\frac{1}{2}$, 1, 3, 6, 12 (shilling), 24 (florin), 30 (half-crown), 60 (crown), and 240 (pound). Show that the change-making greedy algorithm is not optimal for this "natural" coin system.

2. Suppose that for the classroom-scheduling problem we use a different greedy strategy. Instead of selecting the class with the earliest finishing time, we select the last class (that is, the one with the latest starting time) that is nonconflicting with all previously selected classes. Argue that this greedy strategy yields an optimal solution as well.

3. Design an efficient greedy algorithm to schedule n classes in m classrooms in a nonconflicting way so that the number of used classrooms is minimal. As in the problem of one classroom scheduling, each class is given by its start time and its finish time. Prove that your algorithm is optimal.

4. Professor Greedworks drives an automobile from city A to city B along a prescribed route. His car's gas tank, when full, holds enough gas for him to drive n miles, and his map gives the distances between gas stations on his route. The professor wishes to make as few stops as possible for gas along the way. Give an efficient method by which Professor Greedworks can determine at which gas stations he should stop, and prove your strategy yields an optimal

solution.

5. **Art gallery guarding.** Imagine that a long hallway with paintings in an art gallery is represented by a line with points $x_1, x_2, \ldots, x_n$ on the line that specify the positions of paintings in this hallway. Suppose that a single guard can protect all the paintings within distance at most d on both sides of his position. Design an algorithm for finding a placement of guards that uses the minimum number of guards to guard all the paintings with given positions.

6. **Egyptian fractions.** Every positive fraction can be expressed as a sum of different positive fractions with unit numerators. For example,

$$\frac{41}{42} = \frac{1}{2} + \frac{1}{3} + \frac{1}{7}.$$

The name for such representations comes after the ancient Egyptians that used them. Notice that a fraction might have more than one Egyptian representation. Describe a greedy algorithm that, given a fraction, determines an Egyptian representation of the fraction.

7. **Map coloring.** The map coloring problem asks to assign colors to different countries of a given map, where adjacent countries are not allowed to have the same color. The goal is to use as few different colors as possible. Describe an efficient greedy algorithm that colors the map with different colors.

8. **Wire connections.** Suppose that n white dots and n black dots are equally spaced and arbitrarily interleaved on a straight line. We want to connect each white dot and a different black dot by a "wire" so that the total (horizontal) length of all n connections is minimal. (Connecting wires may cross each other.) For example, a solution to the following instance

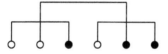

gives the total wire length of $2 + 2 + 3 = 7$. One can easily check that 7 is the minimal length in this instance. Notice however that there are other ways of connecting the dots that yield the minimum. Give a greedy algorithm for this problem. Does the greedy algorithm produces an optimal solution? If so, prove it; if not, provide a counterexample.

9. **The maximum-size independent-set problem on trees.** An *independent set* in a (rooted) tree T is a subset S of the nodes of T such that no node in S is a child or parent of any other node in S. Equivalently, no pair of nodes in S is connected by an edge from T. Give a greedy algorithm to find an independent set of the nodes in T of maximum size. In other words, the algorithm is to find an independent set S in T so that its size $|S|$ is as large as possible. (*Hint:* Show that if x is a leaf in a tree, then there exists a maximum-size independent set that contains x.)

10. Suppose we need to store n programs (or files) of length $\ell_1, \ell_2, \ldots, \ell_n$ on a magnetic tape. We know *a priori* how often each program is used and loaded in the main memory for execution: the program i is requested r_i times, $1 \leq i \leq n$. Each time a program is requested, the tape is wound up to the location of the program, the program is loaded in the main memory, and the tape is rewound to the beginning. (It goes without saying that the tape drive works with constant speed and density.)

Let $r = \sum_{i=1}^{n} r_i$ be the number of all requests for the programs, and $p_i = r_i / r$ be the fraction of all requests asking for the program i. If the programs are held in the order $i_1, i_2, \ldots, i_n$, the average time to load a program is

$$\hat{t} = c \sum_{k=1}^{n} \left(p_{i_k} \sum_{j=1}^{k} \ell_{i_j} \right),$$

where the constant c depends on the recording density and the speed of the tape drive. Our goal is to minimize the average load time $\hat{t}$ of a program.

(a) Give a counterexample showing that it is not necessarily optimal to store the programs in the order of increasing lengths ℓ_i.

(b) Give a counterexample showing that it is not necessarily optimal to store the programs in the order of increasing usage frequencies p_i.

(c) Prove that $\hat{t}$ is minimized if the programs are stored in the order of decreasing values of p_i / ℓ_i (or, equivalently, in increasing order of ℓ_i / p_i).

(d) Describe a greedy algorithm for the problem of optimally arranging programs on a tape. What is the time complexity of your algorithm?

7

Dynamic-Programming Algorithms

The *dynamic programming*[1] technique for algorithm design is similar to the divide-and-conquer technique in that it solves problems by dividing them into subproblems and combining their solutions. However, the dynamic programming is conceptually different from the divide-and-conquer technique, because divide-and-conquer algorithms divide the problem into independent subproblems. In contrast, dynamic programming is applicable when the subproblems are not independent, that is, when subproblems share smaller subproblems. In this case, a divide-and-conquer algorithm does more work than necessary, because it repeatedly solves the common subproblems many times. A dynamic-programming algorithm solves each subproblem just once and saves its solution in a table. Then, every time the subproblem is encountered, we simply look up the answer from the table, avoiding in this way the cost of recomputing the solution. In fact, the subproblem overlap improves the efficiency of a dynamic-programming algorithm.

The underlying idea of dynamic programming is therefore quite simple: avoid computing the same solution many times, usually by maintaining a table of known results and filling the table in as subproblems are solved. The form of a dynamic programming algorithm may vary, but there is the common theme of a table to fill in and an order in which the entries are to be filled. It is sometimes simpler from an implementation point of view to create a table of the solutions to all the subproblems we might ever have to solve. We fill in the table without regard to whether a particular solution to a subproblem is actually needed in the overall solution.

---

[1]The term *dynamic programming* comes from the operations research theory for solving problems in control systems. Thus, the word "programming" in this context, similarly as in the case of *linear programming*, refers to a tabular method, not to writing computer code.

Dynamic programming is a bottom-up technique. We start with the smallest, and thus the simplest subproblems. By combining their solutions, we obtain solutions to subproblems of increasing size until eventually we arrive at a solution to the original problem. On the other hand, divide-and-conquer is a top-down technique: the original problem is attacked head-on in its entirety, and then divided into smaller and smaller subproblems.

The dynamic programming technique is used primarily for optimization problems, where each solution has a value and we wish to find one of the best solutions among many possible such solutions. Often the number of different ways for computing the solution values is exponential, so a brute-force search for the best solution is computationally infeasible for all but the smallest problem sizes. However, we can successfully apply the dynamic programming technique in such situations, if the problem has a certain amount of structure that we can exploit.

The development of a dynamic-programming algorithm roughly follows four steps:

1. Break the global (optimization) problem into similar subproblems and characterize their structure in a simple way.
2. Recursively define the value of an optimal solution to the global problem in terms of the values of optimal subproblem solutions.
3. Compute the values of optimal subproblem solutions in a bottom-up fashion.
4. Construct an optimal solution from the computed intermediate results.

Step 4 may be omitted provided that only the value of an optimal solution is asked for. If we do need to provide an optimal solution itself, we usually maintain additional information in step 3 to enable reconstruction of an optimal solution.

There are few algorithmic techniques that can take problems that seem to require exponential time and produce efficient algorithms to solve them. Dynamic programming is one such technique. In addition, algorithms that result from applications of the dynamic programming technique are usually quite simple—often needing little more than a few lines of code to describe some nested loops for filling in a table. In this chapter, we show some simple examples that exhibit these merits of the dynamic programming technique.

1. Fibonacci numbers

The Fibonacci sequence of numbers is defined by the following recurrence relation:

$$F_0 = 0, F_1 = 1$$
$$F_n = F_{n-1} + F_{n-2}, \quad n \geq 2.$$

Thus, for $n \geq 2$, the n-th Fibonacci number is equal to the sum of previous two Fibonacci numbers. The beginning of the sequence looks like

$$0, 1, 1, 2, 3, 5, 8, 13, 21, 34, 55, \ldots$$

This sequence has numerous applications in art, mathematics, and computer science. De Moivre proved the following formula for the n-th Fibonacci number:

$$F_n = \frac{\phi^n - (1 - \phi)^n}{\sqrt{5}},$$

where the constant ϕ is called the **golden ratio** and has the value

$$\phi = \frac{1 + \sqrt{5}}{2} \approx 1.62.$$

For an approximation of F_n, the term $(1 - \phi)^n$ can be neglected when n is large, since $1 - \phi = (1 - \sqrt{5})/2 \approx -0.62$ and so $|1 - \phi| < 1$. This means that the value of F_n is in the order of ϕ^n, i.e., $F_n = \Theta(\phi^n)$. However, for large values of n, De Moivre's formula is of little help in calculating F_n exactly, since the degree of precision required for the values $\sqrt{5}$ and ϕ increases as n gets larger. (As an aside, it's quite remarkable that these two irrational numbers can be manipulated as in the De Moivre's formula to produce for each n, not only a rational number, but a natural one.)

Consider the problem of computing the value of the n-th Fibonacci number F_n. The first algorithm that solves this problem is obtained directly from the definition of Fibonacci sequence.

```
FIB1(n)
  if n < 2 then
    return n;
  else
    return FIB1(n − 1) + FIB1(n − 2);
```

This algorithm is very inefficient because it recalculates the same value many times. For instance, to calculate FIB1(5) we need values of FIB1(4) and FIB1(3). But FIB1(4) also calls for calculation of FIB1(3). This is illustrated by the tree of recursive calls in Figure 1, where $F(k)$ is the result of the call FIB1(k).

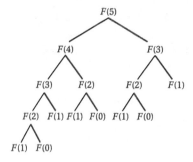

FIGURE 1. Recursive calls made by FIB1(5).

We see that $F(3)$ will be calculated twice, $F(2)$ three times, $F(1)$ five times, and $F(0)$ three times. In fact, the time required to calculate F_n using this algorithm is in the order of the value of F_n itself. To see this, we can write the running time $T_1(n)$ of the recursive FIB1 algorithm as the recurrence equation

$$T_1(n) = \begin{cases} O(1), & \text{if } n = 0 \text{ or } n = 1 \\ T_1(n-1) + T_1(n-2) + O(1), & \text{if } n \geq 2. \end{cases}$$

Since on the right-hand side there are two instances of the function T_1, to solve this recurrence we cannot use the iterative substitution or the master method from Chapter 5. However, since this recurrence looks very much like the definition of Fibonacci numbers, we can apply the guess-and-verify method and show by induction on n that $T_1(n) \geq F_{n+2}$ for all sufficiently large n. Moreover, using the fact $\phi^2 = \phi + 1$ and induction again, it is a simple matter to show that $F_{n+2} \geq \phi^n$ for all $n \in \mathbb{N}$. The running time of the FIB1 algorithm is thus $T_1(n) = \Omega(\phi^n)$. Finally, since $\phi \approx 1.62 > 1.6$, we can conclude $T_1(n) = \Omega(1.6^n)$, which grows exponentially with increasing n.

The second algorithm avoids unnecessary recomputations of the same values using the dynamic programming paradigm. Starting with F_0 and F_1, it calculates all Fibonacci numbers up to F_n by computing the next number in the sequence from previous two. At each step i, $(1 \leq i \leq n)$, it uses two variables to save the values of F_{i-1} and F_i needed for computation of F_{i+1} in the next step.

```
FIB2(n)
  x = 0; y = 1;
  if n < 2 then
    return n;
  else
    for i = 2 to n do
      z = x + y;
      x = y;
      y = z;
    return z;
```

If the running time of the FIB2 algorithm is $T_2(n)$, it is clear that $T_2(n) = O(n)$, which is much, much better than the running time of the first algorithm.

Notice that for this problem it is not necessary to maintain a table of intermediate results, not even an array, but only two additional variables of auxiliary storage—a negligible price to pay for the speed up.

2. Binomial coefficients

Given two nonnegative integers n and m such that $0 \leq m \leq n$, consider the problem of calculating the binomial coefficient $B(n, m) = \binom{n}{m}$ defined by

$$B(n, m) = \frac{n!}{m!(n-m)!}$$

for $n > 0$, and by convention $B(0,0) = 1$. If we make use of the Pascal's formula

$$\binom{n}{m} = \binom{n-1}{m-1} + \binom{n-1}{m}, \quad 0 < m < n$$

we can recursively compute $B(n, m)$ using the following algorithm.

```
BC1(n, m)
    if (m = 0) ∨ (m = n) then
        return 1;
    else
        return BC1(n − 1, m − 1) + BC1(n − 1, m);
```

Alas, the BC1 algorithm for $i < n$ and $j < m$ recalculates many values $B(i, j)$ over and over again. This is illustrated by the tree of recursive calls in Figure 2, where $B(i, j)$ is the result of the call BC1(i, j).

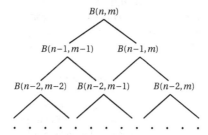

FIGURE 2. Recursive calls made by BC1(n, m).

Since the final result is obtained by adding up a certain number of 1's, the running time of this algorithm is certainly $\Omega\left(\binom{n}{m}\right)$. It is easy to see by induction that $\binom{2n}{n} \geq 2^n$, and so for $m = \lfloor n/2 \rfloor$ the running time of the BC1 algorithm is $\Omega(2^{n/2})$. Therefore, this exponential algorithm is impractical for computing the binomial coefficients, even in case when n is small. We have already met a similar phenomenon in the FIB1 algorithm for calculating the Fibonacci numbers.

If we apply the dynamic programming technique, we can use a table of intermediate results to get a much more efficient algorithm. This table, called Pascal's triangle, is illustrated in Figure 3.

	0	1	2	3	· · ·	$m-1$	m
0	1						
1	1	1					
2	1	2	1				
3	1	3	3	1			
·							
·							
·							
$n-1$	1	·	·	·	· ·	$B(n-1,m-1)$	$B(n-1,m)$
n	1	·	·	·	· ·		$B(n,m)$

FIGURE 3. Pascal's triangle.

The dynamic-programming algorithm BC2(n, m) below returns $B(n, m)$ for the input values $n \geq 0$ and $0 \leq m \leq n$. It builds entire Pascal's triangle with sides of length $n + 1$ and $m + 1$, and then simply returns the requested binomial coefficient.

$$BC2(n, m)$$
$$B(0, 0) = 1;$$
for $i = 1$ **to** n **do**
$\qquad B(i, 0) = 1; B(i, i) = 1;$
$\qquad$ **for** $j = 1$ **to** $i - 1$ **do**
$\qquad\qquad B(i, j) = B(i - 1, j - 1) + B(i - 1, j);$
return $B(n, m);$

The running time of this algorithm is clearly $O(n^2)$, a huge improvement over the BC1 algorithm. In fact, it is not even necessary to keep a table—we could get by with keeping only an array of size m to represent the current row, which we would update from right to left in order to compute the next row.

3. General case of the change-making problem

Consider again the problem of making change, but now using arbitrary coin values. We already met a special case of this problem in which the set of coin values was $\{1, 5, 10, 25\}$. We solved the special-case problem using a greedy algorithm that minimizes the total number of coins used in change. For example, it makes change for a 63-unit amount using two 25-unit coins, one 10-unit coin, and three 1-unit coins. We also noted that, unfortunately, the greedy algorithm is optimal only because of the special values of coins. If we included, say, a 21-unit in the set of coin values, then the greedy algorithm would still give a solution that uses six coins, although the optimal solution uses three coins (all of the 21-unit value).

The question then arises as to how to solve general case of the problem for an arbitrary coin set. We assume there is always a 1-unit coin and that we have unlimited number of coins of all values so that a solution always exists. In general therefore, the problem is to determine the minimal number of coins (as well as their actual values) needed to make change for an amount of $n \geq 0$ units, if the set of coin values is $C = \{c_1, c_2, \dots, c_m\}$, where $c_1 = 1$ and $c_i > 1$ for $i = 2, \dots, m$.

The minimal number of coins, denoted $N(n)$, that makes change for the amount of $n \geq 0$ units can be found using a simple recursive formula

$$N(n) = \begin{cases} 1, & n \in C \\ \min_{1 \leq i < n} \{N(i) + N(n - i)\}, & n \notin C. \end{cases}$$

In other words, if we can make change using exactly one coin, that is the minimum number of coins. Otherwise, for every i such that $1 \leq i < n$, we compute independently the minimum numbers of coins needed to make i units in change and $n - i$ units in change, and sum the numbers up. Finally, of all the $n - 1$ numbers obtained in this way, we then take the smallest one.

As an example, to make 63 units in change using the coins with values from the set $\{1, 5, 10, 21, 25\}$, we would recursively compute the minimum numbers of coins required to make 1 and 62 units in change, and save their sum. Then we would recursively compute the the minimum numbers of coins required to make 2 and 61 units in change, and again save their sum. Repeating this procedure for the amounts of 3 and 60 units, 4 and 59, and so on, 62 and 1 units, we would get 62 sums, of which we finally choose the smallest one. (To obtain actual coin values that make minimal change, ties can be broken arbitrarily.)

While we can slightly improve the algorithm (it is sufficient to compute the sums for each amount i between 1 and $\lceil n/2 \rceil$), there is still to much redundant work. It is easy to convince ourselves that the algorithm runs in exponential time $\Omega(2^n)$. This means that it will not terminate in a reasonable amount of time even when it makes simple 63 units in change.

An alternative recursive solution is to break the given amount into two pieces in each phase: one is taken to be the value of one of the coins, and the other is the rest to the given amount. Thus,

$$N(n) = \begin{cases} 0, & n = 0 \\ \min_{c_j \leq n}\{N(c_j) + N(n - c_j)\}, & n > 0, \end{cases}$$

where the minimum is taken over all coin values $c_j \in C$ for $1 \leq j \leq m$ such that $c_j \leq n$.

Put it another way, since $N(c_j) = 1$ for each such $c_j \in C$,

$$N(n) = \begin{cases} 0, & n = 0 \\ 1 + \min_{c_j \leq n}\{N(n - c_j)\}, & n > 0, \end{cases}$$

where the minimum is again taken over all coin values $c_j \in C$ for $1 \leq j \leq m$ such that $c_j \leq n$.

In the previous example, to make 63 units in change, we now recursively compute the numbers $N(62) = N(63 - 1)$, $N(58) = N(63 - 5)$, $N(53) = N(63 - 10)$, $N(42) = N(63 - 21)$, and $N(38) = N(63 - 25)$. Then we take their minimum, and finally add 1 to obtain $N(63)$. Thus, instead of recursively computing 62 numbers, we get by with only 5 recursive calls, one for each different coin. Once again, a naive recursive algorithm would be still very inefficient because it would recompute subproblem solutions. For instance, to determine $N(62)$, the algorithm would make a further recursive call when it chooses a 10-unit coin and computes $N(52)$. However, it would recompute $N(52)$ when it chooses a 1-unit coin to determine $N(53)$. This redundant work again leads to an exponential running time, which is not hard to see.

The trick is to save answers to the subproblems in an array. We can compute the optimal way to change 1 unit, then 2 units, then 3 units, and so on. Since the solution for larger amounts only depends on the solutions for smaller amounts, we can simply look up the answers from the array when needed instead of recomputing them.

This dynamic programming technique is implemented in the following algorithm that returns the minimum number of coins required to make $n \geq 0$ units in change using the coin values from the set $C = \{c_1 = 1, c_2, \ldots, c_m\}$ given in ascending order.

```
MAKE-CHANGE(n, C)
  N(0) = 0;
  for i = 1 to n do
    〚 Compute N(i) = 1 + min{N(i − c_j)} 〛
                         c_j ≤ i
    M = +∞; j = 1;
    while (j ≤ m) ∧ (c_j ≤ i) do
      if N(i − c_j) < M then
        M = N(i − c_j);
      j = j + 1;
    N(i) = 1 + M;
  return N(n);
```

The running time of this version of the MAKE-CHANGE algorithm is clearly dominated by the two nested loops. Since the inner **while** loop is executed at most m times, the algorithm takes $O(mn)$ time, where m is the number of different denominations of coins, and n is the amount of change we are to make.[2]

As it is written, this version of the MAKE-CHANGE algorithm returns only the minimum number of coins required to make change. Recovering actual coin values in the optimal solution is left as an exercise for the reader.

4. Longest common subsequence

Suppose we have two different versions of a text file and we want to determine which lines are common to the two versions. A useful way to think about this problem is to treat the two files as sequences of symbols, say, the first file as $x = x_1 \cdots x_m$ and the second file as $y = y_1 \cdots y_n$, where x_i represents the i-th line of the first file and y_i represents the i-th line of the second file.

For example, the UNIX command diff compares two text files for their differences. The typical changes that are made to a text file are inserting and deleting a line. A modification of a line can be treated as a deletion followed by an insertion. If we examine two text files in which a small number of such changes have been made when transforming one file into the other, it is usually easy to see which lines correspond to which, and which lines have been deleted and which inserted. The diff command identifies the differences between two files by first finding a longest common subsequence of the two sequences x and y whose elements are the lines of the given text files. The longest common subsequence represents those lines that have not been changed.

A *subsequence* of a sequence of elements is obtained by deleting zero or more elements of the sequence, keeping the remaining elements in their original order. A *common subsequence* of two sequences is a sequence that is a subsequence of both. A *longest common subsequence* (or *LCS* for short) of two sequences is a common subsequence that is not strictly shorter than some other common subsequence of the two sequences.

---

[2]Recall the definition of the O-notation for functions of two variables: $T(m, n) = O(mn)$ if there are positive constants m_0, n_0, and c such that $T(m, n) \leq cmn$ for all $m \geq m_0$ or $n \geq n_0$.

For convenience, throughout this section we assume that sequences of elements are character strings. We can think of characters like a, b, or c, as standing for lines of a text file, or as any other type of sequence elements. However, notice that in general a subsequence in this context is *not* the same as a substring. A substring is a special case of the notion of a subsequence, in which we only allow deletion of a prefix or a suffix of a string.

As an example, for two strings btmatmac and mtamtab, longest common subsequences are, among others, tata and mama. We see that tata is a subsequence of btmatmac because we can take the characters at positions 2, 4, 5, and 7 off the string btmatmac to form tata. Similarly, tata is a subsequence of mtamtab because we can take the characters at positions 2, 3, 5, and 6 off the string mtamtab. By the same token, we can see that mama is also a common subsequence of the two strings. However, we must convince ourselves that tata and mama are *longest* common subsequences; that is, there are no common subsequences of length five or more.

It helps to think of an LCS as a matching between certain characters of the two strings involved. That is, for each character of the LCS, we match the same characters of the two given strings by connecting them with a line, providing that the lines between matched characters do not cross. Figure 4 illustrates these matchings for the strings btmatmac and mtamtab corresponding to the common subsequences tata and mama.

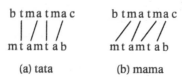

(a) tata   (b) mama

FIGURE 4. Common subsequences as matchings.

We will derive a recursive definition for the length of the LCS of two strings. This definition will let us compute the length easily by constructing a table of lengths of intermediate LCS's. Also, by examining the table we can then discover one of the possible LCS's itself, rather than just its length.

To find the length of an LCS of strings x and y, we calculate the lengths of the LCS's of all pairs of prefixes, one from x and the other from y. A **prefix** is an initial substring of a string. For example, the prefixes of the string btmatmac are ε,[3] b, bt, btm, and so on. Let $x = x_1 \cdots x_m$ and $y = y_1 \cdots y_n$. For each $i = 0, 1, \ldots, m$ and $j = 0, 1, \ldots, n$, we can determine an LCS of the prefix $x_1 \cdots x_i$ of x and the prefix $y_1 \cdots y_j$ of y in the following way.

First of all, if i or j is 0, then one of the prefixes is ε. Hence the only possible common subsequence of the two prefixes is ε, and thus their LCS is also ε.

Second, if both i and j are greater than 0, we consider two cases depending on whether the last symbols of the two strings are equal:

---

[3]ε is the empty string.

1. If $x_i \neq y_j$, then the matching that represents an LCS cannot include both x_i and y_j. Thus an LCS of $x_1 \cdots x_i$ and $y_1 \cdots y_j$ must be either an LCS of $x_1 \cdots x_{i-1}$ and $y_1 \cdots y_j$, or an LCS of $x_1 \cdots x_i$ and $y_1 \cdots y_{j-1}$.

2. If $x_i = y_j$, we can match x_i and y_j, and the matching will not interfere with any other potential matches. Thus an LCS of $x_1 \cdots x_{i-1}$ and $y_1 \cdots y_{j-1}$, concatenated with the symbol $x_i = y_j$, must be also a common subsequence of $x_1 \cdots x_i$ and $y_1 \cdots y_j$. In other words, if l_1 is the length of an LCS of $x_1 \cdots x_i$ and $y_1 \cdots y_j$, and l_2 is the length of an LCS of $x_1 \cdots x_{i-1}$ and $y_1 \cdots y_{j-1}$, then this implies that $l_2 + 1 \leq l_1$. On the other hand, $l_1 \leq l_2 + 1$ holds in general, because $x_1 \cdots x_i$ and $y_1 \cdots y_j$ are one symbol longer than $x_1 \cdots x_{i-1}$ and $y_1 \cdots y_{j-1}$. Therefore, $l_1 = l_2 + 1$.

From these observations we can easily derive a recursive definition for $L(i, j)$, the length of an LCS of the prefixes $x_1 \cdots x_i$ and $y_1 \cdots y_j$:

$$L(i, j) = \begin{cases} 0, & \text{if } i = 0 \text{ or } j = 0 \\ \max\{L(i, j-1), L(i-1, j)\}, & \text{if } i > 0 \text{ and } j > 0 \text{ and } x_i \neq y_j \\ 1 + L(i-1, j-1), & \text{if } i > 0 \text{ and } j > 0 \text{ and } x_i = y_j. \end{cases}$$

Clearly, the length of an LCS for the two entire strings x and y is $L(m, n)$, which we can compute by a simple divide-and-conquer algorithm based on the previous recursive formula. However, the running time of this algorithm would be exponential. For, if the two strings are completely different and we let $k = m + n$, then the running time is described by the recurrence equation $T(k) = 2T(k-1) + \Theta(1)$. The solution to this recurrence is $T(k) = \Theta(2^k)$, which is far too much time to make the simple divide-and-conquer algorithm practical for, say, $m = n = 100$.

The reason this algorithm performs so badly is that it recomputes the lengths of LCS's of the same pairs of prefixes many times. For example, to compute $L(3, 3)$, the algorithm computes $L(2, 3)$ and $L(3, 2)$. But in each of these computations, it computes $L(2, 2)$. We thus do the work of $L(2, 2)$ twice. If we continue to trace the computation, we find that $L(1, 1)$ is computed 6 times, $L(0, 1)$ and $L(1, 0)$ are computed 10 times each, and $L(0, 0)$ is computed 20 times.

We can do much better if we apply the dynamic programming technique to construct a table L that stores the values of $L(i, j)$ for $i = 0, 1, \ldots, m$ and $j = 0, 1, \ldots, n$. Since the computation of $L(i, j)$ involves the values of $L(i, j-1)$, $L(i-1, j)$, and $L(i-1, j-1)$, we can build the table row by row and be assured of finding the needed values in the table when computing $L(i, j)$. This is illustrated in Figure 5.

$L(i-1, j-1)$	$L(i-1, j)$
$L(i, j-1)$	$L(i, j)$

FIGURE 5. Computing the item $L(i, j)$ using already computed items.

The dynamic-programming algorithm LCS(x, y) below computes the length of an LCS of two character arrays $x = [x_1, \ldots, x_m]$ and $y = [y_1, \ldots, y_n]$.

The running time of the LCS algorithm is clearly dominated by the two nested **for** loops, where the outer one iterates m times and the inner one iterates n times. Thus, the LCS algorithm takes $O(mn)$ time.

```
LCS(x, y)
    for j = 0 to n do L(0, j) = 0;
    for i = 1 to m do
    L(i, 0) = 0;
        for j = 1 to n do
        if x(i) ≠ y(j) then
            if L(i − 1, j) ≥ L(i, j − 1) then
                L(i, j) = L(i − 1, j);
            else
                L(i, j) = L(i, j − 1);
        else ⟦ x(i) = y(j) ⟧
            L(i, j) = 1 + L(i − 1, j − 1);
    return L(m, n);
```

If the LCS algorithm returns the entire table L, instead of only the value $L(m, n)$, we can recover one of possible LCS's for the two strings in question. Since the table L gives us the lengths of the LCS's for all pairs of prefixes of the two strings, we will follow a path through the table, beginning at the lower right corner. On this path, we select certain symbols that in the reverse order represent an LCS for the two strings in question.

Suppose that the path from the lower right corner has taken us to the point of row i and column j in the table that corresponds to the pair of symbols x_i and y_j. If $x_i = y_j$, then we treat x_i and y_j as a matched pair of symbols, and include the symbol $x_i = y_j$ in the LCS ahead of all symbols of the LCS found so far. Since $L(i, j)$ in this case was chosen to be $1 + L(i − 1, j − 1)$, we move on our path diagonally left to the upper row, i.e., to row $i − 1$ and column $j − 1$.

It is also possible that $x_i \neq y_j$ at the point (i, j) in the table L. Then $L(i, j)$ must be equal to one of $L(i − 1, j)$ or $L(i, j − 1)$. In the first case, we move one row up along the same column; in the second case, we move one column left along the same row. In case there is a tie, we break it arbitrarily. Figure 6 shows a path that finds an LCS for the example strings btmatmac and mtamtab.

Following the path according to these rules, we eventually arrive at the left or upper edge of the table. At that point, we will have selected symbols in the reverse order of an LCS. We leave it as an exercise to write a more detailed pseudocode that implements this algorithm.

5. The 0-1 knapsack problem

Suppose a student is about to go on a trip carrying a single knapsack. Let W be the maximum total weight of the knapsack capacity. Suppose further that the student has a set of n different useful items that he can potentially take with him, such as a thermos, a few T-shirts, a favorite book, and so on. To simplify our discussion, denote all the items by $1, 2, \ldots, n$ and let $S = \{1, 2, \ldots, n\}$. Let us

7. DYNAMIC-PROGRAMMING ALGORITHMS

		b	t	m	a	t	m	a	c
		0	1	2	3	4	5	6	7	8
	0	0	0	0	0	0	0	0	0	0
m	1	0	0	0	1	1	1	1	1	1
t	2	0	0	1	1	1	2	2	2	2
a	3	0	0	1	1	2	2	3	3	3
m	4	0	0	1	2	2	2	3	3	3
t	5	0	0	1	2	2	3←3	3	3
a	6	0	0	1	2	3	3	3	4←4
b	7	0	1	1	2	3	3	3	4	4

FIGURE 6. A path that finds the LCS mata.

assume that each item $i \in S$ has an integer weight w_i and a value v_i that the student assigns to the item i. (Think of the value of an item as the student's measure of how important it is to take the item with him.)

The student's problem is to maximize the total value of the items that he takes with him, without going over the weight limit W. That is, his objective is to maximize, over all subsets T of S, the sum $\sum_{i \in T} v_i$ subject to $\sum_{i \in T} w_i \leq W$. More formally, the student needs to determine the value M such that

$$M = \max_{T \subseteq S} \left\{ \sum_{i \in T} v_i \,\middle|\, \sum_{i \in T} w_i \leq W \right\}.$$

For example, suppose the student has three items and that the pairs of their values and weights are $(v_1, w_1) = (15,3)$, $(v_2, w_2) = (10,2)$, and $(v_3, w_3) = (10,2)$. Suppose further that $W = 4$. Then the greedy approach based on the item values would yield the total value of 15, because it would select only item 1. However, the optimal solution is the choice of items 2 and 3, giving the total value of 20.

This example is an instance of the *0-1 knapsack problem*. It is is called a "0-1" problem, because each item must be entirely accepted or rejected. There is also a fractional version of this problem, in which we can take a fraction u_i of each item i whose value in the total score contributes $v_i \cdot (u_i / w_i)$.

We can easily solve the 0-1 knapsack problem in $\Theta(2^n)$ time, of course, by enumerating all subsets of S and selecting the one that has highest total value from among all those with total weight not exceeding W. Fortunately, we can derive a dynamic-programming algorithm for the 0-1 knapsack problem that runs much faster than the exponential brute-force algorithm if the number W is relatively small.

As with many dynamic programming problems, one of the hardest parts of designing such an algorithm for the 0-1 knapsack problem is to find a nice characterization for subproblems to be able to derive a recursive definition of the best solution for original problem in terms of the best solutions for subproblems.

To this end, for each $i = 1, 2, \ldots, n$, define the subset S_i of S to contain all the items $1, 2, \ldots, i$, that is, $S_i = \{1, 2, \ldots, i\}$. For convenience, we also let $S_0 = \emptyset$. Call the best subset of S_i to be the one that achieves the maximum value of the objective sum, where the maximum is taken over all subsets of S_i of those items whose total weight does not exceed $W - \sum_{k=i+1}^{n} w_k$. Then we can get the best subset of S if we can somehow gradually build the best subsets of S_i's, starting with the simplest subset S_0 and proceeding in succession.

For every $i = 0, 1, \ldots, n$ and $j = 0, 1, \ldots, W$, let us therefore denote by $M(i, j)$ the maximum total value of a subset of S_i from among all those subsets of S_i of items having the total weight *exactly* j. Thus, for $i = 0$ we have $M(0, j) = 0$ for each $j = 0, 1, \ldots, W$, and for $i \geq 1$ we have

$$M(i, j) = \begin{cases} M(i-1, j), & \text{if } w_i > j \\ \max\{M(i-1, j), M(i-1, j - w_i) + v_i\}, & \text{otherwise.} \end{cases}$$

That is, the value of the best subset of S_i that has total weight j is either the value of the best subset of S_{i-1} that has total weight j, or the value of the best subset of S_{i-1} that has total weight $j - w_i$ plus the value of the item i. Since the best subset of S_i that has total weight j must either contain the item i or not, one of these two choices must be the right choice.

Since the optimal value M we seek is clearly

$$M = \max_{j = 0, 1, \ldots, W} M(n, j),$$

if we compute the values $M(n, j)$ for every $j = 0, 1, \ldots, W$, we can find the optimal value of the 0-1 knapsack problem by taking the largest value among them.

In deriving a dynamic-programming algorithm from this recursive definition, we can make one additional observation that the value $M(i, j)$ is obtained from $M(i - 1, j)$ and possibly $M(i - 1, j - w_i)$. Thus, we can implement this algorithm using only a single array $M(0 \ldots W)$, which we update in a series of iterations indexed by i so that at the end of each iteration $M(j) = M(i, j)$. With this in mind, we give the algorithm KNAPSACK(v, w, W) that takes two arrays $v = [v_1, v_2, \ldots, v_n]$ and $w = [w_1, w_2, \ldots, w_n]$ of the values and weights of n items, as well as the weight limit W, and returns the value M of a maximum-value subset of items having the total weight less than or equal to W.

```
KNAPSACK(v, w, W)
    for j = 0 to W do M(j) = 0;
    for i = 1 to n do
        for j = W downto w(i) do
            if M(j - w(i)) + v(i) > M(j) then
                M(j) = M(j - w(i)) + v(i);
    M = max    M(j);
        j=0,1,...,W
    return M;
```

This algorithm returns only the largest total value, rather than the actual best subset of S. To get the optimal packing set, we should augment the algorithm to remember the choices for each subproblem that lead to the optimal solution. This can be easily done by an array of size n whose i-th element is defined to be 1 if the item i is included in the best subset of S, and 0 otherwise. Details are left as an exercise for the interested reader.

To analyze the KNAPSACK algorithm, we observe that the running time of the first **for** loop in this algorithm is clearly $O(W)$. The running time of the next two nested **for** loops is $O(nW)$, because the outer loop iterates n times and the inner one iterates at most W times. After the loops, we find the optimal value by locating the maximum value in the array $M(0\ldots W)$. Since this last step takes $O(W)$ time, the total running time of the KNAPSACK algorithm is $O(W)+O(nW)+O(W) = O(nW)$.

Notice that the running time of the KNAPSACK algorithm depends on the parameter W that is part of the input. In principle, W takes a value independently of the input item values and weights, and so may be much larger than the number of items n. For example, if $W = 2^n$ then this dynamic-programming algorithm would take $O(n2^n)$-time and would be actually asymptotically slower than the brute-force method. Technically therefore this algorithm is not efficient, since its worst-case running time could be exponential. More precisely, the running time is not a polynomial function of the input size, which is roughly the number of items n *plus* the number W represented in some standard way, say, as a binary number. Since it takes only $O(\log W)$ bits to represent W in binary, even if we count this in the input size, in case $W = 2^n$ the size remains bounded by a polynomial in n while the running time is exponential in n.

It is common to refer to an algorithm such as the dynamic programming KNAPSACK algorithm as being a *pseudo-polynomial time* algorithm, for its running time depends on the magnitude of a number given in the input. In practice however, such algorithms run much faster than any brute-force algorithm, but it is not correct to say they are true efficient algorithms. In fact, there is a theory known as NP-completeness that we study Chapter 12, which shows that it is very unlikely that anyone will ever find a true efficient algorithm for the 0-1 knapsack problem.

Exercises

1. Given the recurrence
$$C(n) = \begin{cases} 1, & \text{if } n = 0 \text{ or } n = 1 \\ C(n-1) + C(n-2) + 1, & \text{if } n \geq 2, \end{cases}$$
prove that $C(n) = F_{n+2} + F_{n-1} - 1 \geq \phi^n$ for $n \geq 2$, where F_n is the n-th Fibonacci number.

2. Let $f(n, m)$ be the function defined for nonnegative integers n and m as follows:
$$f(n, m) = \begin{cases} 0, \text{if } n = 0 \text{ or } m = 0 \text{ or } m \geq n \\ f(n, m-1) + f(n-1, m) + f(n-1, m-1) + 1, \text{otherwise.} \end{cases}$$

(a) Write a recursive algorithm that computes $f(n, m)$ and estimate its running time.

(b) Draw the tree of recursive calls of the recursive algorithm for $f(4, 3)$.

(c) Write a dynamic-programming algorithm that computes $f(n, m)$ and analyze its running time.

3. Modify the dynamic programming MAKE-CHANGE algorithm to return the actual coin values (in addition to the minimal number of coins) that make change for a given amount.

4. For the strings abracadabra and yabbadabbadoo find a longest common subsequence.

5. For the strings savannah and havana

(a) build the table L as in the LCS algorithm;

(b) find all the longest common subsequences.

6. Modify the LCS algorithm to recover a longest common subsequence of two strings.

7. Let $S = \{a, b, c, d, e, f, g\}$ be a set of objects with the value-weight pairs as follows: $a = (12, 4)$, $b = (10, 6)$, $c = (8, 5)$, $d = (11, 7)$, $e = (14, 3)$, $f = (7, 1)$, $g = (9, 6)$. What is an optimal solution to the 0-1 knapsack problem for S assuming we have a sack that can hold objects with total weight of 18?

8. Modify the KNAPSACK algorithm to return the optimal packing set in addition to its maximal value.

9. Suppose you are hosting an Internet auction to sell n widgets. You receive m bids $b_1, b_2, \ldots, b_m$, where each b_i for $i = 1, 2, \ldots, m$ is of the form "I want k_i widgets for d_i dollars." If you want to sell all widgets at the highest price, characterize this optimization problem as a knapsack problem.

10. Suppose we are given an $m \times n$ grid as the one shown in Figure 7.

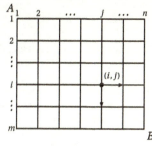

FIGURE 7

We want to get from the point A to the point B by walking along the gridlines. However, at each point of an intersection (i, j), we can move only left to right or top to bottom. Consider all such different paths that start from the point A and go to the point (i, j), and let $p(i, j)$ be their total number.

(a) Derive a recursive definition for $p(i, j)$. (*Hint:* A path from A to (i, j) takes either the edge to the left of (i, j) or the edge above (i, j), but not both.)
(b) Design a recursive algorithm for computing $p(i, j)$.
(c) How many times is $p(2, 2)$ called when the recursive algorithm computes $p(n, n)$? (*Hint:* $\binom{2(n-2)}{n-2}$.)
(d) Design a dynamic-programming algorithm for computing $p(m, n)$ and analyze its running time.

11. Suppose that each edge of the grid in the previous exercise is additionally labeled with a nonnegative number representing its length. Define the length of a path to be the sum of the lengths of the edges along the path. The goal now is to find a shortest path (going left to right or top to bottom) from A to B.

(a) If the length of the horizontal edge entering (i, j) is $h(i, j)$ and the length of the vertical edge entering (i, j) is $v(i, j)$, derive a recursive definition for $\delta(i, j)$, the length of a shortest path from A to (i, j).
(b) Design a dynamic-programming algorithm for computing $\delta(i, j)$ and analyze its running time.

12. Give and analyze a dynamic-programming algorithm that solves the maximum contiguous subsequence sum problem from Section 1.6 in Chapter 2. (*Hint:* For each $i = 1, 2, \ldots, n$, consider contiguous subsequences *ending* exactly at position i. The maximum-sum contiguous subsequence ending at i is either (i) empty, or (ii) the maximum-sum contiguous subsequence ending at $i-1$ followed by the element a_i. Thus, if c_i is the sum of the maximum-sum contiguous subsequence ending at position i, then $c_i = \max\{0, c_{i-1} + a_i\}$ and $c_0 = 0$.)

13. **The set-partition problem.** Given a set $X = \{x_1, x_2, \ldots, x_n\}$ of n positive integers such that $\sum_{i=1}^{n} x_i = M$, design an $O(nM)$-time algorithm for determining whether there is a subset $S \subseteq X$ such that

$$\sum_{x_i \in S} x_i = \sum_{x_i \in X \setminus S} x_i.$$

14. A formal grammar defines a way of generating strings by applying simple substitution rules. A **context-free grammar** G is a 4-tuple $G = (V, T, R, S)$, where individual elements of the tuple have the following meaning: V is a finite set of **variables**, often represented by capital letters; T is a finite set, disjoint from V, of **terminals**, often represented by lowercase letters, numbers, or special characters; R is a finite set of **substitution rules**, also called **productions**, each rule being of the form $A \rightarrow v$ with $A \in V$ and v being a string of variables and terminals; and, $S \in V$ is the **start symbol**. The following is an example of the set R of productions of a context-free grammar $F = \big(\{S, A\}, \{0, 1, \#\}, R, S\big)$:

$$S \rightarrow 0S1$$

$$S \rightarrow A$$
$$A \rightarrow \#$$

If u, v, and w are strings of variables and terminals, and $A \rightarrow v$ is a substitution rule of the grammar, we say that uAw **yields** uvw and write $uAw \Rightarrow uvw$. We say that v is **generated** from u if $u = v$ or if there is a sequence $u_1, u_2, \ldots, u_k$ for $k \geq 0$ such that

$$u \Rightarrow u_1 \Rightarrow u_2 \Rightarrow \cdots \Rightarrow u_k \Rightarrow v.$$

We say that a grammar generates a string of terminals if the string can be generated from the start symbol S. For example, the grammar F generates the string 00#11 with the following sequence of substitutions:

$$S \Rightarrow 0S1 \Rightarrow 00S11 \Rightarrow 00A11 \Rightarrow 00\#11$$

A context-free grammar is in **Chomsky normal form** if every production is of the form $A \rightarrow BC$ or $A \rightarrow a$, where A, B, and C are variables and a is a terminal symbol.

Design an $O(n^3)$-time dynamic-programming algorithm for determining if string $x = x_1 x_2 \cdots x_n$ of terminals can be generated from the start symbol S of a context-free grammar given in Chomsky normal form.

15. **The maximum-weight independent-set problem on trees.** Suppose that each node x in a (rooted) tree T is labeled with a positive number $w(x)$, called the **weight** of the node. Extend the notion of weight for a set of nodes by taking the weight of a set of nodes to be the sum of the weights of the nodes in that set. Recall also from Exercise 9 in Chapter 6 that an independent set in T is a subset S of the nodes of T such that no pair of nodes in S is connected by an edge from T. Design an efficient dynamic-programming algorithm to find a maximum-weight independent set of the nodes in T. In other words, the algorithm is to find an independent set S in T so that its total weight $\sum_{x \in S} w(x)$ is as large as possible. What is the running time of your algorithm?

8

Graph Algorithms

In applications with origins from various scientific disciplines we often need to represent arbitrary relationships among objects. Graphs are natural models for such relationships. For example, cities may be represented by points and lines that connect them may represent roads between the cities. Graphs are a ubiquities data model in computer science, and graph algorithms are fundamental to the field. Although graphs mathematically represent ordinary binary relations, they are always visualized as a set of points connected by lines or arrows. In this regard, the graph data structure is a generalization of the tree data structure we studied in Chapter 4.

Graphs come in several forms. In this chapter we examine directed and undirected graphs, and we first review definitions of the basic notions related to graphs. We then discuss several elementary graph algorithms that are useful in many applications. More advanced graph algorithms will be discussed later in the book.

1. Directed graphs

A *directed graph* (or *digraph* for short) G consists of two sets: a set of vertices V and a set of directed edges E. This is denoted by $G = (V, E)$. A *directed edge* is an ordered pair of vertices, that is, $(u, v) \in E$ for some $u, v \in V$. The vertices are also called *nodes*, and the directed edges are called *arcs*.

Thus, a directed graph G is formally given by two finite sets V and E. Informally, directed graphs are frequently described by diagrams in which we draw small circles for vertices, and for directed edges we draw arrows connecting some of the circles. A directed edge (u, v) is often in the text denoted by $u \rightarrow v$, and in the diagrams depicted by a line leaving the vertex u and ending with an arrowhead pointing into the vertex v. For example, Figure 1 shows a digraph with four vertices and five directed edges.

FIGURE 1. A directed graph.

Formally, for the digraph in Figure 1 we have

$$V = \{u, v, w, x\}, \quad E = \{(u, v), (u, w), (v, x), (w, v), (x, w)\}.$$

The sizes of the sets V and E are $|V| = 4$ and $|E| = 5$.

The vertices of a digraph can be used to represent objects, and the directed edges to represent relationships between the objects. For example, a digraph might represent the flow of control in a computer program, where the vertices represent basic program blocks and the directed edges represent possible transfers of flow of control. Another example is from mathematics, where directed graphs are a handy way of depicting binary relations. If R is a binary relation whose domain is $V \times V$, a graph $G = (V, E)$ represents R if $E = \{(x, y) \in V \times V : xRy\}$. Figure 2 is an illustration of the relation *beats* in the children's game called Scissors-Paper-Stone.

beats	Scissors	Paper	Stone
Scissors	FALSE	TRUE	FALSE
Paper	FALSE	FALSE	TRUE
Stone	TRUE	FALSE	FALSE

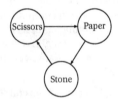

FIGURE 2. The directed graph of a relation.

A *path* in a digraph is a sequence of vertices $v_1, v_2, \ldots, v_k$ such that $v_1 \to v_2$, $v_2 \to v_3, \ldots, v_{k-1} \to v_k$ are directed edges in the digraph. This path is *from* vertex v_1 *to* vertex v_k, *passes through* vertices $v_2, v_3, \ldots, v_{k-1}$, and *ends* at vertex v_k. The *length* of a path π, denoted $\ell(\pi)$, is the number of directed edges on the path. As a special case, a single vertex v by itself denotes a path of length 0 from v to v. In Figure 1, the vertices u, v, x form a path of length 2 from the vertex u to the vertex x.

A path is *simple* if all vertices on the path are distinct. A *cycle* is a path that begins and ends at the same vertex. A cycle is *simple* if its underlying path is simple, excluding the first and the last vertex. In Figure 1, the path w, v, w is a simple cycle of length 3. Notice that any path (cycle) between two distinct vertices in a graph has an underlying simple path (cycle), as suggested in Figure 3.

A directed graph is *semiconnected* if for every pair of vertices u and v there is a path from u to v *or* a path from v to u. A directed graph is called *strongly*

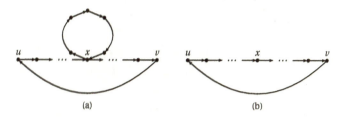

FIGURE 3. A path (cycle) and its underlying simple path (cycle).

connected if for every pair of vertices u and v there is a path from u to v *and* a path from v to u.[1]

A **directed acyclic graph** (or **dag** for short) is a directed graph with no cycles. If we relate dags to trees and arbitrary directed graphs, dags are more general than trees, but less general than arbitrary directed graphs. This relationship is illustrated in Figure 4 with an example of a tree, a dag, and a directed graph with a cycle.

FIGURE 4. Three directed graphs.

The number of arrows leaving a particular vertex is the **outdegree** of that vertex, and the number of arrows pointing into a particular vertex is its **indegree**. We say that the directed edge $u \rightarrow v$ is **from u to v**, and that v is **adjacent** to u.

In computer applications it is often useful to attach data to the vertices and/or directed edges of a digraph. For this purpose we can use a **labeled digraph**, i.e., a digraph in which each directed edge and/or each vertex can have an associated label. A label might be a value of any given data type. In particular, if the labels of directed edges are numbers, then the labeled digraph is often called the **weighted digraph**. These numbers may represent a time, a distance, or any similar numeric quantity that "measures" the relationship between two vertices connected by a directed edge.

Figure 5 shows a labeled digraph in which each directed edge is labeled by a letter that, say, causes transition from one vertex to another.

Figure 6 depicts a weighted digraph whose vertices are cities and whose directed edges are labeled with the cost of the one-way, nonstop air fare between those cities.

---

[1] A directed graph is called **connected** if the corresponding undirected graph is connected. The corresponding undirected graph is obtained by removing directions from all directed edges, and the undirected graph is connected if there is a path (defined in an analogous way) between every pair of vertices.

FIGURE 5. A labeled digraph.

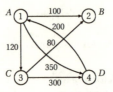

FIGURE 6. A weighted digraph.

A vertex in a labeled digraph can have both a name and a label. Quite frequently, we shall use the vertex label as the name of the vertex. Thus, the numbers in Figure 5 could be interpreted as either vertex names or vertex labels.

1.1. Representation of directed graphs. There are several data structures that we can use to represent a digraph, and the appropriate choice depends on the operations that will be performed on the vertices and directed edges of the digraph. One common representation of a digraph $G = (V, E)$ is the adjacency matrix. If the names of the vertices are enumerated in an arbitrary order so that $V = \{v_1, \ldots, v_n\}$, the **adjacency matrix** for G is an $n \times n$ matrix A such that $A(i, j) = 1$ if there is a directed edge $v_i \rightarrow v_j$, and $A(i, j) = 0$ otherwise. Figure 7 shows an example of a directed graph and its adjacency matrix.

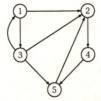

	1	2	3	4	5
1	0	1	1	0	0
2	0	0	0	1	0
3	1	1	0	0	1
4	0	0	0	0	1
5	0	1	0	0	0

FIGURE 7. A directed graph and its adjacency matrix.

In the adjacency-matrix representation, the time required to access any element of the adjacency matrix is constant, that is, it is independent of the number of vertices and edges. Therefore, the adjacency-matrix representation is useful in those graph algorithms in which we frequently need to know whether a given directed edge is present.

For labeled (weighted) digraphs we can use a closely related **labeled (weighted) adjacency-matrix** representation, where $A(i, j)$ is the label of the directed edge $v_i \rightarrow v_j$. If there is no directed edge going from v_i to v_j, then a value that cannot be a legitimate label must be used for $A(i, j)$. Figure 8 shows the labeled adjacency matrix for the digraph in Figure 5. Here, the blank character ␣ represents the absence of a directed edge.

	1	2	3	4
1	␣	a	b	␣
2	a	␣	␣	b
3	b	␣	␣	a
4	␣	b	a	␣

FIGURE 8. The labeled adjacency matrix for the digraph in Figure 5.

Figure 9 shows the weighted adjacency matrix for the digraph in Figure 6. Here, the absence of a directed edge is represented by the symbol ∞, suggesting an "infinite cost."

	1	2	3	4
1	∞	100	120	350
2	∞	∞	∞	∞
3	∞	80	∞	300
4	200	∞	∞	∞

FIGURE 9. The weighted adjacency matrix for the digraph in Figure 6.

The main disadvantage of using an adjacency matrix to represent a digraph is that the matrix requires $\Theta(n^2)$ space even if the digraph is **sparse**, i.e., it has much less than n^2 directed edges. Simply to examine the entire matrix would require $O(n^2)$ time, which would preclude algorithms with $o(n^2)$ running time for manipulating digraphs with only $O(n)$ directed edges.

To avoid this disadvantage, we can take another common approach to represent a digraph by using the **adjacency lists**. An adjacency list for a vertex v is a list of all vertices adjacent to v, in some order. A digraph with the vertex set $V = \{v_1, \ldots, v_n\}$ can be represented by a header array H of size n, where $H(i)$ for $i = 1, \ldots, n$ is a pointer to the adjacency list for vertex v_i. Figure 10 shows the adjacency-lists representation for the digraph in Figure 1.

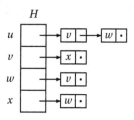

FIGURE 10. The adjacency-lists representation for the digraph in Figure 1.

If directed edges have labels, these labels can be included as part of the elements of the linked lists. Figure 11 shows the adjacency-lists representation for the weighted digraph in Figure 6.

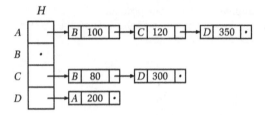

FIGURE 11. The adjacency-lists representation for the weighted digraph in Figure 6.

The adjacency-lists representation for a digraph with n vertices and m directed edges requires space proportional to $n+m$. This is the reason why it is often used for sparse digraphs. However, a potential disadvantage of the adjacency-lists representation is that it may take $\Theta(n)$ time to determine if there is a directed edge $u \to v$, since there can be $\Theta(n)$ vertices in the adjacency list for any vertex u.

2. Undirected graphs

An **undirected graph** (or simply a **graph**) $G = (V, E)$ consists of a finite set of vertices V and a finite set of edges E. It differs from a directed graph in that each edge in E is an *unordered* pair of vertices. If an undirected graph G contains an edge (u, v), the order of vertices u and v in the pair does not matter. So, the pairs (u, v) and (v, u) represent the same edge. Thus, we merely say an edge, not an undirected edge, and in diagrams we represent it by a line, not by a line with two arrowheads on both ends in an analogy with directed edges. Figure 12 shows two examples of undirected graphs.

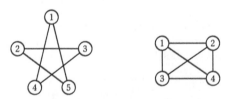

FIGURE 12. Two (undirected) graphs.

Much of the terminology for directed graphs is also applicable to undirected graphs. For example, vertices u and v are **adjacent** if there is an edge (u, v) (or, equivalently, (v, u)) in the graph. We say that an edge (u, v) is **incident** with vertices u and v. The number of edges at a particular vertex is the **degree** of that vertex.

A **path** in a graph is a sequence of vertices connected by edges in the order. The **length** of a path is the number of edges along the path. We postulate that there is a path of length 0 from any vertex to itself. A **simple path** is a path that does not repeat any vertices. A path of length at least 1 is a **cycle** if it starts and ends at the same vertex. A *simple cycle* is one that does not repeat any vertices except for the first and the last.

A graph is *connected* if every two vertices have a path between them. We emphasize that this does not mean that in a connected graph every pair of vertices is joined by an edge.

We say that graph H is a *subgraph* of graph G if the vertices of H are a subset of the vertices of G, and the edges of H are a subset of the edges of G incident with the vertices from H. If we take all the edges of G incident with the vertices from H, we call H an *induced subgraph* of G. Figure 13 shows an example of a graph and one of its induced subgraphs.

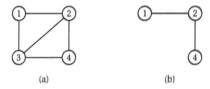

(a)   (b)

FIGURE 13. A graph and one of its induced subgraphs.

A *connected component* of a graph is a maximally connected induced subgraph, that is, a connected induced subgraph that is not a proper subgraph of any other connected induced subgraph.

A connected graph with no cycles is called a *free tree*. A free tree can be made into an ordinary rooted tree if we pick any vertex as the root and orient the edges from the root, then the edges from each child of the root, and so on. A more illustrative way to comprehend this is to imagine that the edges represent strings and that we have lifted the tree by the selected root vertex. Figure 14 shows a disconnected graph consisting of two connected components, where each connected component is a free tree.

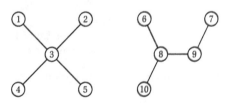

FIGURE 14. Two connected components of a disconnected graph.

2.1. Representation of undirected graphs. The methods of representation for directed graphs can be used to represent undirected graphs if we simply consider an undirected edge (u, v) as two directed edges, one from u to v and the other from v to u. For example, the adjacency-matrix and the adjacency-lists representations for the graph of Figure 13(a) are shown in Figure 15.

Clearly, the adjacency matrix for an undirected graph is symmetric. Similarly, in the adjacency-lists representation, if (u, v) is an edge of an undirected

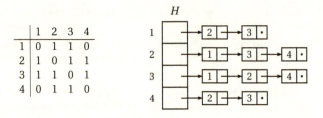

$$
\begin{array}{c|cccc}
 & 1 & 2 & 3 & 4 \\
\hline
1 & 0 & 1 & 1 & 0 \\
2 & 1 & 0 & 1 & 1 \\
3 & 1 & 1 & 0 & 1 \\
4 & 0 & 1 & 1 & 0 \\
\end{array}
$$

FIGURE 15. Representations for the undirected graph of Figure 13(a).

graph, then vertex v is on the list for vertex u and vertex u is on the list for vertex v.

3. Depth-first search

To solve many problems dealing with graphs, we need to visit the vertices and edges of a graph in a systematic fashion. In Chapter 4 we discussed different ways for traversal of trees. Those techniques generalize in a natural way to the two most common methods for graph traversal: depth-first search and breadth-first search. These two search strategies can serve as the basis for many other efficient graph algorithms.

In this section we study depth-first search considering first undirected graphs, and then move on to explore a somewhat more complicated case of directed graphs.

3.1. Depth-first search of undirected graphs. Depth-first search method is a generalization of the preorder traversal of a tree. Suppose we have an undirected graph $G = (V, E)$ in which all vertices are initially marked as unvisited.

Depth-first search works by initially selecting an arbitrary vertex $u \in V$, which is then marked as visited. Then each unvisited vertex v adjacent to u is visited in turn, using depth-first search recursively. Once all vertices that can be reached form u have been visited, the traversal of the graph G is complete providing that G is connected. If some vertices remain unvisited (that is, G is not connected), we select an unvisited vertex as a new starting point, and repeat this process until all vertices of G have been visited.

The basic algorithms DEPTH-FIRST-SEARCH(G) and DFS(u) below implement depth-first search on the input graph $G = (V, E)$ in pseudocode. The DEPTH-FIRST-SEARCH algorithm is a "main" procedure that initializes necessary variables and calls the recursive DFS algorithm appropriately.

```
DEPTH-FIRST-SEARCH(G)
  [ Initially, all vertices are unvisited ]
  for each v ∈ V do c(v) = 0;
  [ Visit each vertex that's left unvisited ]
  for each v ∈ V do
    if c(v) = 0 then
      DFS(v);
  return;
```

```
DFS(u)
  c(u) = 1;   [ u is visited ]
  for each v adjacent to u do
    if c(v) = 0 then
      DFS(v);
  return;
```

The DEPTH-FIRST-SEARCH algorithm uses a global array c to keep track of visited vertices so that $c(v) = 1$ if the vertex v has been visited, and $c(v) = 0$ otherwise. Initially, all vertices are unvisited, and so $c(v) = 0$ for every $v \in V$.[2] The next **for** loop ensures that each connected component of G gets traversed, in case G is a disconnected graph. In particular, while there is some $v \in V$ such that v is still unvisited, the body of the loop selects it and calls DFS(v), which actually traverses (a connected component of) the graph G in depth-first fashion starting from the vertex v.

Each call DFS(u) results on an unvisited vertex u. So, at the very beginning, the DFS algorithm indicates that u is visited, and then explores u's neighbors by examining every edge leaving u. If the algorithm finds a vertex v adjacent to u that is unvisited, it recursively visits v by calling itself on v. After every edge leaving u has been fully examined, u is finished processing, and the algorithm returns.

This technique is called depth-first search because it keeps searching in forward (deeper) direction as long as possible. For example, suppose x is the most recently visited vertex. Depth-first search selects some unexplored edge (x, y) emanating from x. If the vertex y has been visited, the method looks for another unexplored edge emanating from x. If y has not been visited, then the method marks y as visited, and initiates new search starting at y. After completing the search through all paths beginning at y, the search returns to x, the vertex from which y was first visited. The procedure of selecting unexplored edges emanating from x is then repeated until all edges emanating from x have been explored.

In other words, the recursive DFS algorithm tries to initiate as many recursive calls as possible before it ever returns from a call. This recursive chain of calls is only stopped if exploration of the graph cannot continue any further. At this point the recursion "unwinds" so that alternative paths at higher levels can be explored. Figure 16 illustrates a sample run of the DFS algorithm on an undirected graph by showing the nested structure of generated recursive calls. Notice that the results of depth-first search may vary depending upon the order in which the vertices are selected in the **for** loop of the DEPTH-FIRST-SEARCH algorithm, as well as upon the order in which the neighbors of a vertex are visited in the **for** loop of the DFS algorithm.

---

[2]We can think of the elements of the array c as representing the color of a vertex v that can be either black ($c(v) = 1$) or white ($c(v) = 0$). We'll expand this black-and-white coloring scheme later in the section.

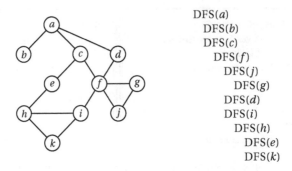

DFS(a)

DFS(b)

DFS(c)

DFS(f)

DFS(j)

DFS(g)

DFS(d)

DFS(i)

DFS(h)

DFS(e)

DFS(k)

FIGURE 16. An undirected graph and a DFS execution from vertex a.

To analyze the running time of the DEPTH-FIRST-SEARCH algorithm, suppose the graph $G = (V, E)$ has n vertices and m edges, i.e., $|V| = n$ and $|E| = m$.[3] The first step of marking all vertices as unvisited requires obviously $O(n)$ time. To estimate the running time of the next **for** loop, we need to calculate the time taken by a call to DFS(v) for each unvisited $v \in V$. Although the DFS algorithm is recursive, in this case we argue differently from the usual way we used when we analyzed a recursive (divide-and-conquer) algorithm.

First we observe that we call DFS(u) exactly once for each vertex $u \in V$. To see why, it is clear that DFS is called at least once for each vertex, because all vertices are initially unvisited and we keep calling DFS in the DEPTH-FIRST-SEARCH algorithm until unvisited vertices are exhausted. To see that DFS is called at most once for each vertex, notice that as soon as we call DFS(u) on a vertex u, we mark the vertex as visited, and we never call again the DFS algorithm on a visited vertex. Thus, if we sum the execution times of all calls, sans their recursive calls, we get the total time spent in all the calls as a group in the **for** loop.

Now, to estimate the running time of the call DFS(u) excluding the time spent in its recursive calls, we notice that it depends on the number of vertices adjacent to the vertex u. If the degree (i.e., the number of neighbors) of the vertex u is d_u, then the body of the **for** loop in the DFS algorithm is executed d_u times. Since in the body of the loop we do not count the time spent for execution of the call DFS(v), the body clearly takes constant time exclusive of this call. Therefore, the total time taken by the call DFS(u), exclusive of its recursive calls, is $O(1 + d_u)$. The additional 1 is needed because d_u might be 0, in which case the call DFS(u) still requires some constant time for execution.

Because we call DFS(u) exactly once for each vertex $u \in V$, the total time spent in all these calls is

$$\sum_{u \in V} O(1 + d_u) = O\left(\sum_{u \in V} (1 + d_u)\right) = O\left(\sum_{u \in V} 1 + \sum_{u \in V} d_u\right) = O(n + 2m) = O(n + m).$$

Finally, we can calculate the running time of the DEPTH-FIRST-SEARCH algorithm as follows. It is the sum of the time taken by the first **for** loop, which is

---

[3]Usually $m \geq n$, but in some graphs there are more vertices than edges.

$O(n)$, the time taken by the second **for** loop exclusive of the DFS calls, which is clearly again $O(n)$, and the time taken by all the DFS calls, which is $O(n + m)$. Thus, we can conclude that the DEPTH-FIRST-SEARCH algorithm takes $O(n + m)$ time, which is linear in the sum of the numbers of vertices and edges.

3.2. Depth-first search tree. Any depth-first search of a connected graph associates a spanning tree to the graph. The search explores some edges to visit unvisited vertices. Since we never visit the same vertex twice, the edges traversed by depth-first search actually form a tree. In fact, we can draw a tree whose parent-child edges are some of the edges of the graph being traversed. If we are in the call DFS(u), and a call to DFS(v) results, then we make v a child of u in the tree. The children of u appear, from left to right, in the order in which the DFS algorithm was called on these children. In this way we obtain a rooted tree whose root is the starting vertex on which the initial call to DFS was made. This tree is called the ***depth-first search tree*** (or the ***dfs tree***) for a given connected graph.

Figure 17 shows the depth-first search tree obtained by depth-first search of the graph of Figure 16 that starts from the vertex a. The solid lines are the traversed edges and the dashed lines are the edges that are not used in the search.

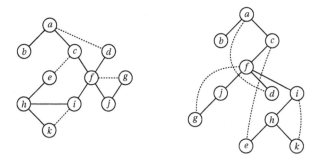

FIGURE 17. The depth-first search tree obtained by the sample DFS run of Figure 16.

If a graph is not connected, depth-first search constructs not a single tree, but rather a collection of trees, or forest—one tree in the collection for each connected component of the graph. When depth-first search generates a sequence of trees, we add trees to the sequence in the left-to-right order. For simplicity however, we generally assume that an undirected graph is connected, since all considerations with obvious modifications apply to a dfs forest as well.

Edges of the graph can be classified according to their relationship with edges of the depth-first search tree constructed during depth-first search. An edge (u, v) such that DFS(u) directly calls DFS(v), or vice versa, is referred to as the ***tree edge***. The solid edges in Figure 17 represent the tree edges. A dashed edge in the same figure connects two vertices that are in the ancestor-descendant relationship in the depth-first search tree. A dashed edge is an edge (u, v) such that neither DFS(u) nor DFS(v) called the other directly, but one called the other indirectly. Since each dashed edge is a "shortcut" that goes from a descendant up the tree to an ancestor, it is called the **back edge**. A useful observation here, which we prove in the next section, is that this is a complete classification of the edges of an undirected graph.

3.3. A closer look at depth-first search. Depth-first search of a graph $G = (V, E)$ imposes a nice structure on the vertices and edges of the graph. To study this structure in more detail, let's consider a more concrete implementation of the depth-first search algorithm. We assume the graph is given by its adjacency-lists representation. That is, for each vertex $v \in V$ there is an adjacency list $L(v)$ that contains all the neighbors of v in G in some order.

For any unvisited vertex v, the call $DFS(v)$ initiates depth-first search from v to unvisited vertices of G. This call is active until after $DFS(v)$ has explored all incident edges with v. By a switch of perspective, we say that a vertex v is **active** if during the course of depth-first search we are in the call of $DFS(v)$. When the recursive call $DFS(v)$ terminates, it returns to the caller procedure, and we never call DFS on v again. We say that the vertex v is **inactive** from the moment $DFS(v)$ terminates.

During depth-first search therefore, vertices change their state going through from being unvisited, then active (i.e., visited but not yet finished processing), to eventually becoming inactive (i.e., visited *and* finished processing). We represent the state of a vertex by assigning a color to it, where white color denotes an unvisited vertex, gray color denotes an active vertex, and black color denotes an inactive vertex.

It is also useful to timestamp each vertex during the search, so that we can keep track on a time line when a vertex changes its state from unvisited to active and from active to inactive. Thus, each vertex v receives two timestamps: the first timestamp records when v becomes active (i.e., the moment when it is first visited and grayed), and the second timestamp records when v becomes inactive (i.e., the moment when the search finishes examining v's adjacency list and blackens v). These timestamps are helpful in reasoning about the behavior of depth-first search.

We use four auxiliary arrays to represent all this information collected during depth-first search. We also use a global counter t to generate time ticks for the timestamping purpose.

An array c stores the color for each vertex $v \in V$: $c(v) =$ WHITE if v is unvisited; $c(v) =$ GRAY if v is active (i.e., v is visited but not yet finished processing); and $c(v) =$ BLACK if v is inactive (i.e., v is visited and finished processing).

Two arrays a and b record the state transition time of each vertex $v \in V$: $a(v)$ stores timestamp when v becomes active, and $b(v)$ stores timestamp when v becomes inactive. If there are n vertices, these timestamps are integers between 1 and $2n$, since there is one activity event and one inactivity event for each of the n vertices.

Finally, to construct the dfs tree (forest), we use an array p such that, for each vertex $v \in V$, $p(v)$ stores parent pointer that points back to the vertex from which v was visited.

This expanded version of the basic depth-first search algorithm is again implemented using a "main" procedure DEPTH-FIRST-SEARCH and a recursive procedure DFS.

DEPTH-FIRST-SEARCH(G)
   $t = 0;$ ⟦ beginning of epoch ⟧
   ⟦ Initially, all vertices are unvisited ⟧
   for each $v \in V$ **do** $c(v) = $ WHITE;
   ⟦ Visit each vertex that's left unvisited ⟧
   for each $v \in V$ **do**
     if $c(v) = $ WHITE **then**
       $p(v) = $ NIL;   ⟦ v is the root of the dfs tree ⟧
       DFS(v);
   return;

As before, the expanded procedure DEPTH-FIRST-SEARCH(G) has two tasks. The first one is to initialize necessary variables (according to the new scheme). Its second task is to select a yet-unvisited vertex v from every connected component of G and to initiate depth-first search of each component by calling the expanded procedure DFS(v).

DFS(u)
   $t = t + 1; a(u) = t;$ ⟦ timestamp the activity event for u ⟧
   $c(u) = $ GRAY;   ⟦ u is active ⟧
   for each v in $L(u)$ **do**
     if $c(v) = $ WHITE **then**
       $p(v) = u;$ ⟦ u is parent of v ⟧
       DFS(v);
   $t = t + 1; b(u) = t;$ ⟦ timestamp the inactivity event for u ⟧
   $c(u) = $ BLACK;   ⟦ u is inactive ⟧
   return;

The expanded DFS algorithm uses the new coloring scheme to paint vertices. In each call of DFS(u), the newly visited (white) vertex u is immediately colored gray to mark it as visited but not yet finished processing. Then, as usual, all u's neighbors are explored, and those of them that are white (unvisited) are recursively visited. Finally, after every edge leaving u has been examined, u is colored black to mark it as finished processing. The algorithm appropriately records the start of activity and inactivity times for u, as well as sets u as the parent of each neighbor v that is recursively visited.

The following sequence of pictures illustrates the progress of this version of the DFS algorithm corresponding to the sample run of Figure 16.

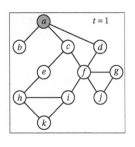
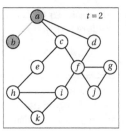
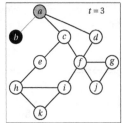

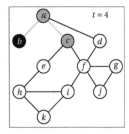
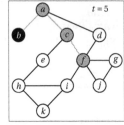
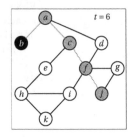

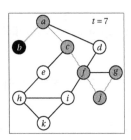
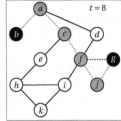
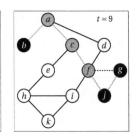

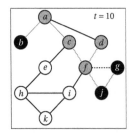
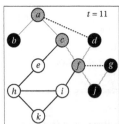
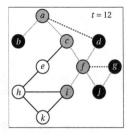

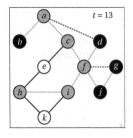
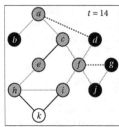
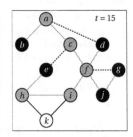

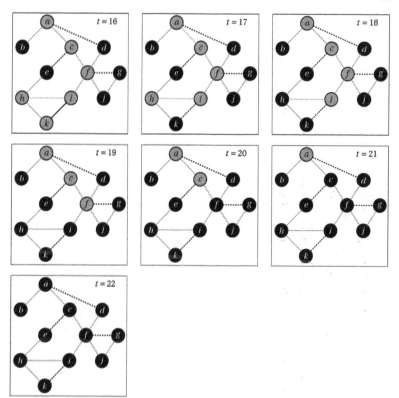

Perhaps the most important property of depth-first search of a graph G is that it induces a tree structure (or forest) on the graph. The dfs tree T produced by depth-first search is just the recursion tree of DFS calls, where a tree edge (u, v) arises if while processing vertex u we call DFS(v) for some neighbor v of u. More precisely, T is determined by the array p, since $u = p(v)$ if and only if DFS(v) is called by DFS(u) when exploring u's neighbors. Observe also that v is descendant of u in T if and only if v is grayed during the time in which u is gray.

Figure 18 illustrates the dfs tree and activity intervals of all vertices corresponding to depth-first search shown in the previous sequence of pictures. Each rectangular node spans the interval given by the start of activity and inactivity times of the corresponding vertex.

The timestamps provide important information about the activity of each vertex. First, for any vertex $v \in V$, it is clear that $a(v) < b(v)$ and, looking at the color of the vertex, we have v is white before time $a(v)$, gray between time $a(v)$ and time $b(v)$, and black thereafter. Second, the following lemma shows that the activity intervals of two vertices are properly nested.

LEMMA (NESTED ACTIVITY LEMMA)   *In any depth-first search of a graph $G = (V, E)$ and for any two vertices $u, v \in V$, the activity intervals of u and v are either nested or disjoint.*

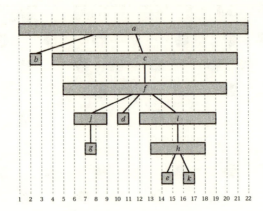

FIGURE 18. The dfs tree and activity interval of each vertex obtained by the sample DFS run of Figure 16.

PROOF: We show the contrapositive: the activity intervals of two vertices cannot properly intersect. Let u and v be two vertices of G, and consider any depth-first search of G. Suppose without loss of generality that, say, $a(u) < a(v)$. This implies that DFS(v) was called after DFS(u). If furthermore we had $b(u) < b(v)$, it would mean that DFS(v) (implicitly) called DFS(u) again so that u can be blackened. This is so because DFS(v) has not yet finished at time $b(u)$ and DFS(u) is the only call that can blacken u. But this is a contradiction with the specification of the algorithm, that is, with the fact that in any depth-first search we never call DFS twice on the same vertex. ∎

If we denote the activity interval of a vertex v by $[a(v), b(v)]$, the Nested Activity Lemma implies the following corollary.

COROLLARY *For any dfs tree T for a graph $G = (V, E)$ and any two vertices $u, v \in V$,*

1. *v is a descendant of u in T if and only if $[a(v), b(v)] \subseteq [a(u), b(u)]$.*
2. *v is a ancestor of u in T if and only if $[a(v), b(v)] \supseteq [a(u), b(u)]$.*
3. *neither u nor v is a descendant or an ancestor of the other in T if and only if $[a(u), b(u)]$ and $[a(v), b(v)]$ are disjoint.*

PROOF: The first claim of the corollary easily follows from the Nested Activity Lemma and

v is a descendant of u in T $\Leftrightarrow$ v is grayed while u is gray during depth-first search

$$\Leftrightarrow a(u) < a(v) < b(v) < b(u)$$

$$\Leftrightarrow [a(v), b(v)] \subseteq [a(u), b(u)].$$

The second and third claim of the corollary are just negation of the first claim, taking into account the Nested Activity Lemma. ∎

We now show more carefully the edge classification property of depth-first search of an undirected graph.

LEMMA (EDGE CLASSIFICATION LEMMA)   *For any dfs tree T produced by depth-first search of an undirected graph G, every edge of G is either a tree edge or a back edge.*

PROOF: Let (u, v) be an arbitrary edge of G. Suppose first that, say, $a(u) < a(v)$, and consider the call DFS(u). Since v is on u's adjacency list, this call eventually examines v. If v was white at that moment, then DFS(u) calls DFS(v), and so v becomes a child of u in T, i.e., (u, v) is a tree edge.

If v was not white at the moment DFS(u) examines v, then DFS(v) was called earlier. Consider then the call DFS(v) that happened before the call DFS(u). Since u is also on v's adjacency list, the call DFS(v) eventually examines u. At that moment, u cannot be white, since $a(u) < a(v)$, and u cannot be black, since otherwise the activity intervals of u and v would properly intersect. Thus, u must be gray, hence v is grayed while u is gray. But this means that v is a descendant of u in T, and so (u, v) is a back edge. The second case $a(u) > a(v)$ is similar, with the roles of u and v reversed. ∎

The proof of this lemma suggests that the expanded DFS algorithm can be modified to classify edges as it explores them. Namely, the type of each edge can be determined according to the color of the other end vertex when the edge is first explored. In particular, assuming $a(u) < a(v)$, i.e., DFS(u) is called before DFS(v), if an edge (u, v) is first explored in the direction from u to v, then (u, v) is a tree edge. Otherwise, if (u, v) is first explored in the direction from v to u, then (u, v) is a back edge.

The following is a modified DFS algorithm that classifies edges of an undirected graph G "on the fly."

```
DFS(u)
    t = t + 1; a(u) = t;   [ timestamp the activity event for u ]
    c(u) = GRAY;           [ u is active ]
    for each v in L(u) do
        if c(v) = WHITE then
            (u, v) is a tree edge;
            p(v) = u;       [ u is parent of v ]
            DFS(v);
        else if c(v) = GRAY then
            (u, v) is a back edge;
        else [ c(v) = BLACK ]
            [ (u, v) = (v, u) is already classified as a back edge ]
    t = t + 1; b(u) = t;   [ timestamp the inactivity event for u ]
    c(u) = BLACK;          [ u is inactive ]
    return;
```

3.4. Depth-first search of directed graphs. Conceptually, depth-first search of directed graphs is essentially the same as depth-first search of undirected graphs, the only difference being in the interpretation of the notion "adjacent vertex." In a directed graph, a vertex v is adjacent to vertex u if there exists a directed edge $u \rightarrow v$. If $u \rightarrow v$ exists but $v \rightarrow u$ does not, then v is adjacent to u but u is not adjacent to v. With this change of interpretation, the basic DEPTH-FIRST-SEARCH and DFS algorithms apply equally well in the case of a directed graph. Moreover, an argument identical to the one for an undirected graph shows that the running time of the DEPTH-FIRST-SEARCH algorithm for a directed graph with n vertices and m edges is also $O(n + m)$. Figure 19 illustrates a sample run of the basic DFS algorithm on a directed graph.

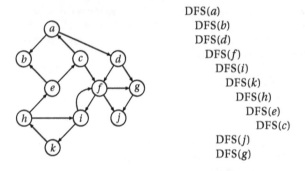

FIGURE 19. A directed graph and a DFS execution from vertex a.

Exactly as for an undirected graph, we can build the depth-first search tree as we depth-first search a directed graph. Specifically, if DFS(u) directly calls DFS(v), we add a tree edge connecting v as a child of u in the tree. Figure 20 shows the resulting depth-first search tree associated with the depth-first search in Figure 19.

Although conceptually similar, depth-first searches of directed and undirected graphs behave quite differently. This is revealed by two properties specific to depth-first search of a digraph.

First, it can produce a forest of the depth-first search trees even if the digraph is connected. For example, had we started the search in Figure 19 from the vertex g, we would get a forest of two or more trees.

Second, in the case of an undirected graph we saw that the nontree edges of the graph necessarily connect some vertex to one of its ancestors. That is, the nontree edges of an undirected graph are always back edges. However, the nontree directed edges in the case of a directed graph G can be of three kinds:

1. **Forward edges**, which are directed edges $u \rightarrow v$ of G such that v is a proper descendant of u, but not a child of u. For example, in Figure 20, $d \rightarrow g$ is a forward edge.

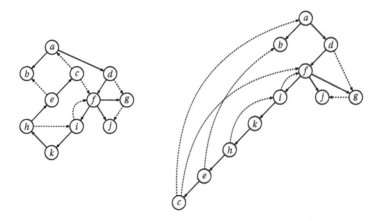

FIGURE 20. The depth-first tree obtained by the sample DFS run in Figure 19.

2. **Back edges**, which are directed edges $u \to v$ of G such that v is an ancestor of u ($u = v$ is allowed). In Figure 20, $c \to f$ is a back edge. Notice that self-loops, which may occur in directed graphs, are considered back edges.

3. **Cross edges**, which are all other directed edges of G. In Figure 20, $g \to j$ is a cross edge. Notice that cross edges may go between vertices in different dfs trees of a forest. Stated another way, cross edges are directed edges $u \to v$ of G such that v is neither an ancestor nor a descendant of u. In general however, given a cross edge $u \to v$, it does not imply that u and v have a common ancestor.

On the other hand, in Figure 20 we see that the cross edge $g \to j$ points from right to left. This is not a coincidence and holds in general. That is, any cross edge always points from right to left, even in a forest of dfs trees.

To see why, take a cross edge $u \to v$, and consider the call DFS(u). It is clear that a node v is a descendant of u in the depth-first search tree if and only if v is added to the tree between the time DFS(u) is invoked and the time DFS(u) is completed. Now, by the time this call considers the edge $u \to v$, it must be the case that v was already visited, because $u \to v$ is not a tree edge. Thus, the node v must have been visited earlier in the call DFS(u) or before DFS(u) was invoked. But the first case is impossible, since otherwise v would be a descendant of u. Therefore, v was added to the depth-first tree before u, and since nodes are added in the left-to-right order, v must be to the left of u.

The expanded form of depth-first search on a directed graph provides also valuable insight into the structure of the graph by considering the colors and timestamps of the vertices. For example, the following sequence of pictures illustrates the progress of the expanded DFS algorithm corresponding to the sample run of Figure 19.

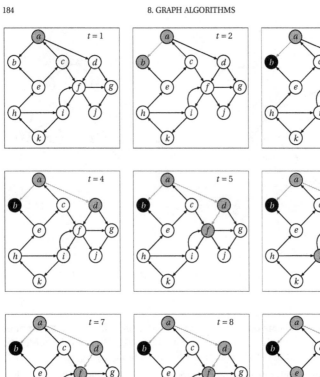

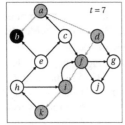
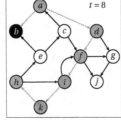
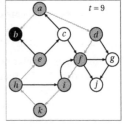

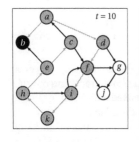
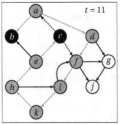
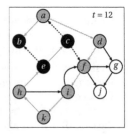

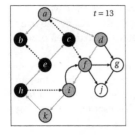
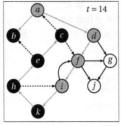
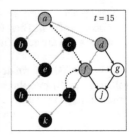

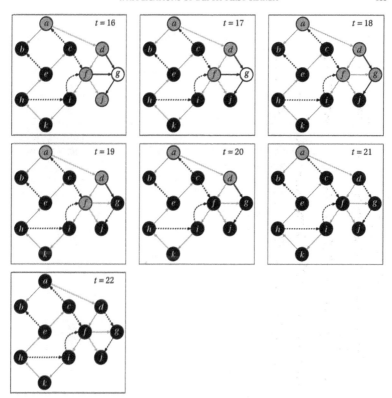

It is easy to observe that the Nested Activity Lemma and its corollary also hold in the case of a directed graph. In addition, we can modify the expanded DFS algorithm to classify directed edges as it explores them. Again, the type of each edge $u \to v$ can be determined according to the color of the vertex v that is reached when the edge is explored:

1. If v is a white vertex, $u \to v$ is clearly a tree edge.
2. If v is a gray vertex, $u \to v$ is a back edge. This follows from the fact that exploration always proceeds from the deepest gray vertex, so a gray vertex that examines another gray vertex reaches an ancestor.
3. If v is a black vertex, $u \to v$ is either a forward or a cross edge. Specifically, if $a(u) > a(v)$ then $u \to v$ is a cross edge, since the activity intervals of u and v are disjoint and $u \to v$ points from right to left. Otherwise, if $a(u) < a(v)$ then $u \to v$ is a forward edge.

A modified DFS algorithm that implements these observations is left as an exercise.

4. Applications of depth-first search

We now present some efficient applications of depth-first search to solving various practical problems related to graphs. We first consider two problems for undirected graphs (finding connected components and articulation points), and then three problems for directed graphs (finding reachable sets, test for acyclicity, and topological sorting). Some additional interesting problems appear in the exercises.

4.1. Connected components. Suppose that a graph represents TV network owned by some company XYZ. Each vertex represents a TV station and each edge represents a bidirectional communication channel between two TV stations. Two TV stations can communicate with each other if and only if they are connected by a channel. Figure 21 shows an example of the TV network in the form of an undirected graph.

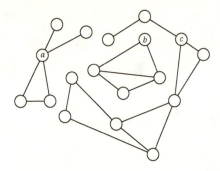

FIGURE 21. TV network represented by a graph.

The company XYZ is preparing for a live broadcast of the Super Bowl. A satellite can broadcast the Super Bowl signal to any number of TV stations. Once a TV station receives the Super Bowl signal from the satellite, it can then broadcast the signal. For example, if the stations a, b, c in Figure 21 receive signal from the satellite, then after broadcasting by these three stations, all other stations in the network receive the Super Bowl signal.

Broadcasting is much cheaper compared to the satellite transmission. Thus, the company XYZ wants to determine the *minimum* number of TV stations that should receive signal from the satellite and yet all TV stations in the network can receive the Super Bowl signal subsequently. What is this number?

As a graph problem, this question clearly translates to finding the number of connected components of the graph that represents a TV network. More precisely, given a graph G, we want to assign numbers to its vertices so that vertices in the same connected component receive the same ordinal number of the corresponding connected component. The key observation here is that if we start depth-first search from some vertex, then all the vertices we will visit belong to the same connected component containing the starting vertex. In other words, connected components of an undirected graph are determined by the spanning trees of the dfs forest.

The following is pseudocode for the algorithm CONNECTED-COMPONENTS(G) that uses this observation to find the connected components of a graph $G =$

(V, E) with n vertices and m edges. It is very similar in structure to the basic DEPTH-FIRST-SEARCH algorithm, except that we now additionally use a global array cc of size n to store assigned vertex numbers, as well as a global variable k to count connected components of G.

CONNECTED-COMPONENTS(G)
for each $v \in V$ **do** $c(v) = 0$;
$k = 0$; $[\![$ number of cc $]\!]$
for each $v \in V$ **do**
if $c(v) = 0$ **then**
$k = k + 1$;
DFS-CC(v);
return cc; $[\![$ vertex cc numbers $]\!]$

DFS-CC(u)
$c(u) = 1$;
$cc(u) = k$;
for each v adjacent to u **do**
if $c(v) = 0$ **then**
DFS-CC(v);
return;

The running time of the CONNECTED-COMPONENTS algorithm is clearly the same asymptotically as the running time of the DEPTH-FIRST-SEARCH algorithm, that is, $O(n + m)$.

4.2. Articulation points. Let $G = (V, E)$ be a connected, undirected graph. A vertex $a \in V$ is said to be an ***articulation point*** if there exist distinct vertices u and v such that every path between u and v contains the vertex a. Put it another way, a is an articulation point of G if removing a and all its incident edges makes the remaining subgraph disconnected. Figure 22 shows an example of a graph and its articulation point a.

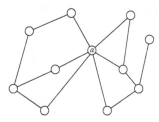

FIGURE 22. An articulation point as a gateway.

The problem of finding articulation points is important in practical applications. For example, we may represent a telecommunication network as a graph in which the vertices are sites that are connected by communication lines. If the graph has no articulation point, it means that the network can continue to function even in the case of failure of one of its sites, either due to the equipment malfunction or an external attack in military situations.

Intuitively, a vertex is an articulation point of the graph G if it is a "gateway" between two disjoint subgraphs in the sense that every path from one subgraph to the other has to pass through that vertex. Not so surprisingly, this condition can be relatively easily detected by running depth-first search on G. However, we have to distinguish two cases depending on the starting vertex of depth-first search.

In the first case, if we start depth-first search from an articulation point, then it will be the root of the depth-first search tree. The root must have at least two children in the depth-first search tree, since after visiting vertices in one subgraph, the search has to backtrack all the way to the articulation point and then visit vertices in the other subgraph. This case is illustrated in Figure 23, where the names of vertices indicate order in which they are visited during depth-first search.

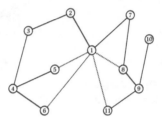

FIGURE 23. Depth-first search that starts from an articulation point.

Thus, the example in Figure 23 suggests that the root is an articulation point if and only if it has two or more children.

In the second case, if we start depth-first search from a non-articulation point vertex located in one subgraph, then the depth-first search tree will cross the articulation point and extend to the other subgraph. This case is illustrated in Figure 24, where depth-first search starts from a vertex in the left subgraph.

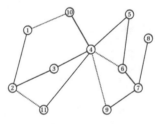

FIGURE 24. Depth-first search that does *not* start from an articulation point.

Consider in Figure 24 a subtree rooted at a child of the articulation point that is contained completely inside the right subgraph. (Notice that not all descendants of the articulation point need to lie inside the right subgraph.) This subtree (that is, some node in the subtree) is not connected via a back edge to any ancestor of the articulation point. Thus, the example in Figure 24 suggests that a vertex u other than the root is an articulation point if and only if there is some child s of u such that the subtree rooted at s is not connected via a back edge to any proper ancestor of u. These observations lead to the following lemma.

LEMMA *Let $G = (V, E)$ be a connected, undirected graph, and let T be a dfs tree for G. Then,*

1. *The root r of the dfs tree T is an articulation point of G if and only if r has at least two children in T.*
2. *A nonroot vertex u of T is an articulation point of G if and only if there is some child s of u in T such that the subtree rooted at s is not connected via a back edge to any proper ancestor of u.*

PROOF: The key observation in the proof is that there are no cross edges of an undirected graph in any depth-first search of G. Now, the first claim is easy to show. In one direction, if the root of T has more than one child, removing the root and all edges incident with it clearly splits G into two or more subgraphs corresponding to the subtrees rooted at the children of the root. Thus, the root is an articulation point.

Conversely, if the root of T has only one child, removing the root and all edges incident with it leaves one subtree rooted at the child of the root, which spans remaining subgraph and so it stays connected. Thus, the root is not an articulation point.

For the second claim, first suppose there is some child s of a nonroot vertex u such that the subtree rooted at s is not connected via a back edge to any proper ancestor of u. If we remove u and all edges incident with it, this will split G into at least two parts—one corresponding to the subtree rooted at s, and the other corresponding to everything outside the subtree. The subtree rooted at s is isolated since no edge connects it with an outside vertex. This is so because a back edge may connect the subtree only with u, which we removed, and there are no cross edges. Thus, the vertex u is an articulation point.

Conversely, if every subtree rooted at each child of u is connected via a back edge to some proper ancestor of u, then removing u will not disconnect G, hence the vertex u is not an articulation point. To see why, let $T_1, T_2, \ldots, T_k$ be the subtrees rooted at the children $s_1, s_2, \ldots, s_k$ of u in T, and let G' be the graph obtained by removing u and all its incident edges from G. To show that there is a path in G' for any two vertices x and y in G', it is easy to see that the most complicated case is if x belongs to one of the subtrees, say T_l, and y belongs to some other, say T_j (see Figure 25). The other possible locations for x and y are handled similarly. Now, if (v, w) is a back edge that connects some w in T_l with a proper ancestor v of u, and (v', w') is a back edge that connects some w' in T_j with a proper ancestor v' of u, then a path between x and y in G' goes from x to w in T_l, then follows the back edge (w, v), then goes from v to v' in T, then follows the back edge (v', w'), and finally arrives at y from w' in T_j. ∎

Given a nonroot vertex u in T as in Figure 25, consider the subtree T_u in T rooted at u. To efficiently use the previous lemma, the idea is to associate each nonroot vertex u in T with the highest vertex v on the path from u to r in T such that v connects T_u via a back edge.

Let B be the set of all back edges induced by depth-first search producing the dfs tree T. Further, for any (root or nonroot) vertex u in T, let B_u be the set

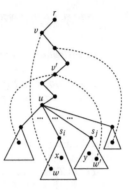

FIGURE 25. A dfs tree and the vertex u that is not an articulation point.

of all back edges with one of their end vertices (or both) in T_u. That is,

$$B_u = \{(v, w) \in B : v \text{ or } w \text{ (or both) in } T_u\}.$$

Next, partition B_u into two subsets: one is B_u' of back edges with only one end vertex in T_u, and the other is B_u'' of back edges with both end vertices in T_u. That is,

$$B_u' = \{(v, w) \in B_u : \text{only } w \text{ is in } T_u\}$$

and

$$B_u'' = \{(v, w) \in B_u : \text{both } v \text{ and } w \text{ are in } T_u\}.$$

By the corollary of the Nested Activity Lemma from Section 3.3, if $(v, w) \in B_u'$, then $a(v) < a(u) \leq a(w)$. (The equality in the second inequality holds if $w = u$.) Thus, if $B_u' \neq \emptyset$, then

$$\min\{a(v): (v, w) \in B_u'\} < a(u).$$

Moreover, the starting time of the activity interval of the highest vertex v in T such that $(v, w) \in B_u'$ is exactly $\min\{a(v): (v, w) \in B_u'\}$.

On the other hand, if $(v, w) \in B_u''$ and, say, v is an ancestor of w, then $a(u) \leq a(v) < a(w)$. (The equality in the first inequality holds if $v = u$.) Thus, if $B_u'' \neq \emptyset$, then $a(v) \geq a(u)$ for every $(v, w) \in B_u''$.

In other words, the mapping $h: V \to \mathbb{N}$ defined by

$$h(u) = \begin{cases} a(u), & B_u' = \emptyset \\ \min\{a(v): (v, w) \in B_u'\}, & B_u' \neq \emptyset \end{cases}$$

can be used to detect the highest vertex v on the path from u to r in T such that v connects T_u via a back edge, since $B_u' \neq \emptyset$ if and only if $h(u) < a(u)$.

One good property of the mapping h is that we can compute it recursively. Given any u in T, it is easy to see that

$$h(u) = \min \begin{cases} a(u), \\ \{h(s): s \text{ is a child of } u\}, \\ \{a(v): (u, v) \in B_u'\}. \end{cases}$$

This is important because we can now easily detect whether a nonroot vertex is an articulation point. Namely, by the previous lemma, a nonroot vertex u of T is an articulation point of G if and only if there is a child s of u such that $h(s) \geq a(u)$.

The following two algorithms incorporate necessary changes based on the previous observations to the expanded depth-first search.

```
ART-POINTS (G)
    t = 0;
    for each v ∈ V do
        c(v) = WHITE;
        ap(v) = FALSE;
    r = any v ∈ V;    [ choose any vertex in G ]
    p(r) = NIL;       [ r is the root of the dfs tree ]
    ch = 0;           [ the root's number of children ]
    DFS-AP(r);        [ call DFS-AP once, since G is connected ]
    return ap;        [ ap gives all articulation points ]
```

```
DFS-AP(u)
    t = t + 1; a(u) = t;
    c(u) = GRAY;
    h(u) = a(u);                    [ initialize h(u) ]
    for each v in L(u) do
        if c(v) = WHITE then        [ v is a child of u ]
            p(v) = u;               [ u is parent of v ]
            DFS-AP(v);
            if p(u) = NIL then      [ u is the root ]
                ch = ch + 1;
                if ch > 1 then      [ the root is an articulation point ]
                    ap(u) = TRUE;
            else                    [ u is a nonroot vertex ]
                if h(v) ≥ a(u) then [ a nonroot vertex is an articulation point ]
                    ap(u) = TRUE;
            h(u) = min{h(u), h(v)}; [ update h(u) with a child ]
        else if c(v) = GRAY then    [ (u, v) is a back edge ]
            h(u) = min{h(u), a(v)}; [ update h(u) with a back edge ]
    t = t + 1; b(u) = t;
    c(u) = BLACK;
    return;
```

We use the global array *ap* to indicate articulation points of the connected, undirected graph G, as well as global variable *ch* to count the number of children of the root of the dfs tree produced by depth-first search. The main algorithm ART-POINTS(G) only initializes these global variables and calls a modified DFS algorithm (DFS-AP) on an arbitrary vertex in G. The DFS-AP algorithm computes h and detects all articulation points accordingly.

Figure 26 shows the pairs $a(u)/h(u)$ next to each vertex u in the dfs tree obtained by the sample run of Figure 16 using the DFS-AP algorithm. Notice that the root a is an articulation point, since it has two children. Also, the vertex f is the only nonroot articulation point, which is detected since f's child j satisfies the condition $h(j) \geq a(f)$ (i.e., $5 \geq 5$).

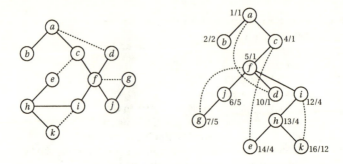

FIGURE 26. DFS-AP computation of articulation points on a sample graph.

Finally, to analyze the running time of the ART-POINTS algorithm, we note that the changes made to the expanded depth-first search increase its running time only by a constant factor. Therefore, for a connected undirected graph G with n vertices and m edges, the algorithm ART-POINTS(G) takes $O(n + m)$ time. In fact, since G is connected, i.e., $n \leq m + 1$, we can express the running time of the ART-POINTS algorithm more succinctly as $O(m)$.

4.3. Reachability problem. A natural question to ask about a directed graph G is, given two vertices u and v, can we get to v from u by following directed edges? Or, more generally, given a vertex u, which vertices can we reach from u? Or, most generally, if we ask the last question for each vertex u in G, then we call it the *reachability problem*. The solution to the reachability problem tells us then for which pairs of vertices u and v there is a path from u to v.

We can devise an algorithm for solving the reachability problem as a simple application of depth-first search. For a vertex u, we mark all vertices as unvisited and call DFS(u). We then examine all the vertices again. Those marked as visited belong to the reachable set of vertices from u, and others do not. If we are interested in another vertex v, we mark all vertices as unvisited again and call DFS(v). After the call, vertices marked as visited belong to the reachable set of vertices from v. Obviously, we can repeat this process for all vertices to get the solution to the reachability problem.

As usual, assume a directed graph G has n vertices and m edges. Since finding the reachable set for one vertex takes $O(n + m)$ time, and we repeat the process n times, the total running time of the algorithm for solving the reachability problem is $O(n(n + m))$.

For the sake of completeness, we present below pseudocode for the algorithm REACHABLE-SETS(G) that in a given directed graph $G = (V, E)$ finds the reachable set for every vertex $v \in V$. The pseudocode is an easy modification of the basic DEPTH-FIRST-SEARCH algorithm, and uses an additional global array rs to indicate the reachable set for each $v \in V$.

REACHABLE-SETS(G)
for each $u \in V$ **do**
  for each $v \in V$ **do** $c(v) = 0$;
  $rs(u) = \emptyset$; $[\![$ u's reachable set $]\!]$
  DFS-RS(u);
return rs; $[\![$ reachable sets $]\!]$
	DFS-RS(u)
	  $c(u) = 1$;
	  for each v adjacent to u **do**
	    if $v = u$ **then** $rs(u) = rs(u) \cup \{u\}$;
	    if $c(v) = 0$ **then**
	      $rs(u) = rs(u) \cup \{v\}$;
	      DFS-RS(v);
	  return;

4.4. Test for acyclicity. We often want to know whether a directed graph contains a cycle. For example, a directed graph may represent the calling structure of a collection of procedures in a large program, where a directed edge goes from a caller to called procedure. A cycle in the calling graph tells us that there is an indirect or a direct recursion, which might not be what we intended.

The four types of the edges of a directed graph G with respect to a particular dfs forest suggest that we can discover a cycle by looking at backward edges. Indeed, if there is a backward edge $u \rightarrow v$ after depth-first search of the graph, there must be a cycle in the graph. The cycle consists of the directed edge $u \rightarrow v$ and the path in the tree from v to u. Figure 27 illustrates this case.

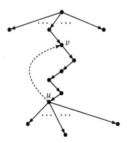

FIGURE 27. A cycle formed by a backward directed edge.

Moreover, those cycles formed by backward directed edges are the only cycles in a digraph. That is, the converse of the above observation is also true: if there is a cycle, there must be a backward directed edge. We prove this by looking at the inactivity timestamps of vertices and with the help of the following two lemmas.

LEMMA   *Let $G = (V, E)$ be a directed graph. Consider any dfs forest T for G, and consider any directed edge $u \to v \in E$. If this edge is a tree, forward, or cross edge with respect to T, then $b(u) > b(v)$. If this edge is a back edge, then $b(u) \le b(v)$.*

PROOF: The proof follows directly from the corollary of the Nested Activity Lemma for directed graphs. For a tree or forward edge $u \to v$, v is a descendant of u in T, and so v's activity interval is contained within u's, implying $b(v) < b(u)$. For a cross edge $u \to v$, the activity intervals of u and v are disjoint. Since the cross edges always point from right to left, DFS(v) was started and finished before the start (and finish) of DFS(u), implying $b(v) < b(u)$. Finally, if $u \to v$ is a back edge with $u \ne v$, v is an ancestor of u in T, and so u's activity interval is contained within v's, implying $b(u) < b(v)$. If the back edge $u \to v$ is a self-loop, i.e., $u = v$, then clearly $b(u) = b(v)$. ∎

Since the tree, back, forward, and cross edges completely partition the set of edges of G, the following corollary is immediate.

COROLLARY   *Given a digraph $G = (V, E)$ and any dfs forest for G, an edge $u \to v \in E$ is a back edge if and only if $b(u) \le b(v)$.*

LEMMA   *Let $G = (V, E)$ be a directed graph, and let T be any dfs forest for G. Then G has a cycle if and only if G has a back edge with respect to T.*

PROOF: ($\Leftarrow$) This direction follows from the opening discussion leading to Figure 27.

($\Rightarrow$) We show the contrapositive. Suppose G has no back edge, and consider any path $v_1 \to v_2 \to \cdots \to v_{k-1} \to v_k$ in G. Any directed edge $v_1 \to v_2$, $v_2 \to v_3$, ..., $v_{k-1} \to v_k$ along the path is a tree, forward, or cross edge in T. Thus, by the previous lemma, $b(v_1) > b(v_2) > \cdots > b(v_{k-1}) > b(v_k)$. For vertices along any path therefore, finish times strictly decrease monotonically. This implies that any path cannot repeat any vertex v, otherwise it would lead to $b(v) > b(v)$, a contradiction. In other words, any path in G cannot close itself, implying there can be no cycle in G. ∎

We can now devise a simple test for acyclicity of a directed graph G. We run first the expanded version of depth-first search on G to compute the inactivity timestamps of all vertices. After that, we examine all the directed edges to check if there is a back edge. If we find one, the graph is cyclic. If there is no back edge, the graph is acyclic.

The following is pseudocode for the algorithm ACYCLICITY-TEST (G) that returns TRUE or FALSE depending on whether the graph G is acyclic or not, respectively. It uses the expanded DEPTH-FIRST-SEARCH algorithm as a subprocedure for computation of the finishing times of all vertices.

ACYCLICITY-TEST (G)
    DEPTH-FIRST-SEARCH (G);
    for each $u \in V$ **do**
        for each v in $L(u)$ **do**
            if $b(u) \le b(v)$ **then** $[\![\ u \to v$ is a back edge $]\!]$
                return FALSE;
    return TRUE;

The algorithm ACYCLICITY-TEST (G) on a digraph G with n vertices and m edges clearly takes $O(n + m)$ time for depth-first search, and another $O(n + m)$ worst-case time for the examination of all directed edges. Thus, the entire acyclicity test also takes $O(n + m)$ time.

Notice that we could modify the expanded DEPTH-FIRST-SEARCH and DFS algorithms to directly detect absence or presence of a back edge in G by considering the color of vertices, and return TRUE or FALSE accordingly. This way we could speed up the algorithm by not running the edge examination phase afterwords. However, we would not get any asymptotic time speed-up, and so we have chosen this more modular version of the acyclicity test. The other version is left as an exercise for the interested reader.

4.5. Topological sort. A directed acyclic graph (dag) is used in many applications to indicate precedences or ordering constraints among events. For example, a complex project is usually divided into a series of smaller tasks, some of which have to be performed in certain specified order so that the project can be brought to completion (e.g. in construction, the first floor must be built before the second floor, but electrical wiring can be done while installing the windows). Another example is a university curriculum that may have courses that require other courses as prerequisites. Directed acyclic graphs are often used to model such situations naturally: a directed edge from task (course) A to task (course) B can represent the relation that A's completion is necessary for B's completion. Figure 28 illustrates a dependency structure of five tasks. In this example, task A_3 can be performed only after tasks A_1 and A_2 have been finished.

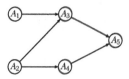

FIGURE 28. A directed acyclic graph of dependencies of five tasks.

Topological sort is an operation of arranging the vertices of a directed acyclic graph $G = (V, E)$ in a list so that if there exists a directed edge $u \to v$, then u precedes v in the list. In other words, it is a process of assigning a linear ordering $v_1, v_2, \ldots, v_n$ to the vertices of G in such a way that $v_i \to v_j \in E$ for some $1 \le i \ne j \le n$ implies $i < j$. Thus, a topological sort of a dag can be viewed as an ordering of its vertices along a horizontal line so that all directed edges go from left to right. For example, a topological sort of the graph of Figure 28 is A_1, A_2, A_3, A_4, A_5, as suggested in Figure 29. Notice that the linear ordering is not unique; another topological sort of the example graph is A_2, A_4, A_1, A_3, A_5.

FIGURE 29. A topological sort of the dag of Figure 28.

A topological sort of a dag G can be easily accomplished by performing depth-first search on G. Since G is acyclic, we know from the last lemma in the previous section that G cannot have a back edge. Thus, if $u \rightarrow v \in E$ (be it a tree, forward, or cross edge), then $b(v) < b(u)$. This suggest that we can get a topological ordering if we list the vertices in reverse order of finishing time. More precisely, define the numbering $v_1, v_2, \ldots, v_n$ of the vertices of G so that $b(v_1) > b(v_2) > \cdots > b(v_n)$. Then $v_i \rightarrow v_j \in E$ for some $1 \leq i \neq j \leq n$ implies $b(v_j) < b(v_i)$, which means $i < j$ in our ordering.

Using these observations to compute a topological ordering of a dag G, we can modify the basic DFS algorithm to add each vertex v at time $b(v)$ (i.e., when it is finished processing) to the end of a linked list. Thus, if $u \rightarrow v \in E$, then $b(v) < b(u)$, hence v will appear before u in the list. However, since a topological ordering requires that u precedes v, we just need to reverse the final list. Another equivalent approach, which we have actually taken in the modified DFS algorithm, is to insert each vertex, as we are finished with it, onto the front of the linked list.

The algorithm TOP-SORT(G) below, which uses a version of the basic DFS algorithm modified in this way, topologically sorts the vertices of a dag G. The TOP-SORT algorithm returns a linked list of vertices whose order defines their final topological ordering.

TOP-SORT(G)	DFS-TS(u)
for each $v \in V$ **do** $c(v) = 0$;	$c(u) = 1$;
$L = \emptyset$; ⟦ L is empty linked list ⟧	**for** each v adjacent to u **do**
for each $v \in V$ **do**	**if** $c(v) = 0$ **then**
if $c(v) = 0$ **then**	DFS-TS(v);
DFS-TS(v);	LIST-HEAD-INSERT(L, u);
return L; ⟦ final topological order ⟧	**return**;

The running time of the TOP-SORT algorithm to topologically sort vertices of a dag G with n vertices and m edges is clearly $O(n + m)$, the same time it takes to perform depth-first search on G.

5. Breadth-first search

Another systematic way of visiting all vertices in a graph is called breadth-first search. It is a generalization of the level-order traversal of a tree. This method is called "breadth-first," because from each visited vertex v we search as broadly as possible by next visiting all the vertices adjacent to v. When depth-first search arrives at v, it next tries to visit some neighbor of v, then a neighbor of this neighbor, and so on. On the other hand, when breadth-first search arrives at v, it first visits all the neighbors of v, and only when this is done, it goes on to visit nodes farther away.

Unlike depth-first search, breadth-first search is not naturally recursive. To better see this, we can easily write a nonrecursive version of the DFS algorithm by simulating recursion with a stack. For example, the following is an iterative implementation of depth-first search.

```
DEPTH-FIRST-SEARCH(G)
  [ Initially, all vertices are unvisited ]
  for each v ∈ V do c(v) = 0;
  [ Visit each vertex left unvisited ]
  for each v ∈ V do
    if c(v) = 0 then
      ITER-DFS(v);
  return;
```

```
ITER-DFS(s)
  c(s) = 1; [ source s is visited ]
  S = ∅;   [ S is empty stack ]
  PUSH(S, s);
  while S ≠ ∅ do
    u = POP(S);
    for each v adjacent to u do
      if c(v) = 0 then
        c(v) = 1; [ v is visited ]
        PUSH(S, v);
  return;
```

Breadth-first search by contrast requires a queue data structure as a guide for the search. In other words, breadth-first search uses a queue to systematically keep track of vertices it is supposed to visit next.

Another stylistic difference is that breadth-first search, due to its iterative nature, *discovers* new vertices as it visits them. That is, when visiting a vertex v, breadth-first search examines all edges leaving v to discover v's neighbors it should visit later. That's why, in this context, we additionally use this term to describe the process of traversing vertices of a graph by exploring its edges.

In pseudocode, breadth-first search on an input graph $G = (V, E)$ is encompassed by two algorithms BREADTH-FIRST-SEARCH and BFS below. Except for using a queue instead of a stack, their structure is identical to that of the corresponding algorithms for the iterative implementation of depth-first search.

The basic BREADTH-FIRST-SEARCH algorithm first initializes a global array c that keeps track of visited (discovered) vertices, and then selects an unvisited vertex v to call a *nonrecursive* algorithm BFS(v) that actually traverses the graph G starting from the source vertex v. This way, each connected component of G gets breadth-first searched, in case G is not a connected graph.

```
BREADTH-FIRST-SEARCH(G)
    ⟦ Initially, all vertices are unvisited ⟧
    for each v ∈ V do c(v) = 0;
    ⟦ Visit each vertex that's left unvisited ⟧
    for each v ∈ V do
        if c(v) = 0 then
            BFS(v);
    return;
```

The basic algorithm BFS(s) maintains a queue of all vertices it is to visit in the breadth-first fashion, starting from the source vertex s. The algorithm iterates as long as the queue is nonempty by dequeuing already discovered vertices and enqueuing their newly discovered neighbors.

```
BFS(s)
    c(s) = 1;               ⟦ source s is discovered ⟧
    Q = ∅;                  ⟦ Q is empty queue ⟧
    ENQUEUE(Q, s);
    while Q ≠ ∅ do
        u = DEQUEUE(Q);     ⟦ u is the next to visit ⟧
        for each v adjacent to u do
            if c(v) = 0 then
                c(v) = 1;   ⟦ neighbor v is discovered ⟧
                ENQUEUE(Q, v);
    return;
```

Observe that the BFS algorithm systematically explores the edges of G to discover every vertex reachable from the source s, and in doing so it expands the frontier between visited and unvisited vertices across the breadth of the frontier.

With an appropriate interpretation of the notion "adjacent vertex," the BFS algorithm can be applied to either directed or undirected graphs. Figure 30 illustrates a sample run of the BFS algorithm on an undirected graph by showing the contents of the queue after each iteration of the **while** loop.

The time complexity of breadth-first search is the same as that of depth-first search. To prove this fact, we can use a similar argument to the one employed for the running time of depth-first search. Namely, each visited vertex by breadth-first search is placed in the queue exactly once, so the **while** loop is executed once for each vertex $u \in V$. The inner **for** loop is executed d_u times, where d_u is the degree (i.e., the number of neighbors) of u. Thus, the **for** loop takes $O(1 + d_u)$ time, where the additional 1 accounts again for the case $d_u = 0$. Therefore, if a graph G has n vertices and m edges, the running time of the algorithm BREADTH-FIRST-SEARCH(G) is:

$$\sum_{u \in V} O(1 + d_u) = O\left(\sum_{u \in V}(1 + d_u)\right) = O\left(\sum_{u \in V} 1 + \sum_{u \in V} d_u\right) = O(n + 2m) = O(n + m).$$

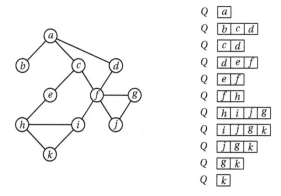

FIGURE 30. An undirected graph and a BFS execution from the vertex a.

5.1. Breadth-first search tree. As for the case of depth-first search, we can also construct a spanning tree (forest) of the graph when we perform breadth-first search. In this case, we consider a graph edge (u, v) as a tree edge if vertex v is first discovered from vertex u in the inner **for** loop of the BFS algorithm.

Figure 31 illustrates the breadth-first search tree associated with the BFS execution shown in Figure 30. As before, we have assumed the search began at vertex a, and we have shown tree edges as solid lines and the other edges as dashed lines.

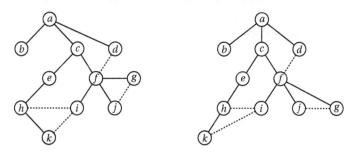

FIGURE 31. A graph and its breadth-first search tree.

It turns out that breadth-first search on a graph induces a similar classification of edges. Specifically, in an undirected graph every edge that is not a tree edge connects two vertices neither of which is an ancestor of the other in the breadth-first search tree. In an analogy with the depth-first search tree, these edges are called *cross edges*.

5.2. A closer look at breadth-first search. To better grasp the essence of breadth-first search, in this section we show an expanded version of the BREADTH-FIRST-SEARCH algorithm, similar in nature to the expanded version of the DEPTH-FIRST-SEARCH algorithm. However, we merge the old BREADTH-FIRST-SEARCH and BFS algorithms into one algorithm, called again BFS, that takes as input a graph and a source vertex form which it performs breadth-first search on the input graph. Although conceptually breadth-first search can proceed from multiple origins, as we saw in the previous sections, the single-source starting point better reflects a typical application of breadth-first search. Namely, as we shall see in the next section, it is usually employed to find shortest-path distances from a given source.

The expanded BFS algorithm keeps track of progress by coloring each vertex white, gray, or black, instead of simply maintaining whether a vertex is discovered or undiscovered. All vertices are initially white, and later become gray and then black. When a white vertex is first discovered, it is colored gray and enqueued for further processing. When a gray vertex is dequeued and finished processing by exploring all its neighbors, it is colored black. White vertices therefore have not been discovered yet, while gray and black vertices are discovered. However, gray vertices are not yet finished processing, while black vertices are finished processing. When a white vertex is first discovered, it is also timestamped so that we can order vertices by their discovery time. This corresponds to the start of activity time for vertices in case of depth-first search. However, the end of activity timestamps are not recorded any more because they don't provide such as a useful information as in the case of depth-first search.

The following is a pseudocode for the expanded BFS algorithm. We assume the adjacency-lists representation of the input graph and use similar data structures to the ones used in the expanded depth-first search algorithm. Additionally, we use an array d to represent the depth (distance) $d(v)$ of each vertex $v \in V$ from the source s. This information will be needed later when we discuss some properties of breadth-first search.

```
BFS(G, s)
    t = 1;                          [ beginning of epoch ]
    [ Initialize all vertices except s ]
    for each v ∈ V \ {s} do
        c(v) = WHITE; d(v) = +∞; p(v) = NIL;
    [ Initialize source s ]
    c(s) = GRAY; a(s) = t; d(s) = 0; p(s) = NIL;
    Q = ∅;                          [ Q is empty queue ]
    ENQUEUE(Q, s);                  [ put initially source s in Q ]
    while Q ≠ ∅ do                  [ visit all vertices ]
        u = DEQUEUE(Q);             [ u is the next to visit ]
        for each v in L(u) do       [ explore u's neighbors ]
            if c(v) = WHITE then    [ if neighbor v is undiscovered ]
                c(v) = GRAY;        [ mark v discovered ]
                t = t + 1; a(v) = t;   [ timestamp discovered vertex ]
                d(v) = d(u) + 1;    [ set the depth (distance) for v ]
                p(v) = u;           [ u is parent of v ]
                ENQUEUE(Q, v);      [ put discovered vertex in the queue ]
        c(u) = BLACK;               [ u is finished ]
    return;
```

The expanded BFS algorithm, after necessary initializations, begins its search from the source vertex s by putting it into the queue Q. This queue contains only gray vertices that have been discovered but not yet finished exploring. In each iteration of the **while** loop, an unfinished vertex u is removed from Q and its edges are explored. In the inner **for** loop, all u's white neighbors that are (first) discovered are added to Q, after properly updating their attributes. Finally, after u's adjacency list is fully examined, u is marked as finished by coloring it black.

Construction of a breadth-first search tree starts with the source vertex s, which is the root of the tree. Whenever a white vertex v is discovered (and colored gray) while exploring the neighbors of a gray vertex u, the vertex v and the edge (u, v) are added to the breadth-first search tree. This is implemented in the algorithm by keeping a pointer $p(v)$ for each vertex v that points back to the parent vertex u. Clearly, since a vertex is discovered only once, this does form a tree. The expanded BFS algorithm also stores value $d(v)$ for each vertex v that represents the depth of v in the bfs tree produced by breadth-first search. In fact, as we shall see in the next section, $d(v)$ is the shortest-path distance from the source vertex to v in the graph.

The following sequence of pictures illustrates the progress of this version of the BFS algorithm on the running example graph of Figure 30. Each picture shows the relevant contents of the queue Q, as well as the discovery time of discovered vertices.

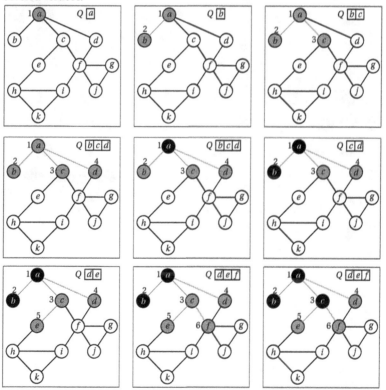

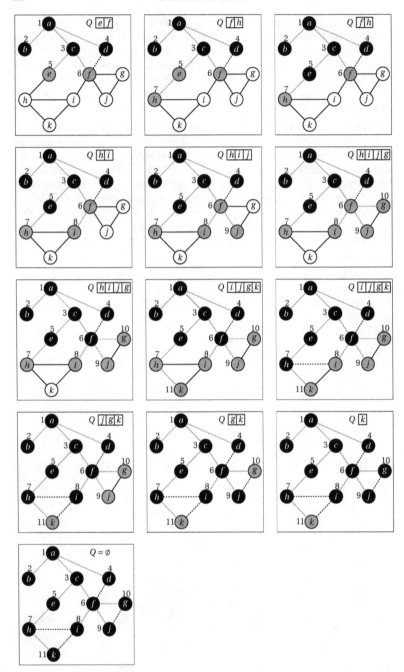

Before discussing more important properties of breadth-first search, we first list some easy observations about the expanded BFS algorithm.

- If (u, v) is an edge and the vertex u is black, then the vertex v is either gray or black. This means that all vertices adjacent to black vertices have been discovered. Gray vertices may have some adjacent white vertices—they represent the frontier between discovered and undiscovered vertices.
- Each vertex is enqueued and hence dequeued exactly once.
- An invariant of the **while** loop is that the queue Q consists of the set of gray vertices.
- The algorithm discovers all vertices that belong to the component of the input graph containing the source vertex s.
- The resulting bfs tree may vary, depending upon the order in which the neighbors of a given vertex are examined in the inner **for** loop.
- If $p(v) = u$ then $a(u) < a(v)$.
- If v is a proper descendant of u in the constructed bfs tree, then $a(u) < a(v)$.

We now investigate more formally other properties of the expanded BFS algorithm, focusing only on undirected graphs and leaving the case of directed graphs as an exercise. The next lemma shows that at all times, the queue Q contains vertices with at most two distinct depth values.

LEMMA (TWO DEPTHS LEMMA)   *Given a graph $G = (V, E)$, at any time during execution of the expanded BFS algorithm, the following invariant is true: if $u_1, u_2, \ldots, u_m$ are vertices in the queue Q, where u_1 is the head and u_m is the tail, then $d(u_1) \le d(u_2) \le \cdots \le d(u_m) \le d(u_1) + 1$.*

PROOF: Clearly, before the source vertex s is enqueued, the condition is vacuously true. Also, before the first iteration of the **while** loop, the condition is true because Q contains only the source vertex s. Suppose now the condition is true at the very beginning of some iteration of the **while** loop and show that it remains true until the next iteration.

To this end, consider the relevant steps that change the contents of the non-empty queue $Q = \langle u_1, u_2, \ldots, u_m \rangle$. First, the head u_1 is removed from the queue, and u_2 becomes the new head. (If the queue becomes empty, the condition holds vacuously.) Since $d(u_1) \le d(u_2)$ before this step, after this step we have $d(u_m) \le d(u_1) + 1 \le d(u_2) + 1$, and the remaining inequalities are unaffected. Thus, the inequalities hold with the new contents $Q = \langle u_2, \ldots, u_m \rangle$.

Next, u_1's newly discovered neighbors $v_1, \ldots, v_k$ are enqueued. (Possibly, there are no such neighbors, but then there is nothing to prove.) Although we should now argue for each enqueue operation separately, for brevity we consider them as a group. That is, the depth value of all these neighbors is updated so that $d(v_1) = \cdots = d(v_k) = d(u_1) + 1$, and the new contents of the queue is

$$Q = \langle u_2, \ldots, u_m, v_1, \ldots, v_k \rangle .$$

Since before this step we have $d(u_m) \le d(u_1) + 1 \le d(u_2) + 1$, this implies that after this step is $d(u_m) \le d(u_1) + 1 = d(v_1) = \cdots = d(v_k) \le d(u_2) + 1$, and the remaining inequalities are unaffected. Thus, the inequalities hold with the new contents $Q = \langle u_2, \ldots, u_m, v_1, \ldots, v_k \rangle$.

The remaining code obviously does not change Q, and so the lemma follows.
∎

We now show that breadth-first search classifies graph edges into tree and cross edges with respect to the produced bfs tree.

LEMMA (EDGE CLASSIFICATION LEMMA) *Let $G = (V, E)$ be a connected graph, and let T be any bfs tree resulting from running the expanded* BFS *algorithm on G. Then any edge $(u, v) \in E$ is either a tree or a cross edge with respect to T. Moreover, if (u, v) is a tree edge and $a(u) < a(v)$, then $d(v) = d(u) + 1$, and if (u, v) is a cross edge and $a(u) < a(v)$, then $d(v) \leq d(u) + 1$.*

PROOF: Take an edge $(u, v) \in E$, and assume without loss of generality that u is discovered before v during breadth-first search. This means $a(u) < a(v)$, and so u is enqueued before v by the way the algorithm works. Consider now the moment when u is removed from the queue Q, and distinguish two cases depending on whether v is in the queue.

In the first case, v is not in Q when u is removed from Q. We show in this case that (u, v) is a tree edge. Now, since v is not in Q and $v \neq u$, the color of v is not gray. The color of v cannot be black either, since it would imply that v had already been dequeued and so that v was enqueued before u, a contradiction. The color of v is therefore white when u is removed from Q. Since v is on u's adjacency list, the edge (u, v) will be examined eventually. Since in the meantime v cannot change its color, (u, v) is added to the bfs tree T by setting $p(v) = u$. Moreover, v's depth value is set to $d(v) = d(u) + 1$, and it is never changed afterwords.

In the second case, v is in Q when u is removed from Q. We show in this case that (u, v) is a cross edge. First off, u cannot be a descendant of v, since $a(u) < a(v)$ and $v \neq u$. Let's show that v cannot be a descendant of u either. Now, since v is in Q, v is a gray vertex, hence v has a parent. (Indeed, v cannot be the root, since v is not the first vertex in Q.) Let w be the parent of v, that is, $p(v) = w$. Then $w \neq u$ because u has not yet been assigned as anyone's parent. For the same reason, w cannot be any vertex in Q. Thus, w must have been dequeued before u, which means w must have been enqueued before u, which means $a(w) < a(u)$. Since $w \neq u$ and $a(w) < a(u)$, we claim that v cannot be a descendant of u. Otherwise, if v were a descendant of u, then v's parent w must have been either u or a proper descendant of u (and hence $a(u) < a(w)$), a contradiction. This shows that (u, v) is a cross edge. Moreover, taking the contents of Q immediately before u is removed from Q, the Two Depths Lemma implies $d(v) \leq d(u) + 1$.

Finally, since the two cases exhaust all possibilities, we can conclude that any graph edge is either a tree or a cross edge. ∎

COROLLARY *In any execution of* BFS(G, s) *on a connected graph $G = (V, E)$, if $(u, v) \in E$ then $|d(u) - d(v)| \leq 1$.*

PROOF: Depending upon which vertex of an edge (u, v) is discovered first, by the Edge Classification Lemma we have $d(v) \leq d(u) + 1$ or $d(u) \leq d(v) + 1$. ∎

Now we can justify why we have referred to the d values as the depth values.

LEMMA   *In any execution of* $\mathrm{BFS}(G,s)$ *on a connected graph* $G = (V,E)$, $d(v)$ *is the depth of vertex* v *in the constructed bfs tree.*

PROOF: The claim is clear for the root vertex s, since $d(s) = 0$. Take any non-source vertex $v \in V \setminus \{s\}$, and suppose its depth in the constructed bfs tree is $k > 0$. Let $s = v_0, v_1, \ldots, v_k = v$ be the unique path in the bfs tree from s to v. This means that for each $i = 1, \ldots, k$, there is a tree edge (v_{i-1}, v_i) and v_{i-1} is the parent of v_i (hence $a(v_{i-1}) < a(v_i)$). Applying the Edge Classification Lemma k times (or inductively) we get

$$d(v) = d(v_k) = d(v_{k-1}) + 1 = d(v_{k-2}) + 2 = \cdots = d(v_0) + k = k. \ \blacksquare$$

Once we know the d values are indeed the depth values, observe that we can interpret the previous corollary as saying that cross edges always go between vertices that are at most one level apart in the bfs tree.

6. Applications of breadth-first search

Typical applications of breadth-first search include two kinds of problems. One is the problem of exploration of certain infinite or huge graphs. Clearly, depth-first search is not applicable in this case, because it would cause an infinite recursion or it would be inefficient because of unnecessary traversal of large portions of these graphs. Yet, breadth-first search can be successfully used for partial exploration of such graphs. We won't pursue this aspect of the breadth-first search applications any further.

The other paradigm of graph algorithms when breadth-first search is usually employed represents the problem of finding shortest-path distances in a weighted graph from a given vertex. Recall that a weighted graph has a value associated with each edge. In this section we discuss a restricted version of the shortest-path problem for unweighted (and undirected) graphs, and the general shortest-path problem for weighted graphs is studied in Chapter 10.

The (undirected) graphs considered in the present section therefore have unnumbered, "equal" edges or, equivalently, all edges have the same weight, say, 1. Given a graph $G = (V,E)$ and two vertices $u, v \in V$, we define the ***shortest-path distance*** (or the ***minimum distance***) from u to v, denoted $\delta(u,v)$, as the minimum number of edges in any path from u to v. By convention, if $u = v$, then $\delta(u,v) = 0$, and if there is no path from u to v, then $\delta(u,v) = +\infty$. A path of length $\delta(u,v)$ from u to v is said to be a ***shortest path*** from u to v. Notice that for unweighted graphs we can define the weight of a path as the number of its constituent edges, and then say that a shortest path is a minimum-weight path.[4]

Given an arbitrary vertex $s \in V$, to find shortest-path distances from s to other vertices of G, we can run the expanded $\mathrm{BFS}(G,s)$ algorithm. Since the algorithm discovers all vertices that belong to the component of the input graph containing the source vertex s, it computes the value $d(v)$ for each vertex v reachable from s, and sets $d(v) = +\infty$ otherwise. Moreover, if v is a vertex in

---

[4]For weighted graphs, the weight of a path is the sum of the weights of its constituent edges.

the component containing s, $d(v)$ is the depth of v in the bfs tree that spans the component. Also knowing that cross edges connect vertices only on the same or consecutive levels of the tree, it is reasonable to expect that the depth value is exactly the shortest-path distance of each vertex from s. While this fact is perhaps intuitively clear, it requires a more careful justification given in the following theorem.

THEOREM *Given a graph $G = (V, E)$ and a vertex $s \in V$, the depth values computed by the call $\mathrm{BFS}(G, s)$ are the shortest-path distances of all vertices from s, that is, $d(v) = \delta(s, v)$ for every $v \in V$.*

PROOF: From the above discussion it follows that we need to prove the claim only for vertices in the component of G that contains the source vertex s. Consider the bfs tree T rooted at s and constructed for this component by the call $\mathrm{BFS}(G, s)$. For any vertex v in T (i.e., any vertex v in G, since T spans G), $d(v)$ is the depth of v in T, and so there is a path T (hence in G) whose length is $d(v)$. This means $d(v) \geq \delta(s, v)$.

To show the opposite inequality, we use induction on the level of T. The base case is clear, since the only vertex on level zero is s, and $d(s) = 0 = \delta(s, s)$. Suppose inductively that for each vertex u on level $k \geq 0$, we have $d(u) \leq \delta(s, u)$. Take a vertex v on level $k + 1$, and consider a shortest path π in the graph G from s to v. By the Edge Classification Lemma and its corollary, all edges along this path connect vertices that are at most one level apart in T. This implies that the path π contains at least one vertex u on level k. Splitting π into two paths, the one from s to u, denoted π_1, and the one from u to v, denoted π_2, we get

$$
\begin{aligned}
\ell(\pi) &= \ell(\pi_1) + \ell(\pi_2) \\
&\geq \ell(\pi_1) + 1 && \text{since } u \neq v \\
&\geq \delta(s, u) + 1 && \text{by definition of } \delta(s, u) \\
&\geq d(u) + 1 && \text{by inductive hypothesis} \\
&= k + 1 && \text{since } u \text{ is on level } k \\
&= d(v) && \text{since } v \text{ is on level } k + 1
\end{aligned}
$$

and so $\delta(s, v) = \ell(\pi) \geq d(v)$, as required. ∎

We conclude with a remark that we can easily recover one of the shortest paths for any vertex v that is reachable from s by following the parent pointers from $p(v)$.

Exercises

1. Consider the (weighted) directed graph in Figure 32.

   (a) How many vertices and edges does the graph have?
   (b) List all simple paths from a to b.
   (c) List all neighbors of the vertex b.
   (d) List all simple cycles from the vertex b.

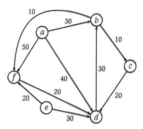

FIGURE 32

2. Answer the same questions from Exercise 1 for the undirected version of the graph in Figure 32. The undirected version is obtained by replacing all directed edges with corresponding undirected edges simply by removing arrows from the directed edges. More precisely, (u, v) is an edge in the undirected version if and only if $u \to v$ is an edge in the directed version.

3. What is the minimum and maximum number of edges of a graph with n vertices if the graph is

   (a) directed?
   (b) undirected?

4. Show that a connected (undirected) graph on n vertices is a free tree if and only if it has $n - 1$ edges.

5. Represent the graph of Figure 32

   (a) by the adjacency lists of its vertices;
   (b) by the adjacency matrix.

6. Answer the same questions from Exercise 5 for the undirected version of the graph in Figure 32.

7. If a digraph is represented by adjacency lists, we may want to add to each vertex a linked list of its predecessors, in addition to the list of its successors. More precisely, for each vertex v we build a linked list of all vertices u such that $u \to v$ is a directed edge in the digraph. Describe an efficient algorithm that performs this task and give its running time.

8. A vertex s of a directed graph $G = (V, E)$ with n vertices is called the **sink** if the indegree of s is $n - 1$ and its outdegree is 0. In other words, for every vertex $v \in V$ different from s, there is an edge $v \to s$ and there is no edge $s \to v$. Describe an $O(n)$-time algorithm that decides whether a directed graph represented by the adjacency matrix contains a sink.

9. Suppose we select the first vertex in the natural order of vertices a, b, c, d, e, f, g, h of the directed graph in Figure 33, whenever we are to choose a vertex among many vertices.

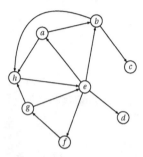

FIGURE 33

   (a) Determine the recursive calls of the basic DFS procedure for the initial call DFS(a), and use indentation to represent their mutual relationship. In case more than one vertex is possible to be visited next, select the one according to the alphabetical order.
   (b) Construct the corresponding depth-first search tree (forest).
   (c) Specify the associated tree, forward, back, and cross edges of the graph.

10. Repeat the same questions from Exercise 9 for the undirected version of the graph in Figure 33.

11. Write a modified expanded DFS algorithm that classifies the edges of a directed graph G "on the fly."

12. Answer the following questions regarding the basic BFS procedure when it is run on the graph in Figure 33, as well as when it is run on the undirected version of the graph.

   (a) Give the contents of the queue of discovered vertices during breadth-first search after the initial call BFS(a) and after each step.
   (b) Construct the corresponding breadth-first search tree (forest).
   (c) Specify the associated tree, forward, back, and cross edges of the graph.

13. Prove that the expanded BFS(G, s) algorithm visits vertices in the order of increasing distance from s. That is, prove that if vertex u is dequeued before vertex v, then $d(u) \le d(v)$.

14. Prove that in a connected undirected graph G there exists a vertex from which the dfs and bfs trees are identical if and only if G is a rooted tree. (These trees are identical if they have the same sets of edges—the order in which the edges are traversed during depth-first and breadth-first search doesn't matter.)

15. A **bridge** of a connected undirected graph G is an edge whose removal disconnects G. For example, the edge (a, b) is a bridge of the graph in Figure 34.

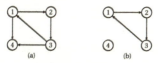

FIGURE 34

(a) Prove that an edge of G is a bridge if and only if it does not lie on any simple cycle of G.

(b) Show how to efficiently compute all the bridges of G using depth-first search.

16. Write an acyclicity test for a directed graph by modifying the expanded DEPTH-FIRST-SEARCH and DFS algorithms to detect absence or presence of a back edge "on the fly."

17. Prove that every dag contains a vertex whose indegree is 0, as well as a vertex whose outdegree is 0. Using this fact, design an efficient algorithm to topologically sort a dag.

18. A simple path in a graph is **Hamiltonian** if it passes through all vertices of the graph. Describe an efficient algorithm that decides whether a dag contains a Hamiltonian path.

19. A **strongly connected component** of a directed graph G is a maximal set of vertices S such that for every pair of vertices $u, v \in S$ there is a path from u to v and a path from v to u in the induced subgraph on S. For example, Figure 35 shows a directed graph on the left and its two strongly connected components on the right.

(a)　　　　(b)

FIGURE 35

Describe an efficient algorithm that finds the strongly connected components of a directed graph. (*Hint:* First run depth-first search on a given graph to compute the finish time for each vertex, and then in decreasing order of the vertex finish times repeat depth-first search on the "reversed" graph, which is obtained by reversing orientation of every directed edge of the original graph.)

20. If $\delta(u, v)$ denotes the length of a shortest path from the vertex u to the vertex v in a graph $G = (V, E)$, the **eccentricity** of the vertex u is defined by

$$\max_{v \in V} \delta(u, v).$$

In other words, the eccentricity of u is the length of a longest shortest path that starts from u. Describe an efficient algorithm that computes the eccentricity of all vertices of a directed graph. (*Hint:* Use breadth-first search.)

21. An undirected graph is ***bipartite*** if its vertices can be partitioned into two disjoint subsets A and B such that each edge (a, b) is incident with vertices from these subsets, that is, $a \in A$ and $b \in B$. Figure 36 shows an example of a bipartite graph.

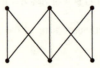

FIGURE 36

(a) Show that every tree is a bipartite graph.
(b) Describe an efficient algorithm that checks whether an undirected graph is bipartite. (*Hint:* Use breadth-first search.)

9

Advanced Data Structures

In Chapters 3 and 4 we studied some elementary data structures to show importance of good organization of data for efficient algorithms. In fact, the field of data structures is among the most fundamental one in computer science and can be studied in its own right. This is witnessed by the fact that most of what computers do is storing, accessing, and manipulating data in one form or another.

In the data structure theory we focus on ways to organize a typically large data set so that information about it can be extracted really fast. An illustrative example are web search engines, which store the contents of the Web using sophisticated indexes so that the engines can quickly return a few pages that are most relevant to a given query (e.g., what documents contain word "algorithmics?"). This example also shows that within the data structures realm we have to work under more stringent time constraints. Namely, in contrast to most algorithmic problems, linear time is too slow—you can't afford to scan through the entire Web to answer a single query—and the goal is typically logarithmic or even constant time per query.

In this chapter we discuss relatively more advanced data structures at a very general level, as well as some of simpler implementations of them. In particular, we look at the dictionary, priority queue, and disjoint-sets data structures.

1. Dictionary

In many computer applications, we need to maintain one dynamic set of elements so that we can efficiently execute three types of operations on the set:

1. Inserting a new element into the set.
2. Finding out whether a given element is currently in the set.
3. Deleting an element from the set.

A typical example is maintenance of, say, an English dictionary as part of a spelling-checker program. In this case we insert a new word from time to time, delete a word that has been entered by a mistake, or look up a string of letters to see whether it is a word.

Because this example is so familiar, a set on which we execute the above operations is called the *dictionary*, no matter what the set is used for. For example, the dictionary data structure may represent a telephone directory, where we add new subscribers, delete the ones who canceled their phones, or check the phone number, if any, of a particular person.

To implement the dictionary data structure, we could organize n elements of a dictionary into a sorted array or a binary search tree. In case of a sorted array, the cost of the find operation is $O(\log n)$ if we use binary search. However, in order to keep the array sorted, insertion becomes an $O(n)$-time operation. Since linear time is not good enough for us, a better approach is to use a binary search tree. For a binary search tree the cost of the insertion, find, and delete operations is $O(\log n)$ on average, as discussed in Chapter 4. Although the binary search tree is fairly easy to implement, this is not quite satisfactory because in the worst case the height of the tree can become $\Theta(n)$. Thus, in the worst case (which is relatively rare, as indicated by the average case analysis, but not uncommon), there is still possibility that we get again linear time operations.

A general approach to remedy these deficiencies is to design height-balanced trees, which are *guaranteed* to have $O(\log n)$ height and consequently $O(\log n)$-time operations. The other approach is to still take chances, but require constant time per operation on average. The simplest example of the former type are AVL trees, which we discuss in the next section. After that we study hash tables, which represent common data structure of the second approach. Of course, there are numerous variations on the theme that try either to reduce the complexities of the operations that keep trees balanced, or to improve the overall efficiency. To name just a few, B-trees, red-black trees, splay trees, treaps, skiplists, and perfect hashing are all examples of different implementations of the dictionary data structure. Moreover, a new kind of cost analysis, the so-called amortized analysis, is used to better describe performance of some of these implementations.

2. AVL trees

Recall from Chapter 4 that a binary search tree has simple operations to insert, delete, or find a key, as well as that the cost of these operations is proportional to the height of the tree. We also showed that the expected height of a randomly built n-node binary search tree is $O(\log n)$, so the BST operations perform quite well on average. However, the worst-case scenario can still give us $O(n)$ time per operation. A pessimistic attitude therefore leads to the idea of keeping the height logarithmic in the tree size at all times, so that the BST operations have guaranteed logarithmic performance.

This is a recurring theme in a wide class of balanced binary search trees, balanced in the sense that the left and right subtrees of each node have not so different heights. Actually, the simplest idea is to require that the left and right subtrees of each node have the same height. This balance condition ensures that the height of the tree is logarithmic, but it's too restrictive because the algorithm for inserting or removing a key becomes prohibitively complex and time consuming. The AVL tree (named after its authors, Adelson-Velskii and Landis) uses a balance condition that is somewhat weaker but still strong enough to guarantee logarithmic height.

An **AVL tree** is a binary search tree with the balance condition, called the **AVL property**, that the heights of the left and right subtrees of each node differ by at most 1. (By definition, the height of an empty tree is −1.) Figure 1 shows two binary search trees—the one on the left satisfies the AVL property and is thus an AVL tree; the tree on the right is not an AVL tree because the grayed nodes are in violation of the AVL property.

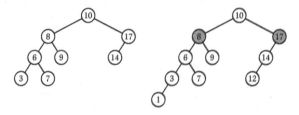

FIGURE 1. Two binary search trees: an AVL tree on the left and a non-AVL tree on the right with unbalanced nodes grayed.

Before discussing how to maintain the AVL balance condition, we should consider the question of whether it is strong enough to guarantee that the height of an AVL tree with n nodes will be $O(\log n)$. The following theorem proves that the minimum number of nodes in an AVL tree is exponential in its height, hence the height of an n-node AVL tree is $O(\log n)$.

THEOREM *An AVL tree of height h has at least ϕ^h nodes, where ϕ is the golden ratio.*

PROOF: We prove that the minimum number of nodes in an AVL tree of height h is at least ϕ^h, hence any AVL tree of height h has at least that many nodes. Let $N(h)$ denote the number of nodes in a size-smallest AVL tree of height h. Clearly, $N(0) = 1$ and $N(1) = 2$. For $h \geq 2$, let T be a size-smallest AVL tree of height h, and consider the two subtrees T_ℓ and T_r of its root. The value $N(h)$ is thus the sum of the T_ℓ's size and the T_r's size, plus 1 for the root. Since T_ℓ and T_r are AVL trees as well, they must themselves be smallest in size for their height. Since the overall tree T has height h, one of the subtrees, say T_ℓ, must have height $h-1$. The other subtree T_r cannot have height less than $h-2$ without violating the AVL balance condition. This implies that T_r must have height $h-1$ or $h-2$. But since T_r must be smallest in size, this rules out the height $h-1$. (A smallest AVL tree of height $h-1$ is at least one node larger (for the root) than a smallest AVL tree of height $h-2$.) Thus, we can write the recurrence

$$N(0) = 1, \quad N(1) = 2$$
$$N(h) = N(h-1) + N(h-2) + 1, \quad h \geq 2$$

for the size of a smallest AVL tree of height h. This recurrence looks very similar to the Fibonacci recurrence, and the guess-and-verify-by-induction method gives us the solution $N(h) = F_{h+3} - 1$, where F_h is the h-th Fibonacci number. It is also a simple matter to show by induction that $F_{h+3} - 1 = F_{h+2} + F_{h+1} - 1 \geq F_{h+2}$ and $F_{h+2} \geq \phi^h$, for $h \geq 0$. Therefore, $N(h) \geq \phi^h$ for $h \geq 0$. ∎

By inverting the claim of the theorem, we can conclude that the height of an AVL tree is at most logarithmic in its size. Namely, if an AVL tree with n nodes has height h, then $n \geq \phi^h$. Hence $h \leq \log_\phi n$, implying $h = O(\log n)$ because logarithms of different bases differ only by a constant factor.

2.1. AVL-tree operations. Because an AVL tree is ordinary binary search tree, key searching in an AVL tree is no different than the search operation in a binary search tree that we discussed in Chapter 4. In this case however, we are guaranteed by the previous theorem that the search will be efficient and take logarithmic time.

Insertions and deletions are more complicated now, because they may destroy the AVL property of some nodes and so we may need to work harder restoring their balance. Luckily, rebalancing itself is purely a local operation, that is, it acts only on nearby nodes taking constant time. As we shall see, this implies that both insertions and deletions stay within the logarithmic worst-case bound. To be fair though, the constants hidden in their asymptotic time are relatively large and, worse, the frequency of having to perform rebalancing per operation is also relatively high. That's why AVL trees are actually of limited practical importance, but they have great historical value for being the first of its kind and for introducing an important balancing idea: *rotations*.

Before discussing how to maintain the balance condition using rotations, observe that each node now must have additional information so that we could detect the source of imbalance during insertions or deletions. The simplest method is to add a new field, say h, to each node that stores the height of the node. In fact, enough additional information a node must have is the difference in heights of its right and left subtrees. This so-called ***balance factor*** for a node x, denoted $b(x)$, is thus $b(x) = h(T_r) - h(T_\ell)$, where T_ℓ and T_r are the x's left and right subtrees, respectively. Clearly, in an AVL tree this number must always be either -1, 0, or $+1$.

In our considerations that follow we will describe the basic operations in terms of the balance factor for two reasons. First, it is more convenient for the discussion. Second, a practical implementation that uses a new field of each node storing its balance factor is more efficient (e.g., it avoids repeated computation of the balance factors and the new field requires only two bits). On the other hand, to make our algorithms as simple as possible, in the pseudocode we will assume the simpler method that uses the height field of each node.

Restructuring of an AVL tree is based on two simple and symmetric rotations around some node: the left and right rotations. The type of rotation performed on a node depends on the nature of the node's height imbalance. Figure 2 shows the transformation of the relevant parts of a tree caused by these rotations.

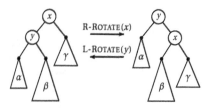

FIGURE 2. Left and right rotations.

In the **right rotation** around x (the **left rotation** is handled symmetrically) we always assume that its left child y is not empty. As we can see from Figure 2,

- x moves down and to the right, into the position of its right child;
- the x's right subtree γ is unchanged, but the y's right subtree β is disconnected from y and attached as the left subtree of x;
- y moves up to take the old x's place, keeping its old left subtree α and taking new x as its new right child.

Observe that the left and right rotations preserve the BST property because the sorted order of nodes $\alpha y \beta x \gamma$ of the affected subtree is not changed after the transformations. Another important observation, used in preserving the AVL property, is that the height of nodes is changed by at most ± 1.

Assuming the pointer-based representation of trees and additional height field h of each node, the tree rotations require just a few pointer changes. We present pseudocode only for the right rotation—the left rotation is symmetrical and is left as an easy exercise. We also give an auxiliary procedure that returns the height of a (possibly empty) node.

HEIGHT(x)	R-ROTATE(x)
if $x = $ NIL **then**	$y = left(x);$
return $-1;$	$left(x) = right(y);$
else	$right(y) = x;$
return $h(x);$	$h(x) = \max\{\text{HEIGHT}(left(x)), \text{HEIGHT}(right(x))\} + 1;$
	$h(y) = \max\{\text{HEIGHT}(left(y)), \text{HEIGHT}(right(y))\} + 1;$
	return $y;$

2.1.1. *Inserting a node.* Insertion in an AVL tree must still be done by preserving the BST property, or else the search order breaks and the search operation will not work correctly. In addition, since we alter the tree by adding a new node, the AVL property may be violated, and in this case we must take a corrective action to keep the tree balanced. Solution to both problems are rotations: we have already seen that rotations preserve the BST property, and we will see how to apply them to fix any height imbalance.

More precisely, the insert operation for AVL trees starts exactly the same as the insertion routine for ordinary binary search trees. That is, we search for the insertion point of the new node in the tree by walking down the path from the root to the new node's correct place as a new leaf. The key observation is that only nodes on this path may have their heights changed, because only those nodes' subtrees are altered. We thus walk this path back from the insertion point of the new leaf up to the root, checking the nodes along the way if any height imbalance has been introduced. If so, we perform a rotation around the node to rebalance it.

To see which rotation fixes a node on the search path, consider the first node x from the insertion point (or the deepest from the root) that needs to be fixed after insertion of a new node. The new node was added to either x's left or x's right subtree and their heights may increase by at most 1. It is thus easy to see that there are principally two symmetric sources of x's height imbalance:

1. New $b(x) = -2$: the tree rooted at x was already heavier on the left (that is, the old $b(x) = -1$, or x's left subtree was one level deeper than x's right subtree), and the new node was added to x's left subtree that caused it to further grow one level deeper.
2. New $b(x) = +2$: the tree rooted at x was already heavier on the right (that is, the old $b(x) = +1$, or x's right subtree was one level deeper than x's left subtree), and the new node was added to x's right subtree that caused it to further grow one level deeper.

Observe that in case 1—case 2 is handled symmetrically—x must have had a nonempty left child y before insertion of the new node. Thus, case 1 breaks further into two subcases 1a and 1b, depending on whether the new node was added to the y's left or right subtree.

Suppose first we have subcase 1a: the new node was added to the y's left subtree. Figure 3 shows this situation and the appropriate transformation that fixes x's height imbalance. The figure also indicates the relevant old and new balance factors.

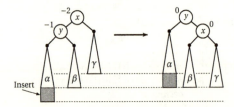

FIGURE 3. Single right rotation around x balances its subtree heavier on the left.

If we analyze subcase 1a carefully, we see that the tree depicted on the left of Figure 3 is the only possible scenario so that x satisfies the AVL property before insertion but violate it afterword. The subtree α has grown one extra level, causing it to be exactly two levels deeper than the subtree γ. The subtree β cannot be at the same level as the (new) subtree α because then x would have been unbalanced before insertion. Moreover, β cannot be at the same level as γ because then y would have been the first node on the path that was in violation of the AVL property.

The operation performed in subcase 1a is single right rotation around x. (In symmetric subcase 2a, when the right subtree of x's right child is too deep, we apply single left rotation around x.) From Figure 3 we see that this rotation evens out the heights of the subtrees of new x and y, hence both new nodes are totally balanced after the transformation. Moreover, the heights of subtrees rooted at the old x (before insertion) and at the new y (after insertion and right rotation) are equal. This implies that no more fixes are necessary or, since x is the deepest unbalanced node, only one right rotation suffices to balance the entire tree in this case.

In the other subcase 1b, when the new node was added to the y's right subtree, it is not enough to apply single right rotation, as suggested in Figure 4.

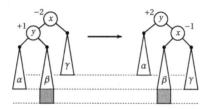

FIGURE 4. Single right rotation around x does not fix x's imbalance when y's right subtree is too high after insertion.

The problem is that the y's right subtree is too high, so the old unbalanced node remains unbalanced (but with balance factor +2) after single right rotation. However, two rotations do suffice. To see how, represent the y's right subtree after insertion as the root z with subtrees β and γ. Since the new node was added to the y's right subtree, the subtree is not empty and so this representation makes sense (β or γ or both may be empty). Subcase 1b with the new picture and the fixed affected subtree are shown in Figure 5. The figure indicates the relevant old and new balance factors, where multiple factors denote the possible values.

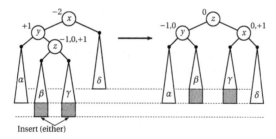

Insert (either)

FIGURE 5. Left-right double rotation fixes x's imbalance when y's right subtree is too high after insertion.

Two single rotations that fix subcase 1b are performed in opposite directions around two different nodes. Specifically, we first do a left rotation around x's left child y (which brings x's left-right grandchild z up one level), and then a right rotation around x (which moves x's new left child z to the root of the resulting subtree). This operation is called a ***double rotation***. (In symmetric subcase 2b we do a right-left double rotation: a right rotation around x's right child followed by a left rotation around x.) The following is pseudocode for the double left-right rotation (and again, the symmetric right-left rotation is left as an exercise).

LR-ROTATE(x)
    $left(x)$ = L-ROTATE($left(x)$);
    return R-ROTATE(x);

Observe again that a double rotation transforms the old subtree rooted at the node that needs to be fixed after insertion into a new subtree of the same height. Since we are looking at the deepest unbalanced node, it follows also in this case that only one double rotation suffices to balance the entire tree.

The recursive algorithm AVL-INSERT(t, x) given below inserts the node x into the AVL subtree rooted at t. (Of course, to insert x into an AVL tree T, we should call AVL-INSERT($root(T), x$).) We assume that the new node x has already been filled in, i.e., $key(x)$ contains its key, $left(x) = right(x) =$ NIL, and $h(x) = 0$. The AVL-INSERT algorithm is similar to the BST-INSERT algorithm for insertion in ordinary binary search trees, extended for the code that fixes any height imbalance by applying appropriate rotation and updating heights as we return from the recursive calls.

AVL-INSERT(t, x)
  if $t =$ NIL **then**
    $t = x$;
  else if $key(x) < key(t)$ **then**
    $left(t)$ = AVL-INSERT($left(t), x$);
    if HEIGHT($right(t)$) − HEIGHT($left(t)$) = −2 **then**   ⟦ case 1 ⟧
      if $key(x) < key(left(t))$ **then**   ⟦ case 1a ⟧
        t = L-ROTATE(t);
      else   ⟦ case 1b ⟧
        t = LR-ROTATE(t);
      $h(t)$ = max{HEIGHT($left(t)$), HEIGHT($right(t)$)} + 1; ⟦ update height ⟧
  else if $key(x) > key(t)$ **then**
    $right(t)$ = AVL-INSERT($right(t), x$);
    ⟦ ...Symmetric rebalancing... ⟧   ⟦ case 2 ⟧
  else
    ⟦ Duplicate insertion is ignored ⟧;
  return t;

This recursive algorithm works by first fixing the deepest unbalanced node, and so it does at most one rotation (single or double). As already mentioned, the height of the affected subtree is the same as it was before insertion, hence no further rotations are required.

2.1.2. Deleting a node. The basic idea for deletion in an AVL tree is the same as for insertion: first delete the node applying the deletion routine for ordinary binary search trees, and then walk back the path from the deletion point up to the root fixing any height imbalances as you go.

Recall that deletion for ordinary binary search trees depends on the number of children of the node being deleted. In case it has no children, the deleted node is replaced by an empty subtree. In case it has exactly one child, the deleted node is replaced by the subtree rooted at the appropriate child. In the most complicated case when the node being deleted has two children, we actually do not remove that node but find its replacement key, and remove the leftmost node in the need-to-be-deleted node's right subtree. Since in the last case we walk back the path from this actual removal point up the tree, we see that the height of the nodes along the way may decrease by at most one.

As for insertion therefore, when we come to an unbalanced node, we can apply an appropriate rotation to restore the AVL property of the node. Again, for deletion there are four cases that we might need to fix but again two of them are mirror-images of the other two. We do not discuss these symmetric cases in much detail, nor do we give the (messier) deletion pseudocode. All this is left as a good exercise for the reader.

Consider a node x from the deletion point that needs to be fixed after removal of a node. Suppose that we have removed a node from the x's left (or symmetrically right) subtree and that as a result this subtree's height has decreased by 1. There are two (sub)cases. First, if the balance factor of x's right child is either 0 or +1, we can apply a single left rotation around x as shown in Figure 6.

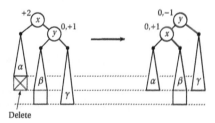

FIGURE 6. Single left rotation for deletion if the x's right subtree is not heavier on the left.

Second, if the balance factor of x's right child is −1, then we need to perform a double rotation: a right rotation around x's right child followed by a left rotation around x. This case is illustrated in Figure 7.

Unlike for insertion though, observe that in both cases we may end up with a shorter transformed subtree (but still by at most one level). So, the height imbalance may propagate up the tree and thus multiple rotations may be needed to balance the entire tree.

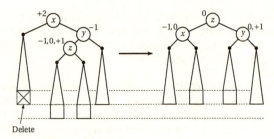

Delete

FIGURE 7. Double right-left rotation for deletion if the x's right
subtree is heavier on the left.

3. Hash tables

Ideally, we would like to construct the dictionary data structure for which
both the insert and find operation takes constant time in the worst case. The
essential problem is that a search of a data structure is always necessary to find
a given element. In other words, to provide constant-time insert and find operations, we need a way to do them *without* performing a search. However, inserting and finding arbitrary elements without a search can only be achieved with
complete a priori knowledge on which elements are to be inserted into the data
structure. Unfortunately, we usually do not have this information in advance to
guarantee the best performance in the worst case.

The hash table is a data structure that is designed specifically with the objective of providing the insert and find operations on its elements in constant time,
but in average case. This objective is achieved at the further expense of a loss of
ordering information among the elements; efficient operations such as finding
minimal (maximal) element or producing a list of all elements in a sorted order
are not supported. In addition, while we still require the ability to remove elements from the data structure, it is not our primary concern to make removal as
efficient as the insert and find operations.

The principal characteristic of the hash table applications is that data needs
to be frequently accessed and the access pattern is either unknown or known
to be random. Typical areas in computer science where hash tables are exten-
sively used include compiler design (symbol tables and associative arrays) and
database systems.

3.1. Basic ideas. A hash table is a generalization of the simpler notion of an
ordinary array. Direct indexing into an ordinary array makes effective use of the
ability to examine an arbitrary position in an array in constant time. The hash
table data structure is based on the idea of using array lookup to speed up computation of an arbitrary mapping. Accessing an array is a very fast operation, so
if we have a mapping whose values can be precomputed and stored in an array,
we can trade space for time to get the mapped values fast.

For example, suppose we are to write an algorithm that computes the function $\pi(n)$ that gives the number of primes less than or equal to n. To solve this
problem, we can write an $O(n \log n)$-time algorithm PI(n), say Erathosten's sieve,

which determines the number of primes less than n. Then, whenever in a program we need the values of the function $\pi(n)$ for different values of the argument n, we would make a call to PI(n) and get the value of $\pi(n)$ in time $O(n \log n)$. However, if we know that the argument n will never exceed some bound, say 1000, we can use another approach. We can precompute all 1000 values of the function $\pi(n)$, store them in an array A, and then return the value $A(n)$ in constant time whenever we need the value of the function $\pi(n)$ for some $n \leq 1000$.

This simple technique of direct addressing works for two reasons. First, it makes sense to use an array only if the domain size of an arbitrary mapping is relatively small. In the previous example, we assumed that possible values for the argument n of $\pi(n)$ were bounded by 1000, so we could use an array of size 1000. Furthermore, we directly used an index into the array to get the value of $\pi(n)$ for some n, because the function $\pi(n)$ maps nonnegative integers. In general, an arbitrary mapping could map nonintegers that we cannot use as an index into the array.

The hash table data structure stores elements of a dynamic set that are distinguished by a special, possibly noninteger field called the *key*. (Notice that we can think of the dynamic set as a mapping from the set of all possible keys to the members of the set.) Without loss of generality, we are assuming that we are to store only keys drawn from some universe U. That is, we identify the keys and the elements they represent; if we do have satellite data, we can always use pointers to elements with underlying keys. Figure 8 shows a hash table with the keys drawn from the set of domestic animals.

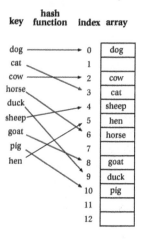

FIGURE 8. A hash table.

The main idea behind hash table is to define an auxiliary function between the set of all possible keys and the positions in an array.[1] The auxiliary function, called the **hash function**, takes a key and returns its **hash value**. Keys vary in

---

[1]This array is often called table in this context, so we will interchangeably use the two terms in this section. Also, the array positions are numbered starting with 0, a convention we will also follow.

type, but hash values are always integers. Instead of using the key as an array index directly, the array index is *computed* from the key. Thus, a hash function converts a key into an integer suitable to index an array where the key is stored. Put it another way, a hash value determines the position of a given key directly from that key.

The algorithmic implementation of hash tables is called **hashing**. Hashing places an additional requirement on the hash function: it should be easy to compute in constant time. Since both computing a hash value and indexing into an array can be performed in constant time, the beauty of hashing is that we can use it to perform constant-time insert and find operations.

3.2. Hash functions. In general, we expect the size of the set of all possible keys U to be relatively large or even unbounded. For example, if the keys are 32-bit integers, then $|U| = 2^{32}$. Similarly, if the keys are character strings of arbitrary length, then the size of U is unbounded.

On the other hand, we also expect that the set K of keys *actually stored* in the hash table to be significantly smaller than U, that is, $|K| \ll |U|$. When the number of keys actually stored is small relative to the total number of possible keys, it seems prudent to use an array of size at least as great as the maximum number of keys to be stored in the array. Figure 9 illustrates the general case.

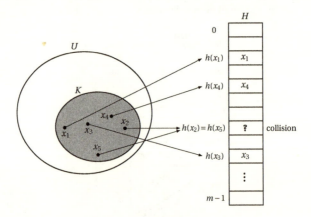

FIGURE 9. Mapping the universe of keys to a hash table of size m.

A hash function h is therefore a mapping $h: U \rightarrow \{0, 1, \ldots, m-1\}$ that maps the set of possible keys to the indices of an array of size m that stores them. Since usually $|U| > m$, the mapping defined by hash function will be a many-to-one mapping. That is, there will exist many pairs of distinct keys x and y such that $h(x) = h(y)$. This situation is called a **collision**. We must be prepared to appropriately deal with collisions in the hash-table operations, because two distinct keys cannot occupy the same location. In the next sections we discuss two techniques for resolving the conflict created by collisions.

The crux of hashing is the choice of a good hash function. In the first place, a good hash function should be avoiding collisions. If the hash function were one-to-one, distinct keys would be hashed (mapped) into distinct array positions, so we could access the keys directly by their array indices. In practice, unless we know something about the keys, we cannot guarantee that there will be no collisions.

If a hash function distributes keys in apparently a random manner through-out the array, collisions are minimized. Thus, a good hash function should be spreading keys evenly. Let p_i be the probability that a hash value is equal to i. A hash function that spreads keys evenly has the property that $p_i = 1/m$ for each $i = 0, 1, \ldots, m-1$. In other words, the hash values computed by the function are uniformly distributed. This theoretically perfect situation is known as **simple uniform hashing**. However, in practice it usually can only be approximated. The requirement that the hash function is to spread the keys uniformly implies that an equal number of keys should hash into each array position. That is, if $U_i = \{x \in U : h(x) = i\}$ is the set of keys that hash to the value i, and U is finite, then $|U_i| = |U|/m$ for each $i = 0, 1, \ldots, m-1$.

The second criterion for selecting a good hash function is that it should be easy to compute. This simply means that the running time of an algorithm for computing the hash function is bounded by a constant.

3.3. Hashing methods. To insert, find, or remove a key, we first pass the key to a hash function in a process called **hashing the key**. This tells us in which array position the key belongs. Generally, most hashing methods assume keys to be nonnegative integers so that they may be easily altered mathematically to make the hash function distribute keys throughout the array more uniformly. When keys are not integers, we can usually encode them into integers without much difficulty.

For example, if the keys are arbitrary character strings, since a character can typically be represented in the 7-bit ASCII code as a number between 0 and 127, we can interpret a string as an integer in the base-128 number system. Thus, one possible interpretation of the string "hash" is

$$\text{'h'} \cdot 128^3 + \text{'a'} \cdot 128^2 + \text{'s'} \cdot 128^1 + \text{'h'} \cdot 128^0 =$$
$$= 104 \cdot 128^3 + 97 \cdot 128^2 + 115 \cdot 128^1 + 104 \cdot 128^0 = 219707880.$$

As another example, if the keys are floating-point numbers, one possibility is to simply reinterpret the bit pattern used to represent the floating-point number as an integer. However, this is only possible when the size of the floating-point type does not exceed the size of the integer type. If this condition is not satisfied, we can extract some part of the floating-point bit pattern. For example, in the case of the IEEE standard 64-bit floating-point format and 32-bit integers, this can be done by shifting the binary representation of the floating-point number 20 bits to the right so as to extract the mantissa of the floating-point number.

Precisely how to encode a set of keys depends a great deal on the charac-teristics of the keys themselves. Therefore, it is important to gain as much of a qualitative understanding of them in a particular application as we can. Then we can exploit some specific knowledge about the keys to reduce the likelihood of a collision.

For example, if the keys are telephone numbers, and we know that the telephone numbers are all likely to be from the same geographic area, then it makes little sense to consider the area codes in the hash function.

Similarly, if we were to hash the identifiers found in a program, we might observe that many have similar prefixes and suffixes since programmers tend to use closely related variables such as `listptr`, `lastptr`, and `leastptr`. A poor way to encode these keys would be any method depending strictly on characters at the beginning and end of the keys, since this would result in many of the same hash values for similar keys. Instead, we might try selecting characters from four positions that seem to be somewhat random, permute them in a way that randomizes them further, and put them into specific bytes of a four-byte integer.

The first step in the hashing process when keys are not integers is therefore to choose an appropriate encoding scheme e such that $e: U \to \mathbb{N}$. The most important goal to remember when selecting such an encoding e is that we are supposed to approximate uniform hashing, that is, to spread keys around the hash table in as a uniform and a random manner as possible.

Once we have a key represented as a nonnegative integer, we further map the resulting integer into one of m array positions. Formally stated, we use a hash function h such that $h: \mathbb{N} \to \{0, 1, \ldots, m-1\}$. The most common ways for selecting a good hash function h like this are the division and multiplication methods.

3.3.1. *Division method.* Perhaps the simplest of all the methods of hashing a nonnegative integer x is by taking the remainder of x divided by m. This is called the **division method** of hashing. In this case, the hash function h is formally given by

$$h(x) = x \pmod{m}.$$

As an example, if the size of the array is $m = 2969$ and we hash the key $x = 219707880$, then its hash value is $h(x) = 219707880 \pmod{2969} = 1880$.

It is tempting to let m be a power of two, i.e., $m = 2^k$ for some integer $k \geq 1$, because division by 2 in binary is just a right shift. In this case, the hash function $h(x) = x \pmod{2^k}$ simply extracts the rightmost, lower-order k bits of the binary representation of x. While this hash function is quite easy to compute, it is not a desirable function because it does not depend on all the bits in the binary representation of x.

When using the division method, we usually choose the value of m to be a prime number not too close to a power of 2. The reason for this is because then the integer division of x by such a prime depends on all the bits of x, not just the rightmost k bits for some small constant k. This clearly increases the likelihood that the keys will be spread out evenly. In the previous example, $m = 2969$ is a good prime number between 2^{11} and 2^{12}.

The division method is extremely simple to implement, since integer division by m and taking the remainder is a basic operation that takes constant time.

A potential disadvantage of the division method is due to the property that consecutive keys map to consecutive hash values. That is, if $h(i) = i$ (mod m), then $h(i + 1) = i + 1$ (mod m), $h(i + 2) = i + 2$ (mod m), and so on. While this ensures that consecutive keys do not collide, it does mean that consecutive array locations will be occupied. We'll see that in certain implementations this can lead to degradation in performance.

3.3.2. *Multiplication method.* The multiplication method avoids the use of division in the hope to speed up the hashing algorithm, since integer division is usually slower than integer multiplication. We can avoid division by making use of the fact that the computer does finite-precision integer arithmetic. For example, if w is the word size of the computer, all arithmetic is done modulo $W = 2^w$. Another advantage of the multiplication method is that the value of m is not critical. We typically choose it to be a power of 2, say $m = 2^k$ for some $k \geq 1$, since then we can easily implement the hash function given by

$$h(x) = \left\lfloor \frac{m}{W} \cdot \left(ax \quad (\text{mod } W)\right) \right\rfloor,$$

where a is a carefully chosen constant.

This looks very complicated to compute, but it actually reduces to one multiplication and shifting, giving a constant-time computation. Namely, since m and W are both powers of 2, the ratio $W/m = 2^{w-k}$ is also a power of 2. Therefore, in order to multiply the term ax (mod W) by m/W, we can divide it by W/m and so we simply shift it to the right by $w - k$ bits. In effect, we are extracting k bits from the middle of the product ax, as suggested in Figure 10.

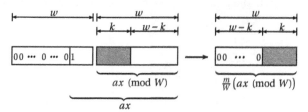

FIGURE 10. Multiplication method when $m = 2^k$.

What is a suitable choice for the constant a? Since the multiplication method only considers a subset of the bits in the middle of the product ax, keys which have a large number of leading or trailing zeros will collide. If we want to avoid this kind of problems, then we should choose an a that has neither leading nor trailing zeros.

Furthermore, if we choose an a to be relatively prime to W, then there exists another number a' such that $a \cdot a' = 1$ (mod W). In other words, a' is the *inverse* of a modulo W, since the product of a and its inverse is one. Such a number has the nice property that if we take a key x and multiply it by a to get ax, we can recover the original key by multiplying the product again by a', because $axa' = aa'x = 1x = x$.

There are many possible constants with the desired properties. One possibility that is experimentally shown to be well-suited for 32-bit arithmetic ($w =$

32) is $a = 2\ 654\ 435\ 769$. This number is relatively prime to $W = 2^{32}$ and has neither many leading nor trailing zeros in the binary representation:

$$(2\ 654\ 435\ 769)_2 = 10011110001101110111100110111001.$$

3.4. Collision resolution. The ideal situation would be to avoid collisions altogether. However, since the number of all possible keys is greater than the size of the array used to store them, there must be at least two keys that have the same hash value. Avoiding collisions altogether is therefore impossible. Thus, while a well-designed, random-looking hash function can minimize the number of collisions, we still need a method for resolving the collisions that do occur.

There are basically two collision resolution techniques called separate chaining and open addressing. When using separate chaining, we put the keys that hash to the same array positions in separate linked lists outside the array. In open addressing, all keys are stored in the hash table itself, and collisions are resolved using various forms of probing different array positions.

To describe these methods of collision resolution, in what follows we refer to the hash function generically, assuming only that its computation takes constant time. Also, for simplicity, in the running examples we use exclusively the division method.

3.5. Chained hash tables. A hash table that uses separate chaining to resolve collisions is simply an array of linked lists. To insert a key into the table, that key is added to one of the linked lists, and the linked list to which it is added is determined by hashing the key. Figure 11 shows an example of a chained hash table.

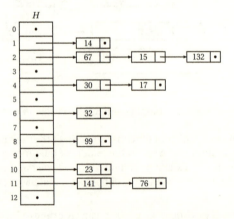

FIGURE 11. Collision resolution by chaining.

A chained hash table therefore fundamentally consists of an array of linked lists. Each list forms a *bucket* into which we place all keys hashing to a specific position in the array. To insert a key, we first hash it to determine which bucket the key belongs to. We then add the key to the front of the appropriate list. To look up or remove a key, we hash it again to find its bucket, then traverse the corresponding list until we find the key we are looking for. Because each bucket is a linked list, a chained hash table is not limited to a fixed number of keys.

The hash-table operations are trivial to implement using the list operations. For example, the following is pseudocode for the insert and find operations—the delete operation is left as an easy exercise.

```
HASH-INSERT(H, x)
    i = h(x);
    LIST-HEAD-INSERT(H(i), x);
    return;
```

```
HASH-SEARCH(H, x)
    i = h(x);
    return LIST-SEARCH(H(i), x);
```

3.5.1. *Analysis of the chained hash-table operations.* Let the chained hash-table contain currently n keys. The worst-case running time for insertion is constant, because the key being inserted is always added to the front of the appropriate list. This holds if the key is not already present in the table, and this condition can be checked, if necessary, at additional cost by performing a search before insertion. The worst-case running time for deletion is essentially the same as that for searching, since in both cases we traverse the corresponding list looking for a key. For searching, the worst-case running time is proportional to the length of the corresponding list. And the length of the list in the worst case can be large when all n keys hash to the same bucket, creating a list of length n. Thus, the worst-case behavior of the find operation is terrible—$O(n)$, or no better than if we used one ordinary linked list for all the keys. Then why have we gone to all the trouble inventing hash tables?

The hash tables are not used for their worst-case performance, instead their power lies in their average (expected) behavior. The essential problem with chained hash tables is that if an excessive number of collisions occur at a specific position, the corresponding bucket becomes larger and larger. Thus, accessing its keys takes more and more time. Even assuming simple uniform hashing, performance degrades significantly if we make the number of buckets in the table small relative to the number of keys we plan to insert. In this situation, all of the buckets become larger and larger.

It is intuitively convincing that performance of the hash table depends on the ratio of the number of keys n actually stored in the hash table to the size of hash table m. This quantity is called the **load factor** of hash table. If we denote it by α, then

$$\alpha = \frac{n}{m}.$$

The load factor of a chained hash table can be less than, equal to, or greater than 1. It indicates the average number of keys we can expect to get in a bucket, assuming simple uniform hashing. To see why, for $i = 0, 1, \ldots, m-1$, denote by n_i the number of keys in the i-th linked list. Then $n = n_0 + n_1 + \cdots + n_{m-1}$, and so the average length of a list (bucket) is

$$\frac{n_0 + n_1 + \cdots + n_{m-1}}{m} = \frac{n}{m} = \alpha.$$

If we are given the load factor α, we can determine the *average* running time of the hash-table operations in terms of α, under the assumption of simple uniform hashing. In fact, the average running time of the insert operation is the same as its worst-case time—$O(1)$. This result does not depend on α, since we always insert to the beginning of an appropriate list. Moreover, the average running time for deletion is essentially the same as the one for searching, since deletion practically consists of searching plus simple pointer manipulation.

It remains therefore to determine the average running time of the find operation. In doing so, we need to make an additional assumption about whether the key that is being sought is in the table or not. If the lookup key is not in the table, the search is said to be unsuccessful. Otherwise, when the key being sought is present in the table, we have a successful search. Nevertheless, our analysis will show that in both case the average running time is $O(\alpha)$.

Let's start with the easier case of an unsuccessful search. In an unsuccessful search for a key x, we check the list $H(i)$ that extends from the position $i = h(x)$. Since we have to scan through the entire list, an unsuccessful search takes time proportional to the length of the list $H(i)$. But since under the simple uniform hashing assumption any list (including an empty list) is equally likely to be checked, we expect that an unsuccessful search takes time proportional to the expected length of a list. And since the expected length of a list is α, it follows that an unsuccessful search takes $O(\alpha)$ time, on average.

A successful search is slightly different, since each list is not equally likely to be searched—the empty lists are ruled out in this case. Thus, we need a more careful argument to show the following result.

THEOREM *In a chained hash table, the average running time of a successful search is $O(\alpha)$, under the assumption of simple uniform hashing and equal probability of any key in the table to be searched for.*

PROOF: Suppose we have n keys stored in a chained hash table of size m. Since any of the n keys is equally likely to be sought in a successful search, the probability that a particular key is the lookup key in a successful search is $1/n$.

In a successful search for any key, before we find the key, we examine all the keys appearing in front of the lookup key in its list. Let X be the random variable of the number of those keys examined during a successful search, not counting the last examination of the lookup key. Clearly, the running time of a successful search is proportional to the value of X. Hence, to determine the average running time of a successful search, we will compute $\mathbb{E}(X)$.

For $i = 1, 2, \ldots, n$, let x_i denote the i-th key inserted into the table. If we are searching for the key x_i, all the keys in front of x_i in the x_i's list hash to the same value equal to $h(x_i)$. Moreover, all those keys were inserted into the table after x_i was inserted, because new keys are placed at the front of the list. Thus, given that the key being searched is x_i, in a successful search we examine those of the keys $x_{i+1}, x_{i+2}, \ldots, x_n$ that hash to the position $h(x_i)$.

For $i = 1, 2, \ldots, n$ and $j = i+1, \ldots, n$, let X_{ij} be the indicator random variable such that

$$X_{ij} = \begin{cases} 1, & \text{if } h(x_j) = h(x_i) \\ 0, & \text{otherwise.} \end{cases}$$

In other words, X_{ij} indicates whether the key x_j, which is inserted into the table after the key x_i, hashes to the same position as x_i. Therefore, given that the key being searched for is x_i, the number of keys appearing in front of x_i in the x_i's list is exactly $\sum_{j=i+1}^{n} X_{ij}$. That is,

$$X \mid \text{the search key is } x_i = \sum_{j=i+1}^{n} X_{ij}.$$

Furthermore, under the assumption of simple uniform hashing, we have

$$\mathbb{P}\{X_{ij} = 1\} = \mathbb{P}\{h(x_j) = h(x_i)\} = \sum_{k=1}^{m} \mathbb{P}\{h(x_j) = k \text{ and } h(x_i) = k\}$$

$$= \sum_{k=1}^{m} \mathbb{P}\{h(x_j) = k\} \cdot \mathbb{P}\{h(x_i) = k\}$$

$$= m \cdot \frac{1}{m} \cdot \frac{1}{m}$$

$$= \frac{1}{m}.$$

Clearly, this implies $\mathbb{E}\{X_{ij}\} = \mathbb{P}\{X_{ij} = 1\} = 1/m$.

Now we are ready to compute the expected number of keys that are examined during a successful search, not counting the last examination of the lookup key. Assuming any key is equally likely to be searched for, and by linearity of expectation, we get

$$\mathbb{E}(X) = \sum_{i=1}^{n} \mathbb{E}(X \mid \text{the search key is } x_i) \cdot \mathbb{P}\{\text{the search key is } x_i\}$$

$$= \sum_{i=1}^{n} \mathbb{E}\left(\sum_{j=i+1}^{n} X_{ij}\right) \cdot \frac{1}{n} = \frac{1}{n} \sum_{i=1}^{n} \sum_{j=i+1}^{n} \mathbb{E}(X_{ij}) = \frac{1}{n} \sum_{i=1}^{n} \sum_{j=i+1}^{n} \frac{1}{m}$$

$$= \frac{1}{nm} \sum_{i=1}^{n} (n-i) = \frac{1}{nm}\left(n^2 - \frac{n(n+1)}{2}\right) = \frac{1}{nm} \frac{n(n-1)}{2}$$

$$= \frac{\alpha}{2} - \frac{\alpha}{2n} = \frac{\alpha}{2}\left(1 - \frac{1}{n}\right)$$

$$\leq \frac{\alpha}{2}. \blacksquare$$

So, while any one find operation can be as bad as $O(n)$, we expect that the running time will be $O(\alpha)$. In fact, if we have a sufficiently good hash function and a reasonable number of keys in the hash table, we can expect that those keys are distributed throughout the table. Therefore, a particular find operation will not be very much worse than the average case.

This analysis tells us also that if the number of available hash-table slots is at least proportional to the number of keys in the table, that is $n = O(m)$, then $\alpha = n/m = O(m)/m = O(1)$. In this case therefore, the hash-table operations take constant time on average. For example, in a chained hash table with $m = 2969$ buckets and a total of $n = 6000$ keys, the load factor of the table is $\alpha = 6000/2969 = 2.02$. Therefore, we can expect to encounter roughly two keys while searching any one bucket. Of course, since uniform hashing is only approximated, in actuality we end up with something more or less than what the load factor is.

Finally, if we know how many keys will be inserted into the hash table *a priori*, then we can choose the table size m which is larger than the maximum number of keys expected. By doing this, we can ensure that $\alpha = n/m \le 1$. That is, a linked list contains no more than one key on average.

3.6. Open-addressed hash tables. In a chained hash table, keys reside in external linked lists extending from each array position. In contrast, in an open-addressed hash table, keys reside in the array itself and open addressing entirely does away with the need for links and chaining. Also, whereas a chained hash table has an inherent means of resolving collisions by inserting a colliding key in the appropriate list, open-addressed hash tables must resolve them in a different way. The way to resolve collisions in an open-addressed hash table is to define the ***probing sequence*** of the array locations for every key, which always leads to the key in question. For example, to insert a key, we probe prescribed array locations until we find an unoccupied one, and insert the key there. To remove or look up a key, we probe the same locations until the key is found or until we encounter an unoccupied location. If we encounter an unoccupied location before finding the key, or if we end up traversing all of the locations, the key is not in the table.

The probing sequence is essentially a sequence of functions $h_0, h_1, \ldots, h_{m-1}$ such that each $h_i \colon \mathbb{N} \to \{0, 1, \ldots, m-1\}$ is a hash function. To insert a key x into the hash table, we examine the array locations

$$h_0(x), h_1(x), h_2(x), \ldots$$

until we find an empty location. Similarly, to find a key x in the hash table, we examine the same sequence of locations in the same order.

The most common probing sequences are of the form

$$h_i(x) = h(x) + g(i) \pmod{m},$$

where $i = 0, 1, \ldots, m-1$. The function $h \colon \mathbb{N} \to \{0, 1, \ldots, m-1\}$ is an ordinary hash function that determines "natural" position $h(x)$ of the key x. It is selected in a way we already discussed, that is, it should be easy to compute and the keys should be distributed uniformly and randomly.

The function $g \colon \{0, 1, \ldots, m-1\} \to \mathbb{N}$ represents the collision resolution strategy. It is required to have two properties. First, $g(0) = 0$ so that the first probe in the sequence is $h_0(x) = h(x) + 0 \pmod{m} = h(x)$. Second, the set of values $\{g(0), g(1), \ldots, g(m-1)\}$, where each value is taken modulo m, must be a permutation of the set of integers between 0 and $m-1$. This second property ensures that the probing sequence eventually probes every possible array location.

The following pseudocode for the insert and find operations of an open-addressed hash table H assumes that each table slot contains either a key or a special value (different from any key) indicating that the slot is empty.

```
HASH-INSERT(H, x)
  k = h(x);  [ natural position ]
  i = 0;
  repeat
    j = k + g(i) (mod m);
    if H(j) = "empty" then
      H(j) = x;
      return j;
    else
      i = i + 1;
  until i = m;
  return "table overflow";
```

```
HASH-SEARCH(H, x)
  k = h(x);  [ natural position ]
  i = 0;
  repeat
    j = k + g(i) (mod m);
    if H(j) = x then
      return j;
    else
      i = i + 1;
  until (H(j) = "empty") ∨ (i = m);
  return "not found";
```

A key removal from an open-addressed hash table is more complicated. The difficulty lies in the fact that when we need to remove a key from some slot j, we cannot simply store the special value in the j-th position marking it as an empty slot. Namely, doing so may make it impossible to find any key during whose insertion we probed the slot j and found it occupied. Any such key x would have been stored in a slot that comes "after" the slot j (in the modulo m sense). So, if in the meantime the slot j is not filled in, HASH-SEARCH(H, x) would encounter an empty slot before it gets to the key x. Thus, it would incorrectly declare that x is not in the table. Some solutions to this problem are discussed in Section 3.6.4, which presents more details of the implementation of an open-addressed hash table.

Regarding performance of the open-addressed hash-table operations, the goal is to minimize the number of probes we have to perform. Exactly how many positions we end up probing depends primarily on two factors: the load factor of the hash table and the degree to which the keys are distributed uniformly and randomly. Note that if n is the number of keys actually stored in the array, and m is the number of array positions into which the keys may be hashed, the load factor $\alpha = n/m$ of an open-addressed hash table is always less than or equal to 1. This follows from the simple fact that an open-addressed hash table cannot contain more keys than the number of positions in the array (that is, $n \leq m$), since no table slot can ever contain more than one key.

3.6.1. *Linear probing.* The simplest collision resolution strategy in open addressing is called the **linear probing**. In linear probing, the function $g(i)$ is a linear function of i in the form $g(i) = ai + b$.

The first required property for the function g that $g(0) = 0$ implies $b = 0$. Then, the second required property for $g(i) = ai$ implies that a and m must be relatively prime. If we know that m will always be a prime number, then any a will do. On the other hand, if we cannot be certain that m is a prime, then a

must be one. Therefore, linear probing sequence that is usually used is

$$h_i(x) = h(x) + i \pmod{m},$$

where $i = 0, 1, \ldots, m - 1$. In other words, given a key, we probe successive locations in the table, starting from the natural location for that key determined by h.

Figure 12 illustrates an example of a hash table using open addressing together with linear probing. For example, consider searching for the key 58. This key hashes to the array position 6 (using the division method for the hash function). The corresponding linear probing sequence begins at position 6 and goes on to positions $7, 8, \ldots$ In this case, the search for the key 58 succeeds after two probes.

0	1	2	3	4	5	6	7	8	9	10	11	12
39			81		18	32	58	86			102	

58

FIGURE 12. Search in an open-addressed hash table with linear probing.

To insert a key x into an open-addressed hash table, an empty location is found by following the same probing sequence that would be used in a search for x. Thus, linear probing finds an empty location by doing a linear search beginning from the array position $h(x)$.

The advantage of linear probing is that it is simple and there are no constraints on m to ensure that all positions will eventually be probed. Unfortunately, linear probing does not approximate uniform hashing very well. In particular, linear probing suffers from a phenomenon known as **primary clustering**. This means that large clusters of consecutive occupied positions begin to develop as the table becomes more and more full. This results in excessive probing, as illustrated in Figure 13.

3.6.2. *Quadratic probing.* An alternative to linear probing that addresses the primary clustering problem is called the **quadratic probing**. In quadratic probing, the function $g(i)$ is a quadratic function of i in the general form $g(i) = ai^2 + bi + c$. Since the first required property that $g(0) = 0$ implies $c = 0$, for quadratic probing in general we have $g(i) = ai^2 + bi$ with $a \neq 0$. However, to make full use of the hash table, the values of a, b, and m are further constrained.

For example, if we take $g(i) = i^2$, as is usually done, it is not clear whether this function has the second required property to range over the entire array. In fact, in general it has not. The following lemma gives the conditions under which this quadratic probing examines different locations.

LEMMA *If the size m of the hash table is a prime number, then the first $\lfloor m/2 \rfloor$ locations probed using $g(i) = i^2$ are distinct.*

PROOF: We argue by contradiction. Given any key x, suppose there are two distinct values i and j such that $0 \leq i < j < \lfloor m/2 \rfloor$ and $h_i(x) = h_j(x)$. This implies $h(x) + i^2 = h(x) + j^2 \pmod{m}$, hence $i^2 = j^2 \pmod{m}$. In other words, $i^2 - j^2 = 0 \pmod{m}$, or $(i - j)(i + j) = 0 \pmod{m}$. Since m is a prime number,

1. insert $x = 32, 18, 81, 39, 86, 102$;   **no collisions**

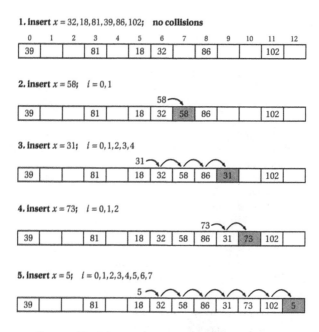

FIGURE 13. Primary clustering with linear probing.

this implies $i - j = 0 \pmod{m}$ or $i + j = 0 \pmod{m}$. Since i and j are distinct, $i - j \neq 0 \pmod{m}$. Furthermore, since both i and j are less than $\lfloor m/2 \rfloor$, the sum $i + j$ is less than m. Thus, $i + j \neq 0 \pmod{m}$ as well. In other words, neither of the possibilities $i - j = 0 \pmod{m}$ nor $i + j = 0 \pmod{m}$ can occur. Since our assumption that among the first $\lfloor m/2 \rfloor$ locations probed there are two of them that are the same leads us to a contradiction, it follows that the first $\lfloor m/2 \rfloor$ probes are all distinct. ∎

Quadratic probing works therefore as long as the table size m is a prime number and there are fewer than $m/2$ keys in the table. This means $n < m/2$, or in terms of the load factor $\alpha = n/m$, this condition is satisfied when $\alpha < 1/2$.

Quadratic probing eliminates the primary clustering phenomenon of linear probing because instead of doing a linear search, it does a quadratic search. For example, it examines the 1-st, 4-th, 9-th, and so on, location from the initial probe position of a key. However, if two keys have the same natural position, then their probing sequences are the same. This fact leads to a milder form of so-called *secondary clustering*.

3.6.3. *Double hashing.* While quadratic probing eliminates the primary clustering problem, it suffers from the secondary clustering. Moreover, in the case $g(i) = i^2$, it places a restriction on the number of keys that can be stored in the table—the table must be less than half full. **Double hashing** is one of the most effective approaches to probing an open-addressed hash table because the resulting probing sequences have many of the characteristics of randomly chosen permutations. Double hashing requires two distinct hash functions h and h'

such that $h: \mathbb{N} \to \{0, 1, \ldots, m-1\}$ and $h': \mathbb{N} \to \{1, 2, \ldots, m-1\}$. The probing sequence is then defined as

$$h_i(x) = h(x) + i\,h'(x) \pmod{m},$$

where $i = 0, 1, \ldots, m-1$. Since in this case the collision resolution function is $g(i) = i\,h'(x)$, clearly $g(0) = 0$ for any $h'(x)$. Hence, this probing sequence satisfies the first required property of starting from the natural position for any key x.

Note that the probing sequence is not predetermined in advance, but depends on each key x as follows. If $h'(x) = 1$, then $h_i(x) = h(x) + i \pmod{m}$, i.e., the probing sequence for the key x is the same as linear probing. If $h'(x) = 2$, then $h_i(x) = h(x) + 2i \pmod{m}$, i.e., the probing sequence examines every subsequent second position starting from the natural position of the key x. In general, if $h'(x) = j$ for $1 \le j \le m-1$, then $h_i(x) = h(x) + ij \pmod{m}$, i.e., the probing sequence examines every array position that is an integer multiple of j modulo m away from the starting position. It is easy to see that, for $i = 0, 1, \ldots, m-1$ and fixed j such that $1 \le j \le m-1$, the product $ij \pmod{m}$ generates a permutation of the set $\{0, 1, \ldots, m-1\}$ if and only if j and m are relatively prime. Thus, since $h'(x)$ can assume any value j between 1 and $m-1$, m must be a prime number to have the probing sequence satisfy the second required property.

Typically, when using double hashing, we choose

$$h(x) = x \pmod{m},$$
$$h'(x) = 1 + \left(x \pmod{m-1}\right).$$

For example, if the size of the hash table is $m = 2969$ (a prime number), and we hash the key $x = 17910$, its natural position is $h(17910) = 17910 \pmod{2969} = 96$. Also, since $h'(17910) = 1 + \left(17910 \pmod{2968}\right) = 103$, we would probe every subsequent 103-th position after the natural position.

The advantage of double hashing is that it generates one of the best forms of probing sequences, producing a good distribution of keys throughout a hash table. However, for double hashing to work at all, the size of the hash table is constrained to be a prime number.

Table 1 summarizes characteristics of the various open-addressing probing sequences.

Probing sequence	Primary clustering	Load limit	Size restriction
Linear probing	yes	none	none
Quadratic probing	no	$\alpha < 1/2$	must be prime
Double hashing	no	none	must be prime

TABLE 1. Characteristics of the open addressing probing sequences.

3.6.4. *Implementation of open-addressed hash tables.* An open-addressed hash table is implemented fundamentally as a single array. Each array element has two fields: *key* and *state*. The former stores the corresponding key, and the latter holds an indicator that marks the underlying array location as either empty, occupied, or deleted. Figure 14 illustrates this approach, where the state of a location is represented by the white (empty), black (occupied), or gray (deleted) color.

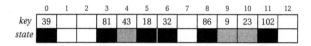

	0	1	2	3	4	5	6	7	8	9	10	11	12
key	39			81	43	18	32		86	9	23	102	
state													

FIGURE 14. Implementation of an open-addressed hash table.

Initially, all array locations are empty. When a key is stored in an array location, the state of that location is changed to occupied. The purpose of the third state, deleted, will be explained when we discuss the operation of removing a key from an open-addressed hash table. We call a location unoccupied if its state is either empty or deleted.

The method for inserting a key into an open-addressed hash table is actually quite simple: find an unoccupied array location and then store the key in that location. To find an unoccupied array location, the array is probed according to a probing sequence.

The find operation used to locate a key follows the same probing sequence that is used during insertion of the key. Therefore, if there is a matching key in the table, the find operation will make exactly the same number of probes to locate the key as were made to store the key into the table in the first place. However, if the lookup key is not in the table, the search immediately terminates when it encounters an empty location. This is because if the target has not been found by the time an empty element is encountered, then the target is not in the table. Any locations marked as deleted are not examined during the search, but they do not terminate the search either.

Removing keys from an open-addressed hash table must be done with some care. A naive approach would be to locate the key to be removed and then change the state of its location to empty. However, that approach does not work, since the find operation stops its search when it encounters an empty location. Therefore, if in the middle of a probing sequence we change the state of a location to empty, all subsequent searches that follow the same probing sequence will stop at the empty location. As a result, subsequent searches for a key may fail even when the key is still in the table.

One way to deal with the problem of removing a key is to have the table end up exactly as it would have appeared had that key never been inserted in the first place. In other words, when a key is removed from the middle of a probing sequence, the keys that follow it in the sequence have to be moved up to fill in the hole. This moving up the keys to positions of their predecessors is further complicated by the fact that we may encounter a key that cannot be moved safely because it hashes to its position. We must then skip this key and find another one behind it in the sequence that can be safely moved up. The details are formidable and in the worst-case this process takes $O(n)$ time. Clearly, we don't want to be removing keys from an open-addressed hash table in this way very often.

Another solution to the deletion problem makes use of the third state, deleted, of each location. Instead of marking some location empty, we mark it deleted when a key is removed. This works because the find operation continues its search past deleted locations. Also, the insert operation is looking for an unoccupied location, i.e., it stops its search when it encounters either an empty or a deleted location. Consequently, the locations marked deleted are available for reuse when insertion is done.

Unfortunately, this technique of marking locations as deleted also has a serious drawback, if a significant number of deletions is made in the table. Once a large number of key removals has been done, it is very likely that there are no locations left that are marked empty. This is because no operation ever marks any location as empty. Consequently, an unsuccessful search will examine all locations in the table, even if only a small fraction of them really contains the keys. For example, if we first fill up the table, and then remove all the keys, every location will be marked as deleted. Since the find operation only stops its search early when an empty location is encountered, and there are no more empty locations, an unsuccessful search must examine all locations. In other words, search times are no longer dependent on the load factor α.

Both solutions to the deletion problem are therefore unsatisfactory in some aspect. So, in applications in which we know *a priori* that a significant number of key removals will be made, chaining is more commonly selected as the collision resolution strategy.

3.6.5. *Analysis of the open-addressed hash-table operations.* It is easy to see that the worst-case time complexity of the open-addressed hash-table operations is as bad as for the chained hash-table operations. If the number of keys actually stored in the table is n, and the probing sequence hits all the keys currently in the table, each of the insert, find, and delete operations can take $O(n)$ time in the worst-case. Thus, open addressing does not bring any improvement in the worst-case over chaining.

The average-case analysis of open addressing is relatively clean, under the assumption of simple uniform hashing, if we ignore the delete operation. We already argued that the delete operation is anyway generally impractical when open addressing is used to resolve collisions. In other words, taking the delete operation completely out of our considerations won't affect much our goal of capturing the average performance of the open-addressed hash tables. Therefore, in the idealized simple uniform hashing model, we assume that the delete operations are not used at all, so we analyze only the insert and find operations. This implies that it is enough to have each table slot be either occupied or empty, that is, the deleted state is superfluous. In fact, observing that for insertion we perform basically an unsuccessful search, it is enough to consider only the two forms of the find operation—successful and unsuccessful search. Moreover, the cost of a successful or unsuccessful search is proportional to the number of probes made. We therefore further simplify our analysis and compute the expected number of probes made in each form of the find operation.

Let's consider first an unsuccessful search, or equivalently, a search for an empty slot in the open-addressed hash-table. The following theorem establishes that we can expect to make about $1/(1 - \alpha)$ probes before we find an empty slot, where α is the load factor.

THEOREM *In an open-addressed hash table, the expected number of probes that is made in an unsuccessful search is approximately $1/(1 - \alpha)$, under the assumption of simple uniform hashing.*

PROOF: Given an open-addressed hash table of size m that contains n keys ($n \le m$), let X be the random variable of the number of probes made in a search for an empty slot in the table. In the best case, we can succeed in finding an empty slot after just one probe; in the worst case, we need $n + 1$ probes if we hit all currently occupied slots. Thus, X ranges between 1 and $n + 1$, and so the expected number of probes is

$$\mathbb{E}(X) = \sum_{k=1}^{n+1} k\mathbb{P}\{X = k\}.$$

Computing exact probability $\mathbb{P}\{X = k\}$ seems like an arduous task, so we apply the usual trick to get around this problem. The expectation sum can be rewritten as

$$\mathbb{E}(X) = \sum_{k=1}^{n+1} k\mathbb{P}\{X = k\}$$

$$= \mathbb{P}\{X = 1\} + \left(\mathbb{P}\{X = 2\} + \mathbb{P}\{X = 2\}\right) + \left(\mathbb{P}\{X = 3\} + \mathbb{P}\{X = 3\} + \mathbb{P}\{X = 3\}\right) +$$

$$+ \cdots + \left(\mathbb{P}\{X = n+1\} + \mathbb{P}\{X = n+1\} + \cdots + \mathbb{P}\{X = n+1\}\right)$$

$$= \sum_{k=1}^{n+1} \mathbb{P}\{X = k\} + \sum_{k=2}^{n+1} \mathbb{P}\{X = k\} + \cdots + \sum_{k=n+1}^{n+1} \mathbb{P}\{X = k\}$$

$$= \mathbb{P}\{X \ge 1\} + \mathbb{P}\{X \ge 2\} + \cdots + \mathbb{P}\{X \ge n+1\}$$

$$= \sum_{k=1}^{n+1} \mathbb{P}\{X \ge k\}.$$

Now, $\mathbb{P}\{X \ge k\}$ is easier to come up with, because the number of probes is greater than or equal to k, $(2 \le k \le n+1)$, if and only if each of the first $k - 1$ probes hits an occupied slot. Thus, under the assumption of simple uniform hashing, we have

$$\mathbb{P}\{X \ge 1\} = 1, \quad \mathbb{P}\{X \ge 2\} = \frac{n}{m}, \quad \mathbb{P}\{X \ge 3\} = \frac{n}{m}\frac{n-1}{m-1},$$

$$\mathbb{P}\{X \ge 4\} = \frac{n}{m}\frac{n-1}{m-1}\frac{n-2}{m-2}, \quad \ldots, \quad \mathbb{P}\{X \ge n+1\} = \frac{n}{m}\frac{n-1}{m-1}\cdots\frac{1}{m-n+1}.$$

Let $p_k = \mathbb{P}\{X \ge k\}$ for $k = 1, 2, \ldots, n+1$. To compute $\mathbb{E}(X) = \sum_{k=1}^{n+1} p_k$, we recursively express p_k's as

$$p_1 = 1, \quad p_{k+1} = p_k \cdot \frac{n+1-k}{m+1-k}, \text{ for } k = 1, 2, \ldots, n.$$

Multiplying the recurrence equation by $m + 1 - k$ for $k = 1, 2, \ldots, n$ and rearranging the terms, we get

$$(m + 1 - k)p_{k+1} = (n + 1 - k)p_k$$

$$(m+1)p_{k+1} - kp_{k+1} = np_k - (k-1)p_k$$
$$(m+1)p_{k+1} = np_k + kp_{k+1} - (k-1)p_k$$

Summing up both sides of the last equation for k ranging from 1 to n yields

(17) $$(m+1)\sum_{k=1}^{n} p_{k+1} = n\sum_{k=1}^{n} p_k + \sum_{k=1}^{n} kp_{k+1} - \sum_{k=1}^{n}(k-1)p_k.$$

Substituting

$$\sum_{k=1}^{n} p_{k+1} = \sum_{k=1}^{n+1} p_k - p_1 = \sum_{k=1}^{n+1} p_k - 1$$

and

$$\sum_{k=1}^{n} kp_{k+1} - \sum_{k=1}^{n}(k-1)p_k = \sum_{k=1}^{n} kp_{k+1} - \sum_{k=1}^{n} kp_{k+1} + np_{n+1} = np_{n+1}$$

in the equation (17) yields

(18) $$(m+1)\sum_{k=1}^{n+1} p_k - (m+1) = n\sum_{k=1}^{n+1} p_k.$$

Solving the equation (18) for $\sum_{k=1}^{n+1} p_k$, we finally get

$$\mathbb{E}(X) = \sum_{k=1}^{n+1} p_k = \frac{m+1}{m+1-n} = \frac{1}{1-\frac{n}{m+1}} \approx \frac{1}{1-\alpha}. \blacksquare$$

This result is actually quite intuitive. The load factor α is the fraction of occupied table slots. Therefore, $1-\alpha$ is the fraction of empty slots, so we would expect to have to probe $1/(1-\alpha)$ slots before finding an empty slot. For example, if the load factor is $3/4$, one fourth of the slots is empty. Therefore, we expect to have to probe four slots before finding an empty slot.

To calculate the expected number of probes in a successful search, we assume that the key being searched for is equally likely to be any of the n keys stored in the table. The main observation is that when a key is initially inserted, we need to find an empty slot in which to place the key. Let x_i denote the i-th key inserted into the table, for $i = 1, 2, \dots, n$. Let $U_m(n)$ denote the expected number of probes it takes to find an empty slot in a table of size m containing n keys. Then, the expected number of probes to find an empty slot into which to place the key x_i is $U_m(i-1)$. But this is exactly the expected number of probes it takes to find the key x_i again.

Let $S_m(n)$ denote the expected number of probes made in a successful search in a table of size m containing n keys. Using $U_m(n) = (m+1)/(m+1-n)$ that we obtained in the previous theorem, and the assumption that any stored key is equally likely with probability $1/n$ to be the lookup key, we can compute $S_m(n)$

as

$$S_m(n) = \frac{1}{n} \sum_{i=0}^{n-1} U_m(i)$$

$$= \frac{1}{n} \sum_{i=0}^{n-1} \frac{m+1}{m+1-i}$$

$$= \frac{m+1}{n} \sum_{i=0}^{n-1} \frac{1}{m+1-i}$$

$$= \frac{m+1}{n} \left(H_{m+1} - H_{m+1-n} \right),$$

where $H_k = \sum_{i=1}^{k} 1/i$ is the k-th harmonic number. Since $H_k \approx \ln k$, we can further approximate the expected number of probes in a successful search as

$$S_m(n) \approx \frac{m+1}{n} \left(\ln(m+1) - \ln(m+1-n) \right)$$

$$= \frac{m+1}{n} \cdot \ln \frac{m+1}{m+1-n}$$

$$\approx \frac{1}{\alpha} \ln \frac{1}{1-\alpha}.$$

For example, if an open-addressed hash table is 75% full ($\alpha = 3/4$), the expected number of probes made in a successful search is approximately 1.85.

4. Priority queue

The priority queue data structure represents a set of elements, each of which has an associated value called the *priority*. For example, the elements of the set could be records, and the priority could be the value of one field in the records. Two distinguishing operations on the priority queue data structure Q are:

1. PQUEUE-INSERT(Q, x) adds a new element x to the set represented by Q;
2. PQUEUE-DELETE(Q) finds in the set an element of highest (or lowest) priority, and removes it from the set returning its priority.

A more concrete application is the way a time-sharing operating system schedules different processes for execution. Namely, all these processes may not have the same priority. At highest priority may be the system processes, such as keyboard or network adapter driver. Then may come user processes that execute user applications in foreground. Below these we may have certain background tasks, such as a tape backup program or a long calculation designated to run with a low priority.

Each process in such a system can be represented by a record that includes an integer field for the ID of the process, and an integer field for the priority of the process. A simplified picture of the scheduling algorithm for a time-sharing operating system is the following. When a new process is spawned in the system, it gets an ID and associated priority. For this process we then execute the insert operation on the priority queue of processes waiting for execution. When a processor becomes idle, or the current executing process has used up its processor

time quantum, the operating system executes the other priority queue operation to get the process of highest priority waiting for execution. This process is
then scheduled next for execution on the processor.

Using priority queues is similar to using queues (where we remove the oldest element) or stacks (where we remove the newest element), but implementing
them efficiently is not so straightforward. An efficient way to implement the priority queue data structure is with a binary heap. We studied the binary heap data
structure in Chapter 4, where we learned the HEAP-INSERT and HEAP-DELETE
operations that take $O(\log n)$ time for a binary heap of n elements.

In practice, priority queues are more complex because we may need to support other operations on them so that they can be properly maintained under all
conditions that may arise in a particular application. These additional desired
operations on a priority queue are:

- Make a priority queue from n given elements;
- Change (decrease or increase) the priority of an arbitrary specified element in the priority queue;
- Delete an arbitrary specified element from the priority queue;
- Merge two priority queues into larger one so that the new priority queue
  contains all the elements of the old priority queues, and the old priority
  queues are (possibly) destroyed.

Recalling the binary heap data structure from Chapter 4, we see that we can
efficiently implement all the additional operations with a binary heap, except
merging two heaps. Namely, to construct a priority queue from given elements,
we can use the HEAP-MAKE procedure that runs in linear time. To change the
priority of an element somewhere in the middle of the heap, we can use the
HEAP-SIFT-DOWN procedure to go down the heap if the priority is increased,
or the HEAP-BUBBLE-UP procedure to move up the heap if the priority is decreased. To delete an arbitrary specified element, we can use similar approach
to the one used for deleting the minimal element. All these operations involve
moving along one path in the heap, perhaps all the way from top to bottom or
from bottom to top in the worst case, and so take logarithmic time. Note carefully that a full implementation of the change-priority and delete operations,
which refer to specific data elements, requires that for each element we maintain a pointer to that element's place in the heap. Details of implementing these
operations are left as an exercise.

However, merging two binary heaps into one seems to be no more efficient
than combining them from scratch, which takes linear time. Since by efficient in
this context we mean at most logarithmic time, efficient implementations of the
insert, find-the-minimum, and merge operations require much more sophisticated tree structure. The difficulty lies in dynamic nature of the operations, so
we need a more flexible tree structure than complete tree that admits an easy
modification. On the other hand, we need to keep such trees sufficiently balanced to ensure logarithmic-time bound for *all* of the three operations. One
example of this sort of trees that we study in the next section is called binomial
heap.

4.1. Binomial heaps. To describe binomial heaps, the starting point is a special kind of general rooted trees, called binomial trees, whose maximum degree of a node is not fixed. A **binomial tree** of order $k \geq 0$, denoted B_k, is defined recursively as follows:

1. B_0 is a single node;
2. For $k \geq 0$, B_k is made of two copies of B_{k-1} that have been linked together so that the root of one B_{k-1} is the new root of B_k, and the root of the other B_{k-1} is a new child of the new root.

An equivalent recursive definition, which is easily verified by induction, is that B_k consists of a (new) root with k children that are copies of $B_0, B_1, \dots, B_{k-1}$. Figure 15 illustrates these two recursive definitions.

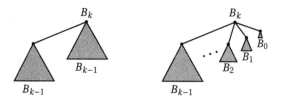

FIGURE 15. Binomial tree B_k defined in two equivalent ways.

An important feature of binomial trees is that they are bushy and shallow. This is suggested in Figure 16, which shows the first five binomial trees.

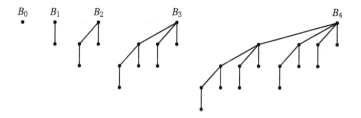

FIGURE 16. Binomial trees B_0, B_1, B_2, B_3, B_4.

Binomial trees have several useful properties. For example, it follows directly from the second definition that the root of B_k has degree k. The following is the list of other properties, which are easy to prove by induction and using the first or second recursive definition of a binomial tree (see Exercise 12).

(i) B_k has 2^k nodes.
(ii) B_k has height k.
(iii) B_k can be obtained from B_{k-1} by adding a new child to every node.
(iv) For every d such that $0 \leq d \leq k$, B_k has $\binom{k}{d}$ nodes at depth d.
(v) For every h such that $0 \leq h < k$, B_k has 2^{k-h-1} nodes with height h and one node (the root) with height k.

For the sake of illustration, let's show the property (iv): the number of nodes in B_k at a given depth is determined by the binomial coefficient. Actually, that's the reason why binomial trees got their name.

LEMMA *For every d such that $0 \le d \le k$, B_k has $\binom{k}{d}$ nodes at depth d.*

PROOF: Let $n_k(d)$ be the number of nodes at depth d in B_k, the binomial tree of order k. We proceed by induction on k using the first definition of B_k. In the base case $k = 0$, B_0 contains a single node and there is only one depth $d = 0$ in the tree. Therefore, B_0 has exactly one node at that depth, that is, $n_0(0) = 1 = \binom{0}{0}$.

Next, by inductive hypothesis assume the claim is true for B_k, and prove it holds for B_{k+1}. First observe that if $d = 0$ or $d = k + 1$, then $n_{k+1}(0) = 1 = \binom{k+1}{0}$ and $n_{k+1}(k + 1) = 1 = \binom{k+1}{k+1}$. Also, if $1 \le d \le k$, according to the first definition of binomial trees, the number of nodes at depth d in B_{k+1} is equal to the number of nodes at depth d in B_k plus the number of nodes at depth $d - 1$ in B_k. Thus, by using the Pascal's formula, we get

$$n_{k+1}(d) = n_k(d) + n_k(d - 1) = \binom{k}{d} + \binom{k}{d-1} = \binom{k+1}{d},$$

which proves the induction step. ∎

A **binomial heap** of a set of elements is implemented not as a single tree but as a collection of heap-ordered binomial trees. That is, a binomial heap consists of a forest of binomial trees with each node in a tree corresponding to one element of the set. Moreover, each binomial tree satisfies the heap property: priority of a node is not larger than priority of its children.

By the property (i) of binomial trees, they only come in sizes that are powers of two. So, the question is how do we implement a binomial heap which holds an arbitrary number of elements n using binomial trees? The answer is related to the binary representation of the number n:

$$n = \sum_{i=0}^{\lfloor \log n \rfloor} b_i 2^i,$$

where $b_i \in \{0, 1\}$ is the i-th bit in the binary representation of n. To make a binomial heap that holds exactly n elements, we use a collection of binomial trees such that B_i belongs to the collection if and only if the i-th bit in the binary representation of n is a one. Clearly, the number of binomial trees in a binomial heap of size n is $O(\log n)$. For example, if $n = 19$ then $(19)_2 = 10011$, and so a 19-element binomial heap is represented by the forest of binomial trees B_4, B_1, and B_0.

4.2. Operations on binomial heaps. To see that certain binomial-heap operations can be performed efficiently, we need to be more specific about how binomial heaps are actually represented. Since binomial trees that belong to a binomial heap are simply general rooted trees with a special shape, we can make use of one of the implementations discussed in Chapter 4.

For the sake of simplicity, we choose a variant of the leftmost-child, right-sibling representation of trees: in each binomial tree, the children of every node are linked into circular doubly-linked list. Moreover, the roots of all binomial trees are grouped together by an array of pointers pointing to them. This array is anyway called the **root list**.

With this representation, every node in a binomial heap points to four other nodes: its parent, its "next" sibling, its "previous" sibling, and its leftmost child (the parent pointers of all the root nodes are NIL). Figure 17 illustrates this representation of the binomial heap $H_{13} = \{B_3, B_2, B_0\}$.

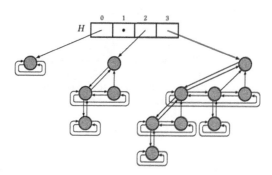

FIGURE 17. Representation of the binomial heap H_{13}, where NIL pointers at tree nodes are omitted for clarity.

The basic operation upon which the binomial-heap operations are based is that of linking two simple binomial heaps B_k together to build a simple binomial heap B_{k+1}. A **simple** binomial heap is just a binomial tree B_k with the heap order. (We hope the same notation B_k for a simple binomial heap and a binomial tree in this context won't cause confusion.)

Given our linked representation of binomial trees, the link operation is easy to implement: the smaller root of the two trees becomes the root of the result, and the larger root becomes the leftmost child of the new root. More precisely, if pointers p and q point to the roots of two simple binomial heaps B_k, then the following procedure BINOM-HEAP-LINK(p, q) builds the simple binomial heap B_{k+1} and returns the pointer to its root.

```
BINOM-HEAP-LINK(p, q)
  if key(p) < key(q) then
    SWAP(p, q);
  [[ key(q) ≤ key(p) ]]
  r = left(q);
  if r ≠ NIL then
    next(p) = r; prev(p) = prev(r);
    next(prev(r)) = p; prev(r) = p;
  left(q) = p; parent(p) = q;
  return q;
```

Since here we only swap the pointers to the original roots if their keys are in wrong order and adjust corresponding links of the original roots, BINOM-HEAP-LINK is a constant-time operation.

4.2.1. *Merging two binomial heaps.* The beauty of binomial heaps is that the merge operation is similar in structure to binary addition. Specifically, the merging of two binomial heaps is done by linking the binomial trees that make up the binomial heaps in the same way in which the bits are combined when adding two binary numbers representing the heap sizes.

For example, consider two binomial heaps H_n and H_m of sizes n and m, respectively. Merging these two heaps results in a heap H_{n+m} that contains $n+m$ elements. If we take, say, $n = 19$ and $m = 14$, since $(19)_2 = 10011$ and $(14)_2 = 1110$, then we need to merge $H_{19} = \{B_4, B_1, B_0\}$ and $H_{14} = \{B_3, B_2, B_1\}$. The sum $19 + 14 = 33$ is calculated in binary as shown on the left of Figure 18.

$$
\begin{array}{cccccc}
  & 1 & 0 & 0 & 1 & 1 \\
+ &   & 1 & 1 & 1 & 0 \\
\hline
1 & 0 & 0 & 0 & 0 & 1
\end{array}
\qquad
\begin{array}{ccccccc}
  & B_4 & \varnothing & \varnothing & B_1 & B_0 \\
+ &   & B_3 & B_2 & B_1 & \varnothing \\
\hline
B_5 & \varnothing & \varnothing & \varnothing & \varnothing & B_0
\end{array}
$$

FIGURE 18. The analogy between the binary addition and the binomial-heap merging.

Recalling that two binomial trees B_k are linked together to obtain the binomial tree B_{k+1}, if we keep the heap order in B_{k+1} and use B_{k+1} as a "carry" for the next position, then the merging of H_{19} and H_{14} is performed by analogy with binary addition, as illustrated on the right of Figure 18.

As Figure 18 shows, final result is the binomial heap $H_{33} = \{B_5, B_0\}$. In other words, the process of merging two binomial heaps mirrors precisely the process of adding two binary numbers. The pseudocode in the following procedure BINOM-HEAP-MERGE(H, G) uses this to merge two binomial heaps given by their associated root lists H and G.

The procedure BINOM-HEAP-MERGE(H, G) merges the heap represented by G *into* the heap represented by H. If the arrays H and G represent the binomial heaps H_n and H_m, respectively, by keeping pointers to the constituent binomial trees, we assume that the sizes n and m of H_n and H_m satisfy $n \geq m$. In addition, we make simplified assumptions that the arrays H and G are sufficiently large and that we don't have to destroy the merged heap H_m.

The **for** loop in the BINOM-HEAP-MERGE algorithm is executed $\lfloor \log n \rfloor$ times because the number of binomial trees in a binomial heap H_n of size n is at most $\lfloor \log n \rfloor$. The running time of the BINOM-HEAP-MERGE algorithm is therefore $O(\log n)$, since the BINOM-HEAP-LINK is a constant-time operation.

```
BINOM-HEAP-MERGE(H, G)
    c = NIL; ⟦ "carry" ⟧
    for i = 0 to ⌊log n⌋ do
        if c = NIL then
            if H(i) ≠ NIL then
                if G(i) ≠ NIL then
                    p = H(i); H(i) = NIL;
                    c = BINOM-HEAP-LINK(p, G(i));
                else
                    H(i) = G(i);
            else
                if H(i) ≠ NIL then
                    if G(i) ≠ NIL then
                        c = BINOM-HEAP-LINK(H(i), G(i));
                    else
                        p = H(i); H(i) = NIL;
                        c = BINOM-HEAP-LINK(p, c);
                else
                    if G(i) ≠ NIL then
                        c = BINOM-HEAP-LINK(c, G(i));
                    else
                        H(i) = c; c = NIL;
        H(i) = c;
    return H;
```

4.2.2. Inserting an element.
A simple way to insert an element into a binomial heap is to represent the new element as a simple binomial heap B_0 and simulate the process of incrementing a binary number. To increment a binary number, we start from the last (zeroth) bit of the number and move from right to left changing 1's to 0's because of the carry produced by doing $(1)_2 + (1)_2 = (10)_2$. We stop when we find the first 0, which we change to 1.

In an analogous way, to add a new element into a binomial heap, we first represent it as a B_0 and make it the carry heap. After that, going from the zeroth index of the associated array, we link heaps that correspond to 1-bits with a carry heap. This is repeated until we encounter the first empty position to put the carry heap. This high-level description of the insert operation is implemented by making use of the merge operation in the following pseudocode.

```
BINOM-HEAP-INSERT(H, x)
    Make G be the array representing x as simple binomial heap B_0;
    BINOM-HEAP-MERGE(H, G);
    return H;
```

Clearly, if we are adding a new element to a binomial heap H_n of size n, in the worst case this may take $\Theta(\log n)$ time because of the carry propagation.

4.2.3. *Deleting the minimum element.* Deleting the minimum element from a binomial heap is a little more complicated. Since a binomial heap is a forest of heap-ordered binomial trees, the minimum element must be at the root of one of the binomial trees. That is, the smallest of the elements at the roots of the binomial trees is the minimum element of the binomial heap.

Let B_k, $0 \le k \le \lfloor \log n \rfloor$, be the binomial tree with the minimum element at the root. Recalling the second definition of the binomial tree B_k as the root node with k subtrees $B_0, B_1, \ldots, B_{k-1}$, the key observation is that the k subtrees form a binomial heap $H_{2^k-1} = \{B_0, B_1, \ldots, B_{k-1}\}$. Since we can easily restructure the children into the heap H_{2^k-1}, using the merge operation we can efficiently combine this binomial heap with the rest of the original heap to complete the operation of deleting the minimum element.

Therefore, to delete the minimum element from a binomial heap, we first scan the root nodes to find the binomial tree with the smallest root and remove that tree from the heap. Then, we consider all subtrees of the root of that tree as a binomial heap and merge that heap back into the original heap. This implementation is given by the following algorithm BINOM-HEAP-DELETE(H).

BINOM-HEAP-DELETE(H)
    Find the index k of the minimum of elements pointed to by the array H;
    $p = left(H(k))$; $H(k) = $ NIL;
    Make G be the array of pointers to elements of the list whose head is p;
    BINOM-HEAP-MERGE(H, G);
    return H;

If the array of pointers H represents binomial heap H_n, the first step of finding the minimum element of H_n takes $O(\log n)$ time, since there are no more than $\lfloor \log n \rfloor$ binomial trees in H_n. If k is index of the pointer to the minimum element, which represents the root of a binomial tree B_k, the second step of initializing the array of pointers G that point to children of the root of B_k takes time $O(k) = O(\log n)$. Finally, the third step of merging H and G also takes time $O(\log n)$, because the corresponding binomial heaps have size at most n. Therefore, the total time of the BINOM-HEAP-DELETE(H) algorithm is $O(\log n)$.

4.2.4. *Making a binomial heap.* Construction of a binomial heap H_n out of n elements can be performed simply by n insert operations on an initially empty binomial heap.[2] Although it seems that this method requires $O(n \log n)$ time, by a closer inspection we can see that it actually takes linear time.

Specifically, the n insert operations on an initially empty heap are analogous to n increment operations on an initially all-zeros binary number. Namely, if adding two bits is executed at a unit cost, the total cost of n binary increments is proportional to the total number of additions of the carry bits generated during the computation. On the other hand, adding the carry bit corresponds exactly to the BINOM-HEAP-LINK operation executed during an insertion of a new element

---

[2]As usual, to represent the binomial heap as an abstract data structure, we should provide common ADS operations like the test for emptiness, fullness, and so on. For example, if we keep separately the length of the root list, then an empty binomial heap has this length equal to zero.

into a binomial heap. Thus, to estimate the cost of the n insert operations on an initially empty heap, it suffices to count the number of carries generated during the n binary increments on an initially all-zeros binary number.

Now, to calculate the total number of generated carries, we first observe that the n binary increments produce the (decimal) numbers $1, 2, \ldots, n$. On roughly one half of the numbers whose 0-th bit is zero (even numbers), the carry is not generated. On the half of the other half (odd) numbers, whose 1-st bit is zero and 0-th bit is one, we get one carry. Continuing argument in this way, we see that the total number of carries is bounded by

$$0 \cdot \frac{n}{2} + 1 \cdot \frac{n}{4} + 2 \cdot \frac{n}{8} + \cdots = \frac{n}{2}.$$

Therefore, because of the correspondence between the carry addition and the constant-time BINOM-HEAP-LINK operation, the n insert operations on an initially empty binomial heap take $O(n)$ time.

5. Disjoint-sets data structure

In some applications we need to group n distinct elements into a collection of disjoint sets. Two important operations are then determining which set a given element belongs to, as well as uniting two given sets from the collection.

For example, consider the problem of finding the connected components of an undirected graph. In Chapter 8 we efficiently solved this problem by performing a depth-first (or a breadth-first) search on the graph. This approach is good when the edges of the graph are "static", that is, not changing over time. Sometimes, however, the edges are added dynamically and we need to maintain the connected components as each edge is added. In this case, the strategy presented in this section may be more efficient than running depth-first search anew each time an edge is added.

Suppose that we want to construct the connected components of an undirected graph $G = (V, E)$ with n vertices and m edges. One way to do this is to begin with a graph G_0 consisting of all the vertices of G and none of the edges. We then examine the edges of G, one at a time, to construct a sequence of graphs $G_1, G_2, \ldots, G_m$, where each G_i for $i = 1, 2, \ldots, m$ consists of all the vertices of G and the first i edges of G considered. Observe that $G_m = G$.

Connected components of the graph G_0 are easy to determine—they are represented by its individual vertices. Next, if $1 \le i \le m$, suppose we have constructed the connected components of the graph G_{i-1}. Then at the i-th step we consider the i-th edge (u, v) of G and distinguish two cases. If the vertices u and v are in same component of G_{i-1}, then G_i has the same set of connected components as G_{i-1}, because the new edge does not connect any vertices that were not already connected.

On the other hand, if u and v are in different components of G_{i-1}, we unite the components containing u and v to get the connected components for G_i. Figure 19 suggests why there is a path from any vertex x in the component of u, to any vertex y in the component of v. Namely, since x and u are from the same component, there is a path π_1 from x to u. Likewise, since v and y are from the same component, there is a path π_2 from v to y. Therefore, traversing the path π_1, then the edge (u, v), and finally the path π_2, we can get from x to y.

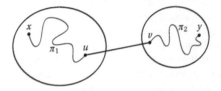

FIGURE 19. Uniting two components containing vertices u and v of the edge (u, v).

Clearly, when we have considered all edges in this manner, we get the connected components of the entire graph G. Figure 20 illustrates this approach to computing connected components on an example graph.

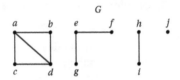

Graph	Edge	Connected components									
G_0		$\{a\}$	$\{b\}$	$\{c\}$	$\{d\}$	$\{e\}$	$\{f\}$	$\{g\}$	$\{h\}$	$\{i\}$	$\{j\}$
G_1	(a,d)	$\{a,d\}$	$\{b\}$	$\{c\}$		$\{e\}$	$\{f\}$	$\{g\}$	$\{h\}$	$\{i\}$	$\{j\}$
G_2	(h,i)	$\{a,d\}$	$\{b\}$	$\{c\}$		$\{e\}$	$\{f\}$	$\{g\}$	$\{h,i\}$		$\{j\}$
G_3	(e,f)	$\{a,d\}$	$\{b\}$	$\{c\}$		$\{e,f\}$		$\{g\}$	$\{h,i\}$		$\{j\}$
G_4	(c,d)	$\{a,c,d\}$	$\{b\}$			$\{e,f\}$		$\{g\}$	$\{h,i\}$		$\{j\}$
G_5	(e,g)	$\{a,c,d\}$	$\{b\}$			$\{e,f,g\}$			$\{h,i\}$		$\{j\}$
G_6	(a,c)	$\{a,c,d\}$	$\{b\}$			$\{e,f,g\}$			$\{h,i\}$		$\{j\}$
G_7	(b,d)	$\{a,b,c,d\}$				$\{e,f,g\}$			$\{h,i\}$		$\{j\}$
G_8	(a,b)	$\{a,b,c,d\}$				$\{e,f,g\}$			$\{h,i\}$		$\{j\}$

FIGURE 20. Constructing connected components of $G_0, G_1, \ldots, G_8$.

If we analyze the previous algorithm described informally, we can see that we need to perform three operations quickly:

1. Initially make a component to be a single vertex.
2. Given a vertex, find its current component.
3. Unite two components into one.

An abstract data structure that efficiently supports these operations is called the ***disjoint-sets data structure***. Generally, a disjoint-sets data structure maintains a collection of k disjoint sets $S_1, S_2, \ldots, S_k$. Each set in the collection is identified by a ***representative***, which is some member of the set. It does not matter which member is used as the representative, as long as this is done in a consistent way. This means that if we ask for the representative of a dynamic set twice without modifying the set between the requests, we get the same answer both times. For example, if the set elements can be ordered, the smallest member of a set is an obvious choice for its representative.

If an element of the sets is represented by an object of some type, the following is a list of operations that we wish to support.

1. *Making a singleton set*: DSET-MAKE(x) creates a new set whose only member (and hence representative) is x. Since the sets are disjoint, we require that x is not already in some other set.

2. *Finding the home set*: DSET-FIND(x) returns a pointer to the representative of the unique set containing x.

3. *Uniting two sets*: DSET-UNION(x, y) unites the dynamic sets that contain x and y, say S_x and S_y, into a new set representing the union of these two sets. Thus, it forms a third set S_z such that $S_z = S_x \cup S_y$. The sets S_x and S_y are assumed to be disjoint, and the representative z of the resulting set is any member of $S_x \cup S_y$. Since we require that the sets in the collection are disjoint, we remove the sets S_x and S_y from the collection after uniting them.

Given these operations, we can easily implement the informal algorithm that constructs the connected components of an undirected graph $G = (V, E)$. In the following pseudocode we assume that the vertices $v_1, v_2, \ldots, v_n$ in V are numbered arbitrarily. We keep track of discovered connected components using a pointer array s indexed by the vertices such that $s_v \neq$ NIL indicates the connected component whose representative is $v \in V$.

CONNECTED-COMPONENTS(G)
   for each vertex $v \in V$ **do** $s_v =$ DSET-MAKE(v);
   for each edge $(u, v) \in E$ **do**
      $p =$ DSET-FIND(u); $q =$ DSET-FIND(v);
      if $p \neq q$ **then**
         $r =$ DSET-UNION(u, v);
         Let p, q, r point to the vertices a, b, c;
         $s_a = s_b =$ NIL; $s_c = r$;
   return s;

We will return to the running time of this algorithm shortly after we consider a particular implementation of the disjoint-sets abstract data structure.

5.1. A tree implementation of the disjoint-sets data structure. An efficient way to implement disjoint sets is by a forest of rooted trees, not necessarily binary, where each node contains one member of a set and each tree represents one set. Moreover, nodes in each tree of this *disjoint-sets forest* point only to parents, and the root of each tree contains the representative of the corresponding set. Figure 21 illustrates this representation for the connected components of the graph of Figure 20.

The three typical disjoint-sets data structure operations are performed as follows. Given an object x, the algorithm DSET-MAKE(x) simply creates a tree

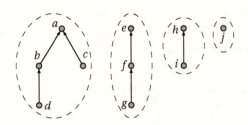

FIGURE 21. Disjoint sets $\{a, b, c, d\}$, $\{e, f, g\}$, $\{h, i\}$, and $\{j\}$.

with just one node containing x. The algorithm DSET-FIND(x) follows the parent pointers from the node x until it reaches the root of the tree, and then returns the root element as the representative of the x's home set. The algorithm DSET-UNION(x, y) merges two trees representing the sets S_x and S_y so that the root with smaller height becomes a child of the root with greater height; ties are broken arbitrarily. This policy of tree merging is called ***union by height***.

To implement a disjoint-sets forest with the union-by-height policy, with each node x we maintain an integer field h that holds the height of x. Each node x also contains a field *key*, which stores the corresponding set element, and a field p, which contains pointer to the parent node of x. The parent field of the root contains NIL. The following is pseudocode for the three disjoint-sets data structure operations.

```
DSET-MAKE(x)
    p(x) = NIL;
    h(x) = 0;
    return x;
```

```
DSET-FIND(x)
    while p(x) ≠ NIL do
        x = p(x);
    return x;
```

```
DSET-UNION(x, y)
    a = DSET-FIND(x);  b = DSET-FIND(y);
    if h(a) > h(b) then
        p(b) = a;
        return a;    ⟦ a is the union's root ⟧
    else
        p(a) = b;
        if h(a) = h(b) then h(b) = h(b) + 1;
        return b;    ⟦ b is the union's root ⟧
```

When a singleton set is created by DSET-MAKE, the initial height of the single node is 0. Each DSET-FIND operation does not change the heights at all. When we apply DSET-UNION operation to two trees, there are two cases depending on the heights of the two roots. If they have unequal heights, we make the root of larger height be the parent of the root of smaller height, but the heights themselves remain unchanged. On the other hand, if the roots have equal heights, we arbitrary choose one of the roots as the parent of the other, and increment its height.

The running time of the DSET-MAKE algorithm is clearly $O(1)$. Since the running times of the other two algorithms depend on the heights of trees, we will show that the heights can grow only logarithmically with the number of nodes in a tree. Therefore, when we follow a path from a node to the root of a tree to find the representative of the corresponding set, we take time proportional to at most the logarithm of the number of nodes in the tree. We derive the logarithmic bound by proving the following theorem.

THEOREM *A tree of height h in a disjoint-sets forest, formed initially by DSET-MAKE and then by (repeated) use of DSET-UNION, has at least 2^h nodes.*

PROOF: We use induction on h. The base case $h = 0$ is obvious, since a tree must be a single node and $2^0 = 1$. Suppose the theorem is true for some $h \geq 0$, and consider a tree T in a disjoint-sets forest of height $h+1$. At some time during the construction of T by the union-by-height operations, the height first reached the value $h+1$. The only way to get a tree of height $h+1$ for the first time is to make the root of some height-h tree T_1 be a child of the root of another height-h tree T_2. Thus, T is T_1 plus T_2 plus perhaps some other trees that were added later, as suggested in Figure 22.

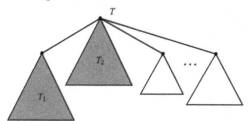

FIGURE 22. A tree of height $h+1$ formed by the union-by-height operations.

By the inductive hypothesis, the trees T_1 and T_2 have at least 2^h nodes each. Since T contains the nodes of T_1, T_2, and perhaps more nodes, T has at least $2^h + 2^h = 2^{h+1}$ nodes. This proves the inductive step, and so the statement of the theorem is true. ∎

We now know that if a tree in a disjoint-sets forest has n nodes and height h, it must be the case that $n \geq 2^h$. Taking logarithms of both sides, we can conclude that $h \leq \log n$. Consequently, when we follow any path from a tree node to its root, we take $O(\log n)$ time.

In other words, the running time of the algorithm DSET-FIND(x) is $O(\log n)$, where $n = |S_x|$. Likewise, the running time of the algorithm DSET-UNION(x, y) is

$$O(\log n_1) + O(\log n_2) + O(1) = O(\log n_1 + \log n_2),$$

where $n_1 = |S_x|$ and $n_2 = |S_y|$. Since

$$\log n_1 + \log n_2 \leq \log (n_1 + n_2) + \log (n_1 + n_2) = 2\log (n_1 + n_2),$$

by letting $n = |S_x \cup S_y| = |S_x| + |S_y| = n_1 + n_2$ we can express this running time more concisely as $O(\log n)$.

Finally, let us estimate the time it takes for the CONNECTED-COMPONENTS algorithm to process a graph G with n nodes and m edges. The first **for** loop clearly takes $O(n)$ time. Body of the second **for** loop is executed m times, and the body takes, as we have just determined, $O(\log n)$ time. Thus, the second **for** loop takes $O(m \log n)$ time, and so the total time of the CONNECTED-COMPONENTS algorithm is $O(n + m \log n)$.

Exercises

1. Consider the recurrence

$$N(h) = N(h-1) + N(h-2) + 1$$

for the size of a smallest AVL tree of height h. Prove that $N(h) > 2^{h/2-1}$, so $h < 2 \log N(h) + 2$. (*Hint:* Use $N(h-1) > N(h-2)$, hence $N(h) > 2N(h-2)$.)

2. Give a pseudocode for the procedures L-ROTATE(x) and RL-ROTATE(x) implementing the single left and double right-left rotations around the node x, respectively.

3. For the AVL tree in Figure 23 show the result of

    (a) inserting the key 14;
    (b) deleting the key 1.

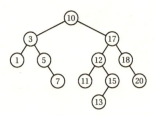

FIGURE 23

4. Show which rotation fixes an unbalanced node in the symmetric cases of those discussed for deletion in an AVL tree.

5. Give a pseudocode for the deletion algorithm in an AVL tree.

6. Illustrate the contents of a hash table after insertion of the keys 5, 28, 19, 15, 20, 33, 12, 17, 10, in this order. Suppose the size of the hash table is 9 and the hash function is $h(x) = x \pmod 9$. Furthermore, assume the collision resolution strategy used is separate chaining.

7. Illustrate the contents of a hash table after insertion of the keys 10, 22, 31, 4, 15, 28, 17, 88, 59, in this order. Suppose the size of the hash table is 11 and the hash function is $h(x) = x \pmod{11}$. Further assume the collision resolution

strategy used is open addressing with

(a) linear probing;
(b) quadratic probing;
(c) double hashing and secondary hash function $h'(x) = 1 + \left(x \pmod{10}\right)$.

8. For each of the two collision resolution strategies determine in the big-O notation the space complexity of the corresponding hash-table representation.

9. In the separate-chaining collision resolution strategy, instead of linked lists, we may use other data structures. For the case of binary search trees, as well as sorted linked lists, answer the following questions.

(a) What is the worst-case time for the insert and find operations?
(b) What is the expected time for the same operations?

10. **Internally chained hash tables.** Consider another implementation of the chained hash table in which all the information is stored logically within a single array. Thus the linked lists of the chained hash table are not separately chained, rather they are linked together using a "pointer" (cursor) field of each element in the array. (This implementation is sometimes referred to as the ***chained scatter table***.) More precisely, in this implementation each element of the hash table contains three fields. One field indicates whether an element is free or occupied. For a free element, the other two fields contain pointers; for an occupied element, the second field contains the key and the third field contains a pointer. The pointer fields are used in the following way:

   • All free elements in the table are chained into a doubly-linked list.
   • All keys that hash to the same location are chained into a singly-linked list. Therefore, an occupied element points to the next occupied element in the common list. (Unless, of course, it is the last element of the list, when the third field of the occupied element contains the NIL pointer.)

   For this implementation of the chained hash table describe in detail the operations of key insertion, key finding, and key removal, so that the expected time of all the operations is $O(1)$.

11. Let $h(x)$ be a strongly collision-resistant and one-way hash function. Informally, for the purpose of this exercise, this means that $h(x)$ never causes a collision and it is computationally difficult to determine its inverse function (i.e., it is difficult to calculate the input to the function given its output). Suppose that Alice and Bob are located in remote places and want to flip a fair coin by exchanging email messages or by communicating over the phone. Device an algorithm (protocol) that allows them to do that without cheating. Argue that both of the properties of $h(x)$ are essential by showing who might cheat if either of the requirements is missing.

12. Formally prove the properties of the binomial tree B_k listed on page 241.

13. Implement common operations for the priority queue data structure. That is, write down a pseudocode for the procedures
- PQUEUE-EMPTY(Q)
- PQUEUE-INSERT(Q,x)
- PQUEUE-DELETE(Q)
- PQUEUE-CHANGE(Q,x,y)
- PQUEUE-UNION(Q_1,Q_2)

if the underlying representation of the priority queue is given by

(a) binary heap;
(b) binomial heap.

14. Illustrate the execution of the CONNECTED-COMPONENTS algorithm for the weighted graph G in Figure 24, if the edges are considered in increasing order of their weight.

$$G$$

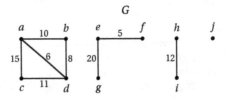

FIGURE 24

10

Weighted Graphs

As mentioned in Chapter 8, the breadth-first search strategy can be used to find a shortest path between two vertices in a connected graph. This approach makes sense in cases where each edge is as good as any other, but there are many situations where this approach is not appropriate.

For example, we might be using a graph to represent a computer network (such as the Internet), and we might be interested in finding the fastest way to route a data packet between two computers. In this case, it is probably not appropriate for all the edges to be equal to each other, because some connections in a computer network are typically much faster than others—say, some edges might represent slow phone-line connections while others might represent high-speed, fiber-optic connections.

Likewise, we may use a graph to represent the cost of dedicated data transmission lines between cities in which some company has branch offices. To build a WAN of its offices, the company might be interested in finding the smallest cost of renting the lines from the phone company. In this case, it is again not appropriate for all the edges to be equal to each other, because some intercity data lines will likely cost much more than others because the cost is proportional to intercity distances.

Thus, it is natural to consider graphs whose edges are not weighted equally. In this chapter we study (directed or undirected) graphs with a numeric label associated with each edge. Recall from Chapter 8 that the label is called the *weight* of an edge, and the corresponding graph is called the *weighted graph*. Edge weights can be any kind of numbers that represent some numerical quantity such as distance, time, connection cost, connection speed, and so on. Given an edge $e = (u, v)$, we denote its weight by $w(e) = w(u, v)$. An example of a weighted graph is shown in Figure 1.

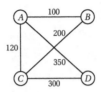

FIGURE 1. A weighted (undirected) graph.

1. Minimal spanning trees

Consider in more detail the example of a hypothetical company that wants to build a network of its offices. Figure 2 shows the graph that represents all possible leased lines between cities in which the company has offices. Vertices of the graph represent corresponding cities, edges of the graph represent the intercity data lines, and edge weights represent the cost of renting corresponding lines. We have assumed that not all connections are feasible due to geographical or physical constraints.

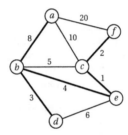

FIGURE 2. A weighted graph of cities and their connections.

The most economical network is shown with thick lines. We see that the solution has the following properties:

1. It has the minimal cost, where the cost of a solution is equal to the sum of its edge weights.
2. It connects indirectly all vertices, since it is required that any two offices can communicate with each other.
3. It does not contain any cycle. Otherwise, we could reduce the cost by discarding any edge in a cycle. Note that connectivity is still maintained.

Put it in more abstract terms, the optimal solution in this example represents a minimal spanning tree. A ***spanning tree*** of an undirected graph G is a tree formed (without a cycle) from the graph edges that connect all the vertices of G. More precisely, the edges of the spanning tree are a subset of the edges of G such that there is a path between any two vertices using only edges in the subset. Thus, the tree edges span the vertices of G, and so the tree is called the spanning tree.

Notice that the trees we are talking about here are not quite the same as the trees of Chapter 4. Rather, in this section, we look at the trees just as undirected graphs with no cycles. In other words, there is no node designated as the root, and there is no notion of children or the order among children. Such unrooted and unordered trees are called *free trees*.

It is easy to see that a spanning tree of G exists if and only if G is connected. The cost of a spanning tree is the sum of the weights of the edges in the tree. A **minimal spanning tree** (or **MST** for short) is a spanning tree of minimum cost. Thus, the sum of the edge weights in a minimal spanning tree is as small as of any spanning tree for a given graph. In short, a minimal spanning tree is a *tree* because it is acyclic, it is *spanning* because it covers every node, and it is *minimal* because of minimum cost.

There are many algorithms to find a minimal spanning tree of an undirected weighted graph. We consider two algorithms for solving the MST problem, which fall into the category of greedy algorithms. As we discussed in Chapter 6, the greedy method builds the solution by iteratively choosing objects that locally optimize some cost function at each step. The greedy choice of the MST algorithms is based on the following fact that we call the **MST property**.

LEMMA (MST PROPERTY) *Let $G = (V, E)$ be a weighted connected graph, and let V_1 and V_2 be a partition of the set of vertices V into two disjoint nonempty subsets. If $(u, v) \in E$ is an edge of minimum weight such that $u \in V_1$ and $v \in V_2$, then there is a minimal spanning tree of G that has (u, v) as one of its edges.*

PROOF: First note that there exists an edge with one endpoint in V_1 and the other in V_2, since V_1 and V_2 are two disjoint nonempty sets and G is connected. Let T be any minimal spanning tree for G. (Such a tree exists since G is connected.) If T includes an edge (u, v) of minimum weight from among of those with $u \in V_1$ and $v \in V_2$, there is nothing to prove. Otherwise, if we add such an edge (u, v) to T, we create exactly one cycle. (This is one of the properties of a tree.) In this cycle that contains the "bridge" edge (u, v) between V_1 and V_2, there must be another edge (u', v') in T such that $u' \in V_1$ and $v' \in V_2$. If not, the cycle could not close, that is, there would be no way to follow the cycle from, say, u and get back to u, without crossing the edge (u, v) for the second time. This situation is illustrated in Figure 3.

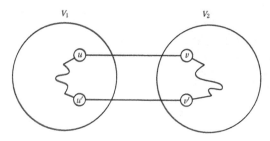

FIGURE 3. Creating a cycle in a minimal spanning tree.

If we now remove the edge (u', v') from T, we will break the cycle and obtain a new tree T' that spans G. Moreover, the cost of T' is no larger than the cost of T, since the weight of (u, v) is no larger than the weight of (u', v'). Therefore, the new spanning tree T' is a minimal spanning tree and includes the minimum-weight edge (u, v) between V_1 and V_2. ∎

The greedy MST algorithms "grow" a minimal spanning tree by selecting its edges one by one in some way that exploits the MST property. Kruskal's and Prim's algorithm are classical representatives of this approach.

1.1. Kruskal's algorithm. Kruskal's algorithm is a simple extension of the algorithm from Chapter 9 for finding connected components of a graph. Basically, the only difference from the CONNECTED-COMPONENTS algorithm is that we are required now to consider edges in the increasing order of their weights.

Given an undirected weighted graph $G = (V, E)$, in Kruskal's algorithm we start with a forest $T = (V, \emptyset)$ consisting only of the vertices of G and having no edges. Therefore, each vertex is initially in its own tree all by itself. The algorithm then examines edges from E in order of increasing weight. If an edge e connects two vertices in two different trees of the forest, then e is added to the set of edges of the forest T, and the two trees connected by e are merged into a single tree. On the other hand, if e connects two vertices that are already in the same tree of the forest T, then e is discarded since it would cause a cycle if added to the forest. Eventually, when the forest T becomes a single tree, i.e., when all vertices are in one component, T is a minimal spanning tree for G. Figure 4 illustrates how Kruskal's algorithm works on the input graph of Figure 2.

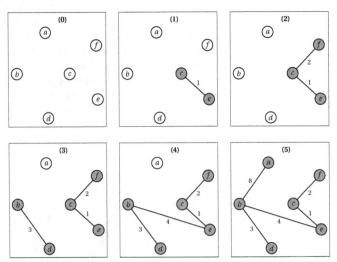

FIGURE 4. The execution of Kruskal's algorithm on the sample graph of Figure 2.

Correctness of Kruskal's algorithm follows from the MST property. To see this by the previous lemma, each time Kruskal's algorithm adds an edge $e = (u, v)$ to the forest T, we can define partitioning of the set of vertices V by letting V_1 be the component containing one end of e (say u) and letting V_2 contain the rest of the vertices in V. This clearly defines a disjoint partitioning of the vertices of

V. More importantly, since we are examining edges in increasing order of their weights, e must be a minimum-weight edge with one vertex in V_1 and the other vertex in V_2. Thus, Kruskal's algorithm always adds a valid MST edge.

Since Kruskal's algorithm maintains disjoint sets of nodes of the trees in the forest T that gets transformed into a minimal spanning tree, we use the disjoint-set data structure from Chapter 9. Also, in the algorithm it is preferable to have the graph represented by an array of edges with associated weights, rather than by the adjacency matrix or adjacency lists. (If not given initially, this representation is easily obtained from either of the other two in linear time.) The following is pseudocode for Kruskal's algorithm that finds a minimal spanning tree of an undirected weighted graph $G = (V, E)$ with n vertices and m edges.

MST-KRUSKAL(G)
   Sort edges in E by increasing weight;
   for each $v \in V$ **do**
      DSET-MAKE(v);
   $T = \emptyset$; ⟦ T contains edges of the MST ⟧
   while $|T| < n - 1$ **do**
      Let $e = (u, v)$ be the smallest edge not yet considered;
      $p = $ DSET-FIND(u);
      $q = $ DSET-FIND(v);
      if $p \neq q$ **then**
         DSET-UNION(p, q);
         $T = T \cup \{(u, v)\}$;
   return T;

We can estimate the running time of the MST-KRUSKAL algorithm as follows. To sort the edges of G we need $O(m \log m)$ time, if we use an efficient sorting algorithm. The **for** loop that initializes n disjoint sets takes an additional $O(n)$ time. The body of the **while** loop is executed at most m times, and each iteration takes $O(\log n)$ time provided that we implement the disjoint-set data structure operations as in Chapter 9. Thus, the **while** loop takes $O(m \log n)$ time. Since for a connected graph is $m \geq n - 1$, it follows that $n = O(m)$. Also, since $m \leq \binom{n}{2} = n(n-1)/2$ is always true, it follows that $\log m = O(\log n)$. Therefore, the total running time of the MST-KRUSKAL algorithm is

$$O(m \log n) + O(m) + O(m \log n) = O(m \log n).$$

A somewhat faster version of the MST-KRUSKAL algorithm would keep the edges of G in a heap, say binary, so that the initial sorting of the edges would not be necessary. Instead, we could initially make the heap out of the edges of G in time $O(m)$ using the HEAP-MAKE algorithm from Chapter 4. In the **while** loop, however, taking the smallest edge not yet considered would not be now a constant-time operation. Rather, it would take $O(\log m) = O(\log n)$ time using the HEAP-DELETE algorithm.

Although this would not improve the asymptotic running time of Kruskal's algorithm, it is particularly advantageous in cases in which a minimal spanning tree is found early when a considerable number of remaining edges do not need to be examined. In such cases, the original version of the MST-KRUSKAL algorithm wastes time sorting all those useless edges.

1.2. Prim's algorithm. In Kruskal's algorithm we greedily choose edges not caring much about their relation to previously chosen edges, except that they do not form a cycle. This approach produces a forest of trees that grow rather haphazardly. On the other hand, in Prim's algorithm we grow a minimal spanning tree "naturally," starting from an arbitrary root vertex. Then, in each iteration, we add a new edge and a new vertex to the tree already constructed, and the algorithm stops when all the vertices have been added.

More precisely, given an undirected weighted graph $G = (V, E)$, in Prim's algorithm we start with a tree $T = (\{s\}, \emptyset)$ consisting only of an arbitrary source vertex s of G and having no edges. Assume inductively that we have constructed a partial tree $T = (V_T, E_T)$ with $V_T \subset V$ and $E_T \subset E$, which is a minimal spanning tree for the induced subgraph of G on the vertices in V_T. Then, in the next iteration, we choose a minimum-weight edge (u, v) connecting some vertex u in T (i.e., $u \in V_T$) and some vertex v outside of T (i.e., $v \in V \setminus V_T$). The edge (u, v) and the vertex v are then added to T, and the process is repeated until all vertices of G are exhausted. Figure 5 illustrates one iteration of Prim's algorithm, where (u, v) is selected as a minimum-weight edge such that $u \in V_T$ and $v \in V \setminus V_T$.

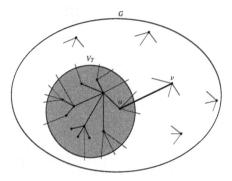

FIGURE 5. One iteration of Prim's algorithm.

Again, the correctness of Prim's algorithm follows from the MST property, because by always choosing the minimum-weight edge joining a vertex inside the partial T and some vertex outside the partial T, we are always assured of adding a valid edge to the minimal spanning tree. Figure 6 shows how Prim's algorithm works on the input graph of Figure 2 if started from the vertex a.

To implement Prim's method, we need to keep track of the set of edges E_T and the set of vertices V_T of the partial tree T. Also, the edges that lead out of the partial tree T must be kept in such a way so that we can quickly choose an edge of the smallest weight. So, in order to be able to efficiently manage the necessary information, we have chosen the following data structures.

- For the set E_T we use its linked-list representation so that adding an edge to E_T takes constant time.

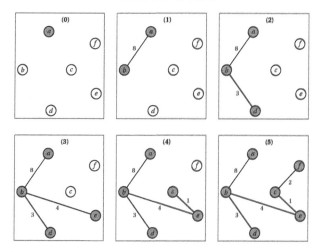

FIGURE 6. The execution of Prim's algorithm on the sample
graph of Figure 2 starting from the vertex a.

- For the set V_T we may use any representation of the set data structure
  in which testing for membership and adding a new element both take
  constant time. A simple bit-array or a hash table representation satisfies
  these requirements.
- To quickly decide which vertex is next to join the partial tree T, we maintain a priority queue of the edges of G. More precisely, each element of
  the priority queue has two fields, where one of them contains an edge
  $e = (u, v)$ and the other contains its weight $w(e)$, but the edge weight is
  the key of each element.

Further assuming the adjacency-lists representation of the graph $G = (V, E)$
with n vertices and m edges, we can now give a pseudocode for Prim's algorithm
that finds a minimal spanning tree of G.

```
MST-PRIM(G)
    s = any v ∈ V;        [[ pick any vertex of G ]]
    V_T = {s}; E_T = ∅;   [[ initialize partial tree ]]
    Q = ∅;                [[ Q is empty priority queue ]]
    for each x ∈ V such that e = (s, x) ∈ E do
        PQUEUE-INSERT(Q, (w(e), (s, x)));
    while |V_T| < n do    [[ or while Q ≠ ∅ ]]
        (u, v) = PQUEUE-DELETE(Q);
        E_T = E_T ∪ {(u, v)};
        V_T = V_T ∪ {v};
        for each x ∈ V \ V_T such that e = (v, x) ∈ E do
            PQUEUE-INSERT(Q, (w(e), (v, x)));
    return T = (V_T, E_T);
```

The running time of the MST-PRIM algorithm is dominated by two loops: the first **for** loop that initializes the priority queue Q, and the second **while** loop that constructs a minimal spanning tree. The **for** loop takes $O(n \log n)$ time, since it executes at most n insertion operations on a priority queue of size at most n. The **while** loop executes at most m removal and insertion operations on the priority queue of size at most m, besides some constant-time work in each iteration. Thus, the **while** loop takes $O(m \log m) = O(m \log n)$ time. The total running time of the MST-PRIM algorithm is therefore

$$O(n \log n) + O(m \log n) = O(m \log n).$$

We see that Prim's and Kruskal's algorithms have asymptotically the same running time. However, if we use the weighted adjacency-matrix representation for the graph G, we can write another version of Prim's algorithm that takes $O(n^2)$ time independently of the number of edges. One simple way to do this is to maintain two arrays. One array c for each vertex in $V \setminus V_T$ gives its currently closest vertex in V_T. That is, for every $x \in V \setminus V_T$, $c(x) = y \in V_T$ such that $w(x, y)$ is the minimum of $w(x, z)$ over all vertices z in V_T. The other array d for each vertex in $V \setminus V_T$ gives the actual weight of its smallest edge to V_T. That is, for every $x \in V \setminus V_T$, $d(x) = w(x, c(x))$.

In each iteration of this version of Prim's algorithm, we first can scan the array d to find the minimum-weight edge between $V \setminus V_T$ and V_T, and then add the corresponding vertex, say $v \in V \setminus V_T$, to V_T, as well as the edge $(v, c(v))$ to E_T. After that we update the arrays d and c, taking into account the fact that v has been added to V_T. We do this by checking for every vertex x in the new $V \setminus V_T$ whether $d(x) > w(x, v)$, and if so replace the old $d(x)$ with $w(x, v)$ and the old $c(x)$ with v. Actual details are left for the interested reader (see Exercise 6). Since finding the minimum value of the array d and updating the arrays d and c each takes $O(n)$ time, it is clear that one iteration of this version of Prim's algorithm cumulatively takes $O(n)$ time. As we need at most n iterations to obtain a spanning tree of the graph G in this way, the running time of this version of Prim's algorithm is therefore $O(n^2)$.

Thus, in case the graph G is dense (i.e., $m = \Theta(n^2)$), this second version is better than Kruskal's algorithm or the first version of Prim's algorithm. However, if the graph G is sparse (i.e., $m = \Theta(n)$), the second version is inferior to Kruskal's algorithm or the first version of Prim's algorithm. Finally, if $m = \Theta(n^2 / \log n)$, all three algorithms take roughly the same time, and the choice of a specific algorithm may depend on some other requirements.

2. Single-source shortest paths

Since in this and the next section we study path-finding problems in a weighted graph G, we switch our terminology and talk about edges as having *length* instead of less natural *weight* in this context. (Of course, length or weight may represent something totally different, like time.) Also, if $e = (u, v)$ is an edge of G, its length is denoted $\ell(e)$ or $\ell(u, v)$ instead of $w(e)$ or $w(u, v)$. We define the **length of a path** as the sum of the lengths of the edges along the path. That is, if a path π consists of vertices $v_1, v_2, \ldots, v_k$, then the length of π, denoted $\ell(\pi)$, is defined as

$$\ell(\pi) = \sum_{i=1}^{k-1} \ell(v_i, v_{i+1}).$$

The **shortest-path distance** (also called the **shortest distance** or the **minimum distance**) from a vertex u to a vertex v in G, denoted $\delta(u, v)$, is the length of a minimum-length path from u to v, if such a path exists. We use the convention that $\delta(u, v) = +\infty$ if there is no path at all from u to v in G, and $\delta(v, v) = 0$ for any vertex v. Notice that even if there is a path from u to v in G, the shortest-path distance from u to v may not be defined well, if there is a cycle in G whose total length is negative. A path of length $\delta(u, v)$ from u to v is said to be a **shortest path** from u to v.

Now consider the problem of finding the shortest path from some vertex s to each other vertex in a weighted graph G. This problem is often called the **single-source shortest paths** problem.[1] To solve this problem we first discuss a simple, yet common, case when all the edge lengths in G are nonnegative (that is, $\ell(e) \geq 0$ for each edge e of G). This implies that there are no negative-length cycles in G, and so the shortest-path distances are well-defined.

2.1. Dijkstra's algorithm. Recall that computing the shortest paths in the special case when all edge lengths are the same can be solved using breadth-first search. The general single-source shortest paths problem can be solved by a greedy method that performs a "weighted" breadth-first search starting at a given vertex. This approach results in an algorithm known as **Dijkstra's algorithm**. To simplify the description of Dijkstra's algorithm, we further assume that the input graph G is undirected and leave it as an easy exercise to the reader to adapt the algorithm presented here for a directed weighted graph.

In the application of the greedy method in Dijkstra's algorithm, the cost function we are trying to optimize—the shortest-path distance, is also the function that we are trying to compute. Although this may seem like a circular reasoning, we can avoid the problem by computing actually an approximation to the distance function, which in the end will be equal to the true shortest distance.

The crux of Dijkstra's algorithm is that we discover the shortest path from a given vertex s to other vertices by finding the closest vertices first. In other words, we determine the shortest paths in the increasing order of these shortest paths. The algorithm works by maintaining a set S of *solved* vertices whose minimum distance from the source vertex s is known. For each vertex v of G, the algorithm also maintains a value $d(v)$ that records the length of the shortest path we have found so far from s to v. More precisely, for a solved vertex v, $d(v)$ is the overall shortest distance from s to v; for an unsolved vertex v, $d(v)$ is the length of the shortest *special* path from s to v. The **special path** is a path that starts at s, goes only through solved vertices, and then at the last edge jumps out of the solved region to v. This situation is illustrated in Figure 7.

---

[1]Seemingly simpler problem of finding a shortest path from a given source vertex to one particular destination vertex appears just as hard in general as the single-source shortest paths problem.

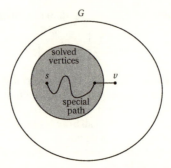

FIGURE 7. Solved vertices and a special path from s to v of Dijkstra's algorithm.

Initially, the set of solved vertices S contains only the source vertex s. Also, $d(s) = 0$ and $d(v) = +\infty$ for each $v \neq s$. In each iteration of the algorithm, we select an unsolved vertex v with the smallest $d(v)$ value of any unsolved vertices. Then we "solve" v by adding it to the set S. Moreover, to account for the fact that v is now solved and that there may be shorter special paths to get to its neighbors, we adjust the value $d(u)$ for each vertex u that is unsolved and adjacent to v. That is, for any such vertex u, a new special path to u follows a shortest path from s to v and ends with the edge (v, u). The new special path to u has the length $d(v) + \ell(u, v)$, and if this path is shorter than the existing special path to u of length $d(u)$, we should update $d(u)$ to the value $d(v) + \ell(u, v)$. Taking by convention $\ell(u, v) = +\infty$ if (u, v) is not an edge of G, this adjustment is therefore done as

$$\textbf{if } d(v) + \ell(u, v) < d(u) \textbf{ then}$$
$$d(u) = d(v) + \ell(u, v);$$

for each unsolved neighbor u of v. Notice that if the newly discovered special path to u is no better than the old special path, then we do not change $d(u)$.

This path-length update operation, which takes an old estimate and checks if it can be improved to get closer to its true value, is known as a ***relaxation***. A metaphor for why we call this a relaxation comes from a spring that is stretched out and then "relaxed" back to its true resting shape. We write the relaxation operation more succinctly as

$$d(u) = \min\{d(u), d(v) + \ell(u, v)\},$$

and an example is shown in Figure 8.

Observe that whenever we set $d(u)$ to a finite value, there is always evidence of a path of that length. Therefore, $d(u) \geq \delta(s, u)$ for every vertex u.

The following is a pseudocode for Dijkstra's algorithm that computes the length $d(v)$ of the shortest path from a source vertex s to each vertex v of a graph $G = (V, E)$.

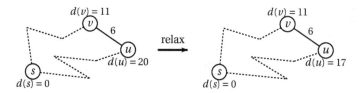

FIGURE 8. Relaxation.

```
SSSP-DIJKSTRA(G, s)
    S = {s};
    d(s) = 0;
    for each v ∈ V such that v ≠ s do
        if (s, v) ∈ E then d(v) = ℓ(s, v); else d(v) = +∞;
    while S ≠ V do
        Choose v ∈ V \ S such that d(v) is a minimum;
        S = S ∪ {v};
        for each u ∈ V \ S such that (u, v) ∈ E do
            d(u) = min{d(u), d(v) + ℓ(u, v)};     ⟦ relaxation ⟧
    return d;
```

Figure 9 illustrates iterations of the SSSP-DIJKSTRA algorithm performed on a sample graph.

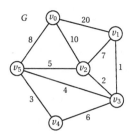

step	v	S	$d(v_0)$	$d(v_1)$	$d(v_2)$	$d(v_3)$	$d(v_4)$	$d(v_5)$
0	–	$\{v_0\}$	0	20	10	$+\infty$	$+\infty$	8
1	v_5	$\{v_0, v_5\}$	0	20	10	12	11	8
2	v_2	$\{v_0, v_2, v_5\}$	0	17	10	12	11	8
3	v_4	$\{v_0, v_2, v_4, v_5\}$	0	17	10	12	11	8
4	v_3	$\{v_0, v_2, v_3, v_4, v_5\}$	0	13	10	12	11	8
5	v_1	$\{v_0, v_1, v_2, v_3, v_4, v_5\}$	0	13	10	12	11	8

FIGURE 9. The execution of Dijkstra's algorithm on the sample graph G for $s = v_0$.

It is probably not immediately clear why Dijkstra's algorithm correctly finds the shortest path from the source vertex s to every other vertex in the graph.

Dijkstra's algorithm is an example in which the greedy choice in each iteration turns out to be the best choice overall. In this case, the locally best approach is to find the distance to the vertex v that lies outside S but has the shortest special path. The greedy method works correctly, i.e., there cannot be a shorter nonspecial path from s to v, since the edge lengths are nonnegative. This is formally established by the following theorem.

THEOREM *In the* SSSP-DIJKSTRA *algorithm, the invariant of the* **while** *loop is:*

(a) *If $v \in S$, then $d(v)$ is the length of a shortest path from s to v, and any shortest path includes only vertices from S;*

(b) *If $v \notin S$, then $d(v)$ is the length of a shortest special path from s to v.*

PROOF: It is clear that these two conditions hold at the outset of the **while** loop in the SSSP-DIJKSTRA algorithm. Namely, the set of solved vertices S contains initially only the source vertex s. Then, (a) the shortest path from s to itself has length 0; and (b) a special path from s to any other vertex v, i.e., a path wholly within S except v, follows the single edge (s, v), if it exists. Thus, the d values are correctly initialized before the **while** loop.

Suppose inductively that (a) and (b) hold after $k - 1 \geq 0$ iterations. Then $|S| = k$, because each iteration adds exactly one vertex to S. Call this set the old S. Consider the k-th iteration and let v be the vertex solved in the k-th iteration. We need to show that (a) and (b) hold for the new S, which is the old S plus the solved vertex v.

To show condition (a) for the new S, because the condition holds for the old S by the inductive hypothesis, we only need to show that $d(v)$ is the minimum distance from s to the k-th solved vertex v. Now, since v was not in the old S, by part (b) of the inductive hypothesis, $d(v)$ is the length of a shortest special path through the old S. Therefore,

(19) $$d(v) \geq \delta(s, v).$$

To show that $d(v)$ is actually equal to the minimum distance from s to v, consider any path π from s to v. As suggested in Figure 10, this path must leave for the first time the old S to go to some vertex u, and then perhaps wander in and out of the old S several times before ultimately arriving at v.

If we denote the path from s to u by π_1, and the path from u to v by π_2, then $\ell(\pi) = \ell(\pi_1) + \ell(\pi_2) \geq \ell(\pi_1)$, because there are no negative-length edges in the graph.[2] Also, $\ell(\pi_1) \geq d(u)$ by part (b) of the inductive hypothesis, because the path π_1 is wholly included except u in the old S. Furthermore, $d(u) \geq d(v)$ by the choice of v in the **while** loop. Thus,

$$\ell(\pi) = \ell(\pi_1) + \ell(\pi_2) \geq \ell(\pi_1) \geq d(u) \geq d(v),$$

which implies $\delta(s, v) \geq d(v)$. Together with (19) we can therefore conclude $d(v) = \delta(s, v)$. Moreover, any shortest path from s to v includes only vertices from the

---

[2]The fact that all edge lengths are nonnegative is vital; the argument wouldn't work without it, and Dijkstra's algorithm would be actually incorrect.

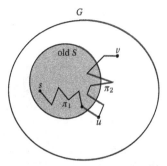

FIGURE 10. Arbitrary path from s to v whose first part to u is special.

new S, since a shortest path is a special path through the old S. This completely proves part (a) of the invariant.

To show (b) for the new S, consider any vertex u outside the new S. On a shortest special path from s to u, there must be some vertex that immediately precedes u. This vertex could be either the vertex v solved in the k-th iteration or some other vertex w, as suggested in Figure 11.

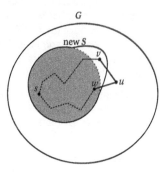

FIGURE 11. The shortest special path from s to u.

If the predecessor of u is v, then the length of the shortest special path is $d(v) + \ell(u, v)$. Alternatively, if the predecessor of u is some $w \neq v$, then v cannot appear on a shortest special path from s to w. This follows from part (a) of the inductive hypothesis, since w is in the old S and any shortest path to w includes only vertices from the old S. Therefore, by part (b) of the inductive hypothesis, the length of the shortest special path in the second case is $d(u)$.

If we recall that the algorithm performs the relaxation operation for each neighbor u of v that lies outside the new S, we see that both of the previous cases are covered in the algorithm. This completely proves part (b) of the invariant. ∎

To finish the proof that the SSSP-DIJKSTRA algorithm works correctly, we observe that after execution of the **while** loop we have $S = V$. This means that all vertices are solved at the end, that is, for any $v \in V$, $d(v)$ is the length of a shortest path from s to v.

2.1.1. *Analysis of the* SSSP-DIJKSTRA *algorithm.* If we use the weighted adjacency-matrix representation for the input graph G, the linked-list representation for the sets S and $V \smallsetminus S$, and the array representation for the shortest-path distances d, then it is easy to determine the running time of SSSP-DIJKSTRA algorithm. The initialization phase clearly takes $O(n)$ time. In the body of the **while** loop, choosing v that minimizes d takes $O(n)$ time, adding v to S (and removing it from $V \setminus S$) takes constant time, and the relaxation procedure on each edge connecting v and its neighbor u outside S takes $O(n)$ time. Thus, the total running time of the body is $O(n)$. Since the body of the **while** loop is executed $n-1$ times, the main loop takes $O(n^2)$ time. The total time of this version of the SSSP-DIJKSTRA algorithm is therefore $O(n^2)$.

If we use the adjacency-lists representation for the graph G, we can apply the idea similar to that of Prim's algorithm. This time we use a priority queue of the vertices of G ordered by their d values. Thus, if we use the binary min-heap implementation of the priority queue, initialization of the heap takes $O(n)$ time. Selecting a vertex v that minimizes the d values now consists of removing the root from the heap, which takes $O(\log n)$ time. The relaxation procedure now consists of checking each edge connecting v and its neighbor u outside S to see whether $d(v) + \ell(u, v) < d(u)$. If so, we must modify $d(u)$ and bubble u up the heap, which again takes $O(\log n)$ time. This happens at most once for each edge of the graph. Therefore, we have to remove the root from the heap $n-1$ times, and to perform at most m bubble-up operations, giving a total running time $O((n + m)\log n)$. If the graph is connected, then $m \geq n - 1$, and so the running time of this version of the SSSP-DIJKSTRA algorithm is $O(m \log n)$.

The adjacency-matrix representation for the graph G is therefore preferable if the graph is dense, while it is better to use the adjacency-lists representation and a priority queue if the graph is sparse.

2.2. Single-source shortest paths in directed acyclic graphs. As we already noted, Dijkstra's algorithm can be easily adapted to work for directed graphs. However, we can solve the single-source shortest paths problem faster if the digraph has no cycles, that is, if it is a weighted directed acyclic graph (dag).

Recall from Chapter 8 that a topological sort of a dag $G = (V, E)$ is a listing of its vertices $v_1, v_2, \ldots, v_n$ such that if (v_i, v_j) is an edge of G, then $i < j$. You may also recall that topological sort of G involves performing depth-first search on G and listing each vertex during the search in the reverse order of its finishing time. The time needed for such a topological sort of a dag with n vertices and m edges is $O(n + m)$.

Given a topological ordering of a weighted dag, we can compute all shortest paths from a given vertex s in $O(n + m)$ time. The idea is to visit the vertices of G according to the topological ordering, relaxing the outgoing edges with each visit. This approach is implemented in the algorithm SSSP-DAG(G, s) below that computes the array d such that $d(v)$ is the minimum distance from the source vertex s to each vertex $v \in V$.

```
SSSP-DAG (G, s)
  ⟦ Initialize array d ⟧
  d(s) = 0;
  for each v ∈ V such that v ≠ s do
    d(v) = +∞;
  ⟦ Get list L of vertices in topological order ⟧
  TOP-SORT(G);
  ⟦ Skip vertices in L up to s ⟧
  repeat
    v = LIST-HEAD-DELETE(L);
  until v = s;
  ⟦ Relax outgoing edges of the remaining vertices in L ⟧
  while L ≠ ∅ do
    for each u ∈ V such that v → u ∈ E do
      d(u) = min{d(u), d(v) + l(v, u)};      ⟦ relaxation ⟧
    v = LIST-HEAD-DELETE(L);
  return d;
```

Figure 12 illustrates the execution of the SSSP-DAG algorithm on a sample graph.

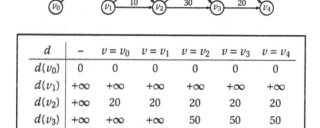

d	$-$	$v = v_0$	$v = v_1$	$v = v_2$	$v = v_3$	$v = v_4$
$d(v_0)$	0	0	0	0	0	0
$d(v_1)$	$+\infty$	$+\infty$	$+\infty$	$+\infty$	$+\infty$	$+\infty$
$d(v_2)$	$+\infty$	20	20	20	20	20
$d(v_3)$	$+\infty$	$+\infty$	$+\infty$	50	50	50
$d(v_4)$	$+\infty$	$+\infty$	$+\infty$	30	30	30

FIGURE 12. The execution of the SSSP-DAG algorithm on the dag G for $s = v_0$.

Correctness of the SSSP-DAG algorithm follows from the following theorem.

THEOREM *In the* SSSP-DAG *algorithm, the invariant of the* **while** *loop is: for each vertex v not on the list L, $d(v)$ is the minimum distance from s to v.*

PROOF: We use induction on the number of iterations k of the **while** loop. We first note that before the very beginning of the **while** loop, the list L (which may be empty) contains only vertices to the right of s in a topological ordering. (The **repeat** loop searches for s on the list removing all vertices up to and including s.) Thus, if the **while** loop is not executed at all ($k = 0$), each vertex not on the list L is before s in the topological ordering, and so it is not reachable from

s; that is, its minimum distance from s is $+\infty$. Since the algorithm initially sets $d(s) = 0$ and $d(v) = +\infty$ for each $v \neq s$, the invariant is true in the base case of induction.

Suppose now that the invariant is true after $k \geq 0$ iterations and consider the $(k+1)$-st iteration. Each iteration of the **while** loop removes one vertex from the list L according to the topological ordering. For $i = 1, \ldots, k, k+1$, let v_i be the vertex removed from L in the i-th iteration. Observe that the $(k+1)$-st iteration relaxes only outgoing edges from the vertex v_k to the vertices that come to the right of v_k in the topological ordering. Thus, for each vertex $v \in \{v_1, \ldots, v_k\}$ not on the list L, $d(v)$ is not changed and so remains the minimum distance from s to v by the inductive hypothesis. It is left to show that $d(v_{k+1})$ is the minimum distance from s to v_{k+1}.

Consider the vertex v_{k+1}, and first suppose that v_{k+1} is not reachable from s. Since initially $d(v_{k+1}) = +\infty$, if $d(v_{k+1}) < +\infty$ after the $k+1$ iterations, then in one of these iterations the value $d(v_{k+1})$ is lowered. This implies that there would be a vertex v_j with $j \leq k$ such that $d(v_j) < +\infty$ and $v_j \to v_{k+1} \in E$. But this is impossible, since then by the inductive hypothesis v_j would be reachable from s, and so would be v_{k+1}. Therefore, $d(v_{k+1}) = +\infty$ never gets lowered in the $k+1$ iterations.

Suppose next the other case when v_{k+1} is reachable from s. Then, there is a shortest path from s to v_{k+1}. Let v_j be the penultimate vertex on a shortest path from s to v_{k+1}. Since the vertices are numbered according to a topological ordering, we have that $j \leq k$. Thus, $d(v_j)$ is correct by the inductive hypothesis, and in the $(j+1)$-st iteration we relax each outgoing edge from v_j, including the edge $v_j \to v_{k+1}$ on the shortest path from s to v_{k+1}. Therefore, $d(v_{k+1})$ is the minimum distance from s to v_{k+1}. ∎

Assuming a dag G with n vertices and m edges is represented by adjacency lists, we can determine the running time of the SSSP-DAG algorithm by estimating the running times of its three main steps: initialization of the array d, topological sorting of G, and sequential processing of all vertices of G according to their topological order. The first step clearly takes $O(n)$ time. The second step, as we recalled, requires $O(n+m)$ time. If $v_1, v_2, \ldots, v_n$ is a topological ordering of the vertices of G, the third step in the worst-case takes $\sum_{i=1}^{n} O(1+\alpha_i)$ time, where α_i is outdegree of v_i and the additional 1 accounts for constant time needed to process each vertex. Thus, the last step takes time equal to

$$\sum_{i=1}^{n} O(1+\alpha_i) = O\left(\sum_{i=1}^{n}(1+\alpha_i)\right) = O\left(n + \sum_{i=1}^{n}\alpha_i\right) = O(n+m).$$

The total running time of the SSSP-DAG algorithm is therefore $O(n+m)$.

3. All-pairs shortest paths

Suppose we wish to compute the minimum distance between every pair of vertices in a weighted graph G with n vertices and m edges. This problem is called the ***all-pairs shortest paths*** problem. Of course, if G has no negative-length edges, then we could run Dijkstra's algorithm in turn from each vertex of G. This approach would take $O(n(n+m)\log n)$ time, assuming G is represented by adjacency lists. In the worst case when the graph is dense ($m = \Theta(n^2)$), this bound could be as large as $O(n^3 \log n)$. In the other extreme when the graph is sparse ($m = \Theta(n)$), this bound could be as small as $O(n^2 \log n)$.

In this section, we consider an algorithm for solving the all-pairs shortest paths problem in $O(n^3)$ time, even if the graph contains negative-length edges (but not negative-length cycles). This algorithm is called **Floyd's algorithm** and it works with an adjacency matrix, rather than adjacency lists.

3.1. Floyd's algorithm. The general all-pairs shortest paths problem can be solved by a dynamic-programming algorithm due to Floyd. Although it solves an apparently harder problem, it is conceptually simpler then the Dijkstra's greedy approach.

Let $G = (V, E)$ be a weighted graph, and let the vertices of G be arbitrarily numbered so that $V = \{v_1, v_2, \ldots, v_n\}$. Floyd's algorithm is based on the idea to split the computation of the shortest path between two vertices v_i and v_j into computation of two shortest paths, one from v_i to v_k and the other from v_k to v_j, where v_k is an intermediate vertex on a path from v_i to v_j. More precisely, we compute the minimum distance from v_i to v_j using only intermediate vertices in the set $V_k = \{v_1, v_2, \ldots, v_k\}$, for each $k = 1, 2, \ldots, n$. If this computation is straightforward for the sequence of sets $V_1, V_2, \ldots, V_n = V$, then we will be able to ultimately compute the desired solution relatively efficiently.

For convenience, let $V_0 = \emptyset$. The minimum distance $D(i, j, k)$ from v_i to v_j using intermediate vertices on the paths between them only from the set V_k is computed for $k = 0, 1, 2, \ldots, n$ as follows. Initially, for $k = 0$,

$$D(i, j, 0) = \begin{cases} 0, & \text{if } i = j \\ \ell(v_i, v_j), & \text{if } (v_i, v_j) \in E \\ +\infty, & \text{otherwise} \end{cases}$$

and for $k = 1, 2, \ldots, n$,

$$D(i, j, k) = \min \{D(i, j, k-1), \ D(i, k, k-1) + D(k, j, k-1)\}.$$

In words, the shortest path going from v_i to v_j and passing only through the vertices in V_k is equal to the shorter of two possible shortest paths.

The first possibility is simply the shortest path from v_i to v_j which actually never passes through vertex v_k. This gives $D(i, j, k-1)$, the distance of the short-est path from v_i to v_j using only vertices in V_{k-1}.

The second possibility is the shortest path from v_i to v_j which does pass (once) through vertex v_k. This gives $D(i, k, k-1) + D(k, j, k-1)$, the sum of the distance of the shortest path from v_i to v_k using vertices in V_{k-1} and the distance of the shortest path from v_k to v_j using vertices in V_{k-1}.

Moreover, there is no other shorter path from v_i to v_j using vertices in V_k than these two. If there was such a shorter path and it excluded v_k, then it would violate the definition of $D(i, j, k-1)$, and if there was such a shorter path and it included v_k, then it would violate the definition of $D(i, k, k-1)$ or $D(k, j, k-1)$. In fact, this argument is still valid even if there are negative-length edges in G, just as long as there are no negative-length cycles in the graph.

Figure 13 illustrates the computation of $D(i, j, k)$ according to the previous formula. In that figure, the end points v_i and v_j may be in or out of the set V_k represented by the shaded area.

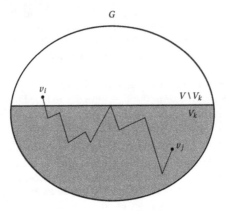

FIGURE 13. Computation of $D(i, j, k)$ in Floyd's algorithm.

As any dynamic-programming algorithm, Floyd's algorithm represents an easy implementation of the recursive formula for computation of the minimum distance for every pair of vertices of the graph G with n vertices and m edges.

```
APSP-FLOYD (G)
    for i = 1 to n do
        for j = 1 to n do
            if i = j then D(i, j, 0) = 0;
            else if (v_i, v_j) ∈ E then D(i, j, 0) = ℓ(v_i, v_j);
            else D(i, j, 0) = +∞;
    for k = 1 to n do
        for i = 1 to n do
            for j = 1 to n do
                D(i, j, k) = min {D(i, j, k − 1), D(i, k, k − 1) + D(k, j, k − 1)};
    return D;
```

Note that Floyd's algorithm computes the three-dimensional matrix $D(i, j, k)$. However, we can regard it as a computation of the sequence of matrices $D_k(i, j)$ for $k = 1, 2, \ldots, n$. The matrix D_k is computed in the k-th iteration of the outermost loop of the triply nested **for** loops.

In addition, in the case when each edge has an associated nonnegative length, values in the k-th row and the k-th column of D_k are the same as those of D_{k-1}, since $D_{k-1}(k, k)$ is always zero. More precisely, for every (fixed) $k = 1, 2, \ldots, n$, we have $D_k(i, k) = D_{k-1}(i, k)$ for each $i = 1, 2, \ldots, n$ and $D_k(k, j) = D_{k-1}(k, j)$ for each $j = 1, 2, \ldots, n$. Since no value with either index equal to k changes during the k-th iteration, it is possible to compute the new value $D_k(i, j)$ as the minimum of the old value $D_{k-1}(i, j)$ and the sum of the new values $D_k(i, k)$ and $D_k(k, j)$, rather than the sum of the old values $D_{k-1}(i, k)$ and $D_{k-1}(k, j)$.

This means that we can get by with only one copy of the two-dimensional $n \times n$ matrix D, although at first sight a three-dimensional $n \times n \times n$ matrix seems necessary. This version of Floyd's algorithm is an easy modification of the previous version, because we only need to get rid of the third index of D. Details are left as an exercise for the reader.

Figure 14 shows the sequence of distance matrices D_k obtained when the modified APSP-FLOYD algorithm is run on the sample graph of Figure 2.

D_0	a	b	c	d	e	f
a	0	8	10	$+\infty$	$+\infty$	20
b	8	0	5	3	4	$+\infty$
c	10	5	0	$+\infty$	1	2
d	$+\infty$	3	$+\infty$	0	6	$+\infty$
e	$+\infty$	4	1	6	0	$+\infty$
f	20	$+\infty$	2	$+\infty$	$+\infty$	0

D_1	a	b	c	d	e	f
a	0	8	10	$+\infty$	$+\infty$	20
b	8	0	5	3	4	28
c	10	5	0	$+\infty$	1	2
d	$+\infty$	3	$+\infty$	0	6	$+\infty$
e	$+\infty$	4	1	6	0	$+\infty$
f	20	28	2	$+\infty$	$+\infty$	0

D_2	a	b	c	d	e	f
a	0	8	10	11	12	20
b	8	0	5	3	4	28
c	10	5	0	8	1	2
d	11	3	8	0	6	31
e	12	4	1	6	0	32
f	20	28	2	31	32	0

D_3	a	b	c	d	e	f
a	0	8	10	11	11	12
b	8	0	5	3	4	7
c	10	5	0	8	1	2
d	11	3	8	0	6	10
e	11	4	1	6	0	3
f	12	7	2	10	3	0

D_4	a	b	c	d	e	f
a	0	8	10	11	11	12
b	8	0	5	3	4	7
c	10	5	0	8	1	2
d	11	3	8	0	6	10
e	11	4	1	6	0	3
f	12	7	2	10	3	0

D_5	a	b	c	d	e	f
a	0	8	10	11	11	12
b	8	0	5	3	4	7
c	10	5	0	7	1	2
d	11	3	7	0	6	9
e	11	4	1	6	0	3
f	12	7	2	9	3	0

D_6	a	b	c	d	e	f
a	0	8	10	11	11	12
b	8	0	5	3	4	7
c	10	5	0	7	1	2
d	11	3	7	0	6	9
e	11	4	1	6	0	3
f	12	7	2	9	3	0

FIGURE 14. The execution of the modified APSP-FLOYD algorithm on the sample graph of Figure 2.

If the graph G is represented by the weighted adjacency matrix, the running time of the APSP-FLOYD algorithm is clearly $O(n^3)$. This simply follows from the fact that the time complexity is dominated by the running time of the triply nested **for** loops.

To prove correctness of the algorithm, one can easily show by induction on k that after k iterations of the outermost loop of the triply nested **for** loops,

$D(i, j, k)$ contains the length of the shortest path from vertex v_i to vertex v_j that passes only through vertices from the set V_k.

Exercises

1. Determine all spanning trees for the graph in Figure 15.

FIGURE 15

2. Show that a graph is connected if and only if it has a spanning tree.

3. Prove that any minimal spanning tree contains a minimum-weight edge of a weighted graph.

4. For the weighted graph in Figure 16 illustrate the execution of

   (a) Kruskal's algorithm;
   (b) Prim's algorithm starting from the vertex b;
   (c) Dijkstra's algorithm if the source vertex is a;
   (d) Floyd's algorithm.

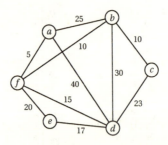

FIGURE 16

5. Give a detailed pseudocode for the second version of Kruskal's algorithm that maintains a (binary or binomial) heap of the edges of a graph.

6. Give a detailed pseudocode for the second version of Prim's algorithm that uses the weighted adjacency-matrix representation of a graph with n vertices and takes $O(n^2)$ time.

7. Give a more detailed pseudocode for the SSSP-DIJKSTRA algorithm if an input graph is represented by

   (a) the adjacency matrix;

(b) the adjacency lists.

8. Give a counterexample of a graph with negative edges for which Dijkstra's algorithm works incorrectly.

9. Give a pseudocode for the version of Floyd's algorithm that uses only one copy of the two-dimensional matrix D.

10. Formally prove correctness of the APSP-FLOYD algorithm.

11. **Warshall's algorithm.** Given an ordinary (non-weighted) directed graph by the adjacency matrix A, we want to compute another matrix B such that $B(i, j) = 1$ if there is a *path* from the vertex v_i to the vertex v_j, and $B(i, j) = 0$ otherwise. Modify Floyd's algorithm to compute the matrix B and determine time complexity of the modified algorithm.

11

Randomized Algorithms

In this chapter we discuss randomized algorithms that make random choices in unit time as they process the input. To make a random choice, these algorithms are allowed to sample a random element from some combinatorial structure, such as a set or a graph. Typically, these algorithms may choose a random element from a given set, pick a random edge from a graph with specified properties, or generate a random permutation among all possible permutations of some length. Besides, these operations are considered basic in the sense that they can be performed in one time step.

Generally speaking, we call an algorithm *randomized* if its behavior is determined not only by its input but also by outcomes of some random decisions. These outcomes are usually simulated by values produced by a deterministic algorithm that returns numbers that are statistically indistinguishable from pure random numbers. Such numbers that "look" random are called pseudorandom numbers, and we assume that each one can be generated in one unit of time.

It is important to note that the randomized algorithms is not the same concept as the probabilistic analysis of algorithms. In probabilistic analysis, deterministic algorithms are analyzed assuming random input. For example, in sorting one might analyze the deterministic quicksort algorithm, where the choice of the pivot element is fixed, on an input which is a random permutation of n elements. On the other hand, randomized algorithms use random outcomes to guide their execution, which amounts to picking a deterministic algorithm from a suite of algorithms. Moreover, randomized algorithms give guarantees for the worst-case input.

Why might it be useful to design algorithms that behave randomly? Two main advantages of randomized algorithms are speed and simplicity. For many problems, randomized algorithms run faster than the best known deterministic algorithms. Moreover, many randomized algorithms are simpler to describe and implement than deterministic algorithms of comparable performance. (Surprisingly, as a side effect of simplicity, randomized algorithms are often more reliable than their deterministic counterparts.) For many problems therefore, a randomized algorithm is the simplest algorithm available, or the fastest, or both.

Before we illustrate some basic ideas of the design and analysis of randomized algorithms, we consider the problem of generating random samples that may be required in randomized algorithms. Namely, there is a question of generating randomness within the inherently deterministic computers that must implement randomized algorithms. To answer this question, we discuss the uniform pseudorandom number generators in the next section. However, we won't study the more general problem of converting the uniform (pseudo) random numbers into some other type of the probability distribution.

1. Random numbers

Simulating stochastic processes on the computer requires a source of randomness. For example, suppose we need to simulate rolling of a fair die. We could do this by taking the number of seconds in the current time, integer dividing it by 10, and finally adding 1. One problem with this method is that it does not work well if we need to repeat this experiment many times. We would obtain the same result of multiple rolling of a die for 10 seconds, which is a very long time for computers and hardly a random-looking an experiment. Thus, this is not very practical for computer applications, especially when millions of random numbers are required as is often the case. What we really need is a long sequence of numbers with the same properties as a random sequence. In this section, we discuss two methods that are used for generating sequences of numbers which try to *imitate* truly random numbers.

In practice, such numbers that appear to be random are generated on the computer by completely deterministic algorithms and therefore cannot possibly be random. In other words, if started at different times or on different computers with the same initial state of its internal data, the sequence of numbers produced by these algorithms will be exactly the same. This is the reason why these numbers are called ***pseudorandom numbers***, and the algorithms that produce them are called ***pseudorandom number generators.*** However, by a slight abuse of language we will interchangeably use the term "random" and "pseudorandom", unless the distinction is important.

True randomness is impossible to obtain on a deterministic computer, so we settle for less and seek a pseudorandom number generator such that the sequence it generates satisfies most of statistical tests of randomness. All pseudorandom number generators fail some statistical tests, but the bad generators fail simple tests and the good generators fail only complicated tests. Since there is no universal collection of tests that can guarantee, when passed, that a given generator is fully reliable, pseudorandom number generators should be constructed based on a sound mathematical analysis of their structural properties.

Suppose we want random integer numbers between 0 and some positive integer m, uniformly distributed. In the ***uniform distribution***, all numbers in the specified range are equally likely to occur. To produce random real numbers uniformly distributed over the interval $[0, 1]$, one can simply divide these random integers by m. Most other distributions (normal, chi-square, geometric, Poisson) can be derived from the uniform distribution by applying appropriate transformations to the uniform random numbers. That's why we restrict our attention only to the *uniform* random number generators.

1.1. Linear congruential generator. One of the simplest and most used uniform generators is the ***linear congruential generator*** (or ***LCG*** for short). It is by no means the best generator, but it is suitable for use in applications where a good approximation to a random sequence is acceptable. Given an initial value x_0, $(0 < x_0 < m)$, LCG computes a sequence of integers $x_1, x_2, \ldots$ between 0 and $m - 1$ according to the recursive formula

$$x_{i+1} = ax_i \quad (\text{mod } m),$$

for $i = 0, 1, \ldots$, where a, $(1 < a < m)$, and m are carefully chosen positive integers. Actually, this is called a ***multiplicative linear congruential generator***, because a linear congruential generator in full generality has an additional integer constant b and is given by $x_{i+1} = ax_i + b$ (mod m). Since $b = 0$ is almost always used in practice, we call this special case an LCG.

The number x_0 that starts the sequence is known as the ***seed***. If $x_0 = 0$, then the sequence is not random since it is all zeros. If m is a prime, then x_i is never 0. For example, if $m = 11$, $a = 3$, and the seed $x_0 = 1$, then the numbers generated are

$$3, 9, 5, 4, 1, 3, 9, \ldots .$$

First time when we generate a number for the second time (3 in this case), we have a repeating subsequence. In our example, the length of the subsequence that repeats itself is 5 numbers; this is known as the ***period*** of the entire sequence. In general, because there are m possible values, the longest possible period is m. If m is a prime, hence 0 is never produced, the maximum possible period is $m - 1$. For example, if $m = 11$, $a = 7$, and the seed $x_0 = 1$, then the numbers generated are

$$7, 5, 2, 3, 10, 4, 6, 9, 8, 1, 7, 5, \ldots .$$

In this case, the length of the subsequence that repeats itself is $m - 1 = 10$ numbers, which is as good as possible.

If m is a prime, several choices of a give a full period of $m - 1$ (independently of the seed). This type of LCG is called a ***full-period linear congruential generator***. To implement a full-period LCG using 32-bit integer arithmetic, m is chosen to be a large 31-bit prime so that the period is sufficiently large for most applications. A signed 32-bit integer number can lie between -2^{31} and $2^{31} - 1$. Fortunately, the largest possible positive integer $2^{31} - 1 = 2\,147\,483\,647$ is a prime number, and so it makes an excellent choice for the value of m. For this particular prime, some choices of a, such as $a = 16\,807$ or $a = 48\,271$, are recommended because the corresponding linear congruential generators satisfy the following three desirable criteria for good generators:

1. An LCG should be a full-period LCG;
2. The generated sequence should pass a wide variety of statistical tests for randomness;

3. It should be possible to efficiently compute the sequence using 32-bit arithmetic without overflow.

We should emphasize that these generators are very fragile, and tiny changes can completely break their "randomness". For example, we might take an LCG of the form

$$x_{i+1} = 48271 x_i + 1 \pmod{2^{31} - 1},$$

guessing that it would be somehow more random. However, since

$$x_{i+1} = 48271 \cdot 179424105 + 1 \pmod{2^{31} - 1} = 179424105,$$

if the seed is $179\,424\,105$, the generator gets stuck at period 1! Thus, unless we really know better, we should use a well studied formula.

One could argue that the third criterion above is not important, and that an overflow actually contributes to the randomness. However, it is unacceptable because we would no longer have the guarantee of a full period. Unfortunately, given the current pseudorandom number s, a naive approach to compute the next one as an assignment of the form

```
s = (a * s) mod m;
```

is incorrect on most machines with 32-bit integers, since repeated multiplications are certain to overflow. It turns out that computing $x_{i+1} = ax_i \pmod{m}$ without overflow can be done by a slight reordering of the computation.

To show this, let q and r be quotient and remainder of m divided by a, that is, $q = \lfloor m/a \rfloor$ and $r = m \pmod{a} = m - aq = m - a\lfloor m/a \rfloor$. Then,

$$
\begin{aligned}
x_{i+1} &= ax_i \pmod{m} \\
&= ax_i - m\lfloor ax_i/m \rfloor \\
&= ax_i - m\lfloor x_i/q \rfloor + m\lfloor x_i/q \rfloor - m\lfloor ax_i/m \rfloor \\
&= ax_i - m\lfloor x_i/q \rfloor + m(\lfloor x_i/q \rfloor - \lfloor ax_i/m \rfloor).
\end{aligned}
$$

To simplify notation, for fixed values of m and a, we let $\delta(x_i) = \lfloor x_i/q \rfloor - \lfloor ax_i/m \rfloor$. Thus,

$$
\begin{aligned}
x_{i+1} &= ax_i - m\lfloor x_i/q \rfloor + m\delta(x_i) \\
&= a\big(x_i \pmod{q} + q\lfloor x_i/q \rfloor\big) - m\lfloor x_i/q \rfloor + m\delta(x_i) \\
&= a\big(x_i \pmod{q}\big) + aq\lfloor x_i/q \rfloor - m\lfloor x_i/q \rfloor + m\delta(x_i) \\
&= a\big(x_i \pmod{q}\big) + (aq - m)\lfloor x_i/q \rfloor + m\delta(x_i).
\end{aligned}
$$

Since $r = m - aq$, we finally get

(20) $$x_{i+1} = a\big(x_i \pmod{q}\big) - r\lfloor x_i/q \rfloor + m\delta(x_i).$$

The three terms on the right-hand side of the reordered equation (20) have several nice properties:

(i) The first term can always be computed without overflow. To see why, we observe that $0 \le a\big(x_i \pmod{q}\big) < m$. This follows from the fact that $0 \le x_i \pmod{q} < q$, hence $0 \le a\big(x_i \pmod{q}\big) < aq = m - r \le m$.

(ii) If $r < q$, the second term can be computed without overflow. For, if $r < q$, or equivalently $r/q < 1$, then $0 \leq r \lfloor x_i/q \rfloor \leq r \cdot x_i/q = (r/q) \cdot x_i < 1 \cdot m = m$.

(iii) Finally, if $r < q$, then the factor $\delta(x_i)$ of the third term is either 0 or 1. More precisely, let $d(x_i)$ be the result of the subtraction of the first two terms: $d(x_i) = a(x_i \pmod q) - r \lfloor x_i/q \rfloor$. Then,

$$\delta(x_i) = \begin{cases} 0, & \text{if } d(x_i) \geq 0 \\ 1, & \text{if } d(x_i) < 0. \end{cases}$$

To see this, first observe that $-m < d(x_i) < m$ because $0 \leq a(x_i \pmod q) < m$ and $0 \leq r \lfloor x_i/q \rfloor < m$. Now, since $0 \leq x_{i+1} < m$, it follows that

$$-m + m\delta(x_i) < d(x_i) + m\delta(x_i) = x_{i+1} < m,$$

or equivalently $\delta(x_i) < 2$.
On the other hand,

$$0 \leq x_{i+1} = d(x_i) + m\delta(x_i) < m + m\delta(x_i) = m(1 + \delta(x_i)),$$

or equivalently $1 + \delta(x_i) > 0$, that is, $\delta(x_i) > -1$. Combining this with $\delta(x_i) < 2$, we get $-1 < \delta(x_i) < 2$. Since $\delta(x_i)$ is an integer, it follows that $\delta(x_i)$ is either 0 or 1. Therefore, if $d(x_i) \geq 0$, it must be the case that $\delta(x_i) = 0$, since $\delta(x_i) = 1$ would imply $x_{i+1} = d(x_i) + m\delta(x_i) \geq m$, a contradiction. Similarly, if $d(x_i) < 0$, it must be the case that $\delta(x_i) = 1$, since $\delta(x_i) = 0$ would imply $x_{i+1} = d(x_i) + m\delta(x_i) < 0$, a contradiction.

For example, for the particular values of $m = 2^{31} - 1$ and $a = 16807$, we have $q = 127773$ and $r = 2836$. Consequently $r < q$ and $-m < d(x_i) < m$, so $d(x_i)$ fits into a 32-bit signed integer. Moreover, it is not even necessary to compute $\delta(x_i)$, since a simple test of the result of the subtraction of the first two terms determines whether the third term is 0 or m. Namely, if $d(x_i) \geq 0$, then $x_{i+1} = d(x_i)$; if $d(x_i) < 0$, then $x_{i+1} = d(x_i) + m$. Therefore, computation of x_{i+1} according to the equation (20) can be done without overflow.

Using this approach, we give the algorithm LCG(s) below that takes a pseudorandom number s and returns the next one. The algorithm implements the linear congruential generator and uses appropriate global constants m and a, as well as q and r such that $r < q$.

```
LCG(s)
    d = a(s (mod q)) - r ⌊s/q⌋;
    if d ≥ 0 then
        s = d;
    else
        s = d + m;
    return s;
```

1.2. Mersenne twister. The *Mersenne Twister* (or *MT* for short) is a uniform pseudorandom number generator that has become quite popular recently. It is fast, robust, and has provably a huge period of $2^{19937} - 1$. This period explains the origin of the name because $2^{19937} - 1$ is a Mersenne prime.[1] In addition, MT generates pseudorandom numbers that have very good uniformity properties and have passed the most stringent statistical tests. All this is achieved with a slight increase in complexity—MT has several parameters compared to two (a and m) of LCG, as well as in additional memory requirements—MT needs 624 memory words of working area compared to just one word that LCG uses.

MT generates a sequence of 0-1 vectors of length w, which are considered to be uniform pseudorandom integers between 0 and $2^w - 1$. Thus, these vectors are identified with machine words of length w, with the least significant bit at the far right. If the row 0-1 vectors are denoted by bold letters, such as $\mathbf{x}$ and $\mathbf{y}$, then the algorithm is based on the linear recurrence relation

$$(21) \qquad \mathbf{x}_{k+n} = \mathbf{x}_{k+m} + \mathbf{x}_{k+1} \begin{pmatrix} 0 & 0 \\ 0 & \mathbf{I}_r \end{pmatrix} \mathbf{A} + \mathbf{x}_k \begin{pmatrix} \mathbf{I}_{w-r} & 0 \\ 0 & 0 \end{pmatrix} \mathbf{A}, \quad k = 0, 1, 2, \ldots,$$

where $\mathbf{I}_r$ and $\mathbf{I}_{w-r}$ are identity matrices of size r and $w - r$, respectively; $\mathbf{0}$ is the zero matrix of appropriate size; and $\mathbf{A}$ is a 0-1 matrix of size $w \times w$. Before we explain the parameters of the recurrence, first note that we need n initial vectors (seeds) $\mathbf{x}_0, \mathbf{x}_1, \ldots, \mathbf{x}_{n-1}$. Then putting $k = 0, 1, 2, \ldots$ in this order, we obtain other vectors of the sequence $\mathbf{x}_n, \mathbf{x}_{n+1}, \mathbf{x}_{n+2}, \ldots$.

Now, the defining recurrence (21) uses several carefully chosen constants: the integer n, which is the degree of the recurrence, the integer m such that $0 \le m < n$, the integer r such that $0 \le r < w$, and the 0-1 matrix $\mathbf{A}$ of size $w \times w$.

Although the recurrence (21) appears complicated, we can rewrite it as

$$(22) \qquad \mathbf{x}_{k+n} = \mathbf{x}_{k+m} \oplus \left((x_{w-1}^k, \ldots, x_r^k, x_{r-1}^{k+1}, \ldots, x_0^{k+1}) \cdot \mathbf{A} \right), \quad k = 0, 1, 2, \ldots,$$

where $\oplus$ is the bitwise EXCLUSIVE-OR operation, and

$$(x_{w-1}^k, \ldots, x_r^k, x_{r-1}^{k+1}, \ldots, x_0^{k+1})$$

represents the vector obtained by taking the upper $w - r$ bits of the vector $\mathbf{x}_k$ and the lower r bits of the vector $\mathbf{x}_{k+1}$. Notice that this vector can be easily computed by masking out the corresponding parts of the vectors $\mathbf{x}_k$ and $\mathbf{x}_{k+1}$, and then applying the bitwise OR operation on them.

The matrix $\mathbf{A}$ is chosen to be in a form that allows fast multiplication from the left. The form is

$$\mathbf{A} = \begin{pmatrix} 0 & 1 & 0 & \cdots & 0 \\ 0 & 0 & 1 & \cdots & 0 \\ \vdots & \vdots & \vdots & \vdots & \vdots \\ 0 & 0 & 0 & \cdots & 1 \\ a_{w-1} & a_{w-2} & a_{w-3} & \cdots & a_0 \end{pmatrix},$$

---

[1] Mersenne primes are prime numbers of the form $2^p - 1$, where p is a positive integer. However, notice that $2^p - 1$ is a prime only if p is a prime.

where $\mathbf{a} = (a_{w-1}, a_{w-2}, \ldots, a_0)$ is a suitably chosen constant vector. Using the notation $\mathbf{x} = (x_{w-1}, x_{w-2}, \ldots, x_0)$, as well as $\mathbf{x} \gg i$ for the operation that shifts $\mathbf{x}$ right for i bits, we have

$$\mathbf{xA} = \begin{cases} \mathbf{x} \gg 1, & \text{if } x_0 = 0 \\ (\mathbf{x} \gg 1) \oplus \mathbf{a}, & \text{if } x_0 = 1. \end{cases}$$

Therefore, computation of the entire recurrence (22) can be done using only the bitshift, bitwise EXCLUSIVE-OR, bitwise AND, and bitwise OR operations.

To obtain good uniformity properties of generated vectors, each vector computed according to the recurrence (22) is finally multiplied from the right by an invertible 0-1 matrix $\mathbf{T}$ of size $w \times w$. This so-called *tempering* matrix $\mathbf{T}$ is implicitly defined so that, given a vector $\mathbf{x}$, the product $\mathbf{y} = \mathbf{xT}$ is computed using the following successive transformations:

$$\mathbf{y} = \mathbf{x} \oplus (\mathbf{x} \gg u)$$
$$\mathbf{y} = \mathbf{y} \oplus \big((\mathbf{y} \ll s) \otimes \mathbf{b}\big)$$
$$\mathbf{y} = \mathbf{y} \oplus \big((\mathbf{y} \ll t) \otimes \mathbf{c}\big)$$
$$\mathbf{y} = \mathbf{y} \oplus (\mathbf{y} \gg l)$$

Here, the parameters u, s, t, and l are integers between 1 and $w - 1$, $\mathbf{b}$ and $\mathbf{c}$ are constant vectors, $\mathbf{x} \ll i$ denotes the operation that shifts $\mathbf{x}$ left for i bits, and $\otimes$ denotes the bitwise AND operation.

By the general theory of linear recurrences, the MT sequence (21) attains the maximal period $2^{nw-r} - 1$ under some conditions that affect the choice of the parameters. Also, the randomness for the whole period is independent of the choice of initial vectors, as long as at lest one of the seeds is different from zero.

In summary, MT depends on two sets of parameters:

1. *Period parameters* determine the period of the generating sequence. These are integers w (word size), n (degree of recursion), m (middle element), and r (partition point of the word), as well as one binary vector $\mathbf{a}$ (matrix $\mathbf{A}$).
2. *Tempering parameters* determine quality of the uniformity of the generating sequence. These are integers u, s, t, and l, as well as two binary vectors $\mathbf{b}$ and $\mathbf{c}$.

Table 1 lists the parameter values, as suggested by the authors of MT (see [38]), which provide for the full period $2^{nw-r} - 1$ and good uniformity properties. Values for the binary vectors $\mathbf{a}$, $\mathbf{b}$, and $\mathbf{c}$ are given in the hexadecimal notation.

Period parameters					Tempering parameters					
w	n	m	r	$\mathbf{a}$	u	s	t	l	$\mathbf{b}$	$\mathbf{c}$
32	624	397	31	9908B0DF	11	7	15	18	9D2C5680	EFC60000

TABLE 1. Parameters for Mersenne Twister (MT19937).

The following pseudocode for MT assumes a global array of n binary vectors (unsigned integers) of word size. These global vectors, denoted by $\mathbf{x}(0), \mathbf{x}(1), \ldots, \mathbf{x}(n-1)$, initially represent the nonzero seed vectors. Then, after execution of one iteration of the MT algorithm, the vectors are shifted in a circular manner so that the new $\mathbf{x}(0)$ becomes the old $\mathbf{x}(1)$, the new $\mathbf{x}(1)$ becomes the old $\mathbf{x}(2)$, and so on, the new $\mathbf{x}(n-2)$ becomes the old $\mathbf{x}(n-1)$. Finally, the new $\mathbf{x}(n-1)$ gets the currently computed vector.

```
MT(x(0), x(1), ..., x(n-1))
    ⟦ Concatenate upper w − r bits of x(0) and lower r bits of x(1) ⟧
    y = (x^0_{w-1}, ..., x^0_r, x^1_{r-1}, ..., x^1_0);
    ⟦ Multiply y from the right by the matrix A ⟧
    if x^1_0 = 0 then
        y = y ≫ 1;
    else
        y = (y ≫ 1) ⊕ a;
    ⟦ Add bitwise modulo 2 the middle element x_m to the result ⟧
    y = x_m ⊕ y;
    ⟦ Multiply the result by the matrix T from the right ⟧
    y = y ⊕ (y ≫ u);
    y = y ⊕ ((y ≪ s) ⊗ b);
    y = y ⊕ ((y ≪ t) ⊗ c);
    y = y ⊕ (y ≫ l);
    ⟦ Shift the global vectors in circular manner ⟧
    for i = 1 to n − 1 do
        x(i − 1) = x(i);
    x(n − 1) = y;
    ⟦ Return the next pseudorandom number ⟧
    return y;
```

2. Las Vegas algorithms

One paradigm of randomized algorithms is called *Las Vegas algorithms*. They always output correct answer, but they are efficient only in the expected sense of the running time. In other words, in the unlucky worst-case they can still perform badly, or can even run forever and give no answer, but their expected running time is provably good. Moreover, sometimes we can make even stronger statement that their running time is small with high probability.[2] Thus, a short informal description of the Las Vegas algorithms is that they are *always correct and probably fast*.

---

[2] The running time $T(n)$ of a randomized algorithm is $O(f(n))$ *with high probability* if $\mathbb{P}\big(T(n) \leq cf(n)\big) \geq 1 - 1/n^d$ for some positive constants c and d.

2.1. Randomized quick sort. Recall from Chapter 5 that the quick sort algorithm consists of two phases: partitioning of the original array into two sections around the pivot element, and then recursively sorting the two sections on either side of the pivot. We learned that the key step in the first phase is the choice of a good pivot element, which should be close to the median value of the array. We also discussed two algorithms for the choice of the pivot when we take the first and the last element of the array.

Let us now consider partitioning of an array A of size n when we choose one of its elements as the pivot at random. The general form of the algorithm, which we repeat here, to quick sort the array is not changed.

RAND-QUICK-SORT (A, l, r)
  if $l < r$ **then**
    $k =$ PARTITION (A, l, r);
    RAND-QUICK-SORT $(A, l, k - 1)$;
    RAND-QUICK-SORT $(A, k + 1, r)$;
  return A;

The only difference is the way we choose the pivot element in the PARTITION algorithm. Instead of choosing the first or the last (or the middle) element, in each recursive step we pick an index m uniformly at random between the lower and upper bounds l and r of the corresponding subarray. For this random choice we suppose that we have available the function $random(i, j)$ that returns a uniform (pseudo) random integer from the set of positive integers $\{i, i+1, \ldots, j\}$. (The function $random$ may be implemented, for example, by using either one of the two generators from the previous section, see Exercise 2.) We then promote the corresponding element $A(m)$ to be the pivot a, and by comparing each array element with a, we partition the array into the section S_1 of elements smaller than or equal to a and the section S_2 of elements larger than a.

PARTITION (A, l, r)
  $m = random(l, r)$;
  $a = A(m)$;
  $i = l - 1;\ j = r + 1$;
  while $i < j$ **do**
    $i = i + 1;\ j = j - 1$;
    while $(A(i) \le a) \wedge (i < j)$ **do** $i = i + 1$;
    while $(A(j) > a) \wedge (i < j)$ **do** $j = j - 1$;
    if $i < j$ **then** $swap(A(i), A(j))$);
  return j;

To analyze the RAND-QUICK-SORT algorithm, we measure its running time in terms of the number of comparisons it performs, because this determines the total number of basic operations in any reasonable implementation. An execution of the call RAND-QUICK-SORT $(A, 1, n)$ is viewed as a binary tree T whose each node is labeled with a distinct element of A. The root of the tree is labeled with the pivot element chosen in the PARTITION algorithm for the call RAND-QUICK-SORT$(A, 1, n)$, the left subtree of the root contains the elements of S_1, and the right subtree of the root contains the elements of S_2. This idea is illustrated in Figure 1.

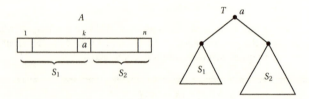

FIGURE 1. The array A and the execution tree T after the first partitioning around the pivot.

The two subtrees S_1 and S_2 of T are recursively determined by execution of the respective calls RAND-QUICK-SORT$(A, 1, k-1)$ and RAND-QUICK-SORT$(A, k+1, n)$. For example, the root of the subtree S_1 is the pivot element chosen in the PARTITION algorithm for the call RAND-QUICK-SORT$(A, 1, k-1)$, and the root of the subtree S_2 is the pivot element chosen in the PARTITION algorithm for the call RAND-QUICK-SORT$(A, k+1, n)$. This process is then repeated, mirroring the recursive execution of the subsequent calls until the tree T is completely built. Clearly, different executions of the procedure RAND-QUICK-SORT$(A, 1, n)$ on the same array A may generate different trees in this way. In fact, each T is a (random) binary search tree whose inorder traversal lists the elements of the array A in a sorted order.

We want to determine the *expected* number of comparisons during an execution of the RAND-QUICK-SORT algorithm. Observe that the root of the corresponding tree T is compared to every element in the two subtrees, but no comparison is performed between an element of the left subtree and an element of the right subtree. Thus, there is a comparison between any two elements of the array if and only if one of these elements is an ancestor of the other in the tree T.

Imagine that we have sorted the array A, and for each $i = 1, 2, \dots, n$ denote by a_i the i-th element of the array A in the sorted order. Thus, a_1 is the smallest, a_2 the second smallest, and so on, a_n is the largest element. Define the random variable X_{ij} that indicates whether a_i and a_j are compared during an execution of the call RAND-QUICK-SORT$(A, 1, n)$. More precisely,

$$X_{ij} = \begin{cases} 1, & \text{if } a_i \text{ and } a_j \text{ are compared during execution} \\ 0, & \text{otherwise.} \end{cases}$$

In other words, X_{ij} is a count of comparisons between a_i and a_j, and so the total number of comparisons is the random variable Y given by

$$Y = \sum_{i=1}^{n-1} \sum_{j=i+1}^{n} X_{ij}.$$

Since we are interested in the expected number of comparisons, we need to determine the expected value of the random variable Y. But by the linearity of expectation this is simplified to determining the expectation of the indicator random variables X_{ij}, that is,

$$\mathbb{E}(Y) = \mathbb{E}\left(\sum_{i=1}^{n-1}\sum_{j=i+1}^{n} X_{ij}\right) = \sum_{i=1}^{n-1}\sum_{j=i+1}^{n} \mathbb{E}(X_{ij}).$$

If p_{ij} denotes the probability that the elements a_i and a_j are compared during an execution, then

$$\mathbb{E}(X_{ij}) = 1 \cdot p_{ij} + 0 \cdot (1 - p_{ij}) = p_{ij},$$

and so

$$\mathbb{E}(Y) = \sum_{i=1}^{n-1}\sum_{j=i+1}^{n} p_{ij}.$$

Thus, the computation of the expected number of comparisons is reduced to the problem of determining the probabilities p_{ij}. Now, to determine these probabilities, for $i \neq j$ consider the sequence $a_i, a_{i+1}, \ldots, a_j$ of the array elements between a_i and a_j, inclusive, in the sorted order. We can make two simple observations about this sequence:

1. The elements a_i and a_j are compared during an execution if and only if a_i or a_j is selected first as a pivot among the elements $a_i, a_{i+1}, \ldots, a_j$ during the execution. To see why, first suppose that, say, a_i is selected first among $a_i, a_{i+1}, \ldots, a_j$ during the execution. Then the right subtree of the node a_i in the corresponding execution tree must contain all the elements $a_{i+1}, \ldots, a_j$. For otherwise, if one of these elements, say a_j, is outside the right subtree, take the first common ancestor x on the paths from a_i and a_j to the root. Since a_i is before a_j in the sorted order, it must be the case that a_i belongs to the left subtree and a_j to the right subtree of x. But then x is between a_i and a_j in the sorted order, and must have been chosen before a_i during the execution, a contradiction. Thus, a_j is a descendant of a_i in the execution tree, hence they are compared during the execution. (In the second case, if a_j is selected first among $a_i, a_{i+1}, \ldots, a_j$ during the execution, then the elements $a_i, a_{i+1}, \ldots, a_{j-1}$ must belong to the left subtree of the node a_j.) Conversely, if a_k for some $i < k < j$ is selected first among $a_i, a_{i+1}, \ldots, a_j$, then a_i and a_j belong to the different subtrees of the node a_k, which means that a_i and a_j are not compared.

2. Any of the elements $a_i, a_{i+1}, \ldots, a_j$ is equally likely to be chosen the first as a pivot during an execution. Therefore, the probability that a_i or a_j is selected first among the elements $a_i, a_{i+1}, \ldots, a_j$ is exactly $2/(j - i + 1)$. So, by the first observation, it follows that $p_{ij} = 2/(j - i + 1)$.

Therefore,

$$\mathbb{E}(Y) = \sum_{i=1}^{n-1}\sum_{j=i+1}^{n} p_{ij} = \sum_{i=1}^{n-1}\sum_{j=i+1}^{n} \frac{2}{j - i + 1}$$

$$= 2\sum_{i=1}^{n-1}\left(\frac{1}{2} + \frac{1}{3} + \cdots + \frac{1}{n - i + 1}\right)$$

$$\leq 2 \sum_{i=1}^{n-1} H_n$$
$$\leq 2n H_n,$$

where $H_n = \sum_{i=1}^{n} 1/i$ is the n-th harmonic number. Since $H_n = \Theta(\log n)$, we have thus established $\mathbb{E}(Y) = O(n \log n)$.

Since the running time is proportional to the total number of comparisons, we can conclude that the *expected* running time of the RAND-QUICK-SORT algorithm is $O(n \log n)$.

It is worth pointing out that the expected running time of the RAND-QUICK-SORT algorithm holds for every input. That is, the $O(n \log n)$ expectation depends only on the random choices made by the algorithm, and *not* on any assumptions about the distribution of the input array elements. The execution of a randomized algorithm can vary even on a fixed input, and so its running time represents a random variable. It is important to distinguish this from the probabilistic analysis of an algorithm, in which we assume some distribution on the input and analyze an algorithm that may itself be deterministic.

2.2. Flipping a biased coin to simulate a fair coin. Suppose we have a possibly biased coin so that each toss produces heads with unknown probability p ($0 < p < 1$) and tails with complementary probability $q = 1 - p$. Assume that each toss of the coin is independent from previous tosses—the probability of getting heads at any given toss is exactly p, regardless of previous outcomes. Unfortunately, we do not know the value of p. Can we design a simple procedure by which we can use this coin to generate a perfectly unbiased coin toss?

Clearly, flipping a biased coin only once will not do, because the outcomes are not equally probable. However, if we toss a biased coin twice, the probability of changing from a head to a tail, or from a tail to a head, will be the same. That is, the probability of getting a head-tail combination is pq (probability of a head *and* a tail), and the probability of a tail-head combination is qp (probability of a tail *and* a head).

This suggests a simple algorithm: flip the coin twice until you get two different values. If you observed a head-tail combination, output 0 ("fair heads"). If you observed a tail-head combination, output 1 ("fair tails").

```
COIN-FLIP(p)
    repeat
        Flip the biased coin twice;
        if a head-tail combination then
            return 0;
        else if a tail-head combination then
            return 1;
        ⟦ Else a head-head or a tail-tail, do nothing ⟧
    until forever;
```

The output of the COIN-FLIP algorithm is clearly a random variable, denote it by X, whose outcome is 0 or 1. Because

$$\mathbb{P}(X = 0) = \mathbb{P}(X = 1) = pq + (p^2 + q^2)pq + (p^2 + q^2)^2 pq + \cdots$$

$$= pq \sum_{i=0}^{\infty} (p^2 + q^2)^i = pq \cdot \frac{1}{1 - (p^2 + q^2)}$$

$$= pq \cdot \frac{1}{2pq}$$

$$= \frac{1}{2},$$

the two outcomes are equally probable, so this correctly simulates a fair coin. It is also clear that the expected running time of the COIN-FLIP algorithm is proportional to the expected number of repetitions of the **repeat** loop, which is

$$\sum_{i=1}^{\infty} i \cdot (p^2 + q^2)^{i-1} 2pq = 2pq \sum_{i=1}^{\infty} i \cdot (p^2 + q^2)^{i-1} = 2pq \cdot \frac{1}{\left(1 - (p^2 + q^2)\right)^2} = \frac{1}{2pq}.$$

Therefore, for a "reasonable" value of p (e.g., $1/4 \le p \le 3/4$) the expected running time is bounded by a small constant, although in theory the algorithm can run forever. In other words, the COIN-FLIP algorithm is always correct and only expectedly fast, which is typical behavior of a Las Vegas algorithm.

2.3. Searching a sorted compact list. A *sorted compact list* represents a totally ordered set X of n elements by a sorted doubly-linked list L. In turn, all elements of the list L are stored using only the first n positions in the multiple-array representation of L. Such a compact representation in storage is often desirable in a paged, virtual-memory environment. Figure 2 shows the set $X = \{1, 2, 3, 4, 5, 6, 7, 8, 9\}$ represented by a sorted compact list.

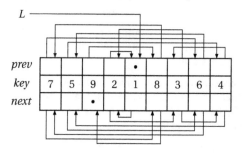

FIGURE 2. Sorted-compact-list representation of the set $X = \{1, 2, 3, 4, 5, 6, 7, 8, 9\}$.

Consider first the problem of finding the smallest (or the largest) element in some set X represented by a sorted compact list L. To make this problem non-trivial, assume we don't have access to the header element pointing to the first element of the list L. In that case, the links that determine sorted order of the list elements are irrelevant and we cannot do better than treating the list as an ordinary, unordered array. Thus, searching for the smallest element deterministically takes $\Theta(n)$ time on average. On the other hand, consider the following randomized algorithm[3]:

RAND-MIN-SCL(L)
1. Take a random sample $S \subseteq X$ of size $\lfloor \sqrt{n} \rfloor$;
2. Find the smallest element $s_0 \in S$ in the sample;
3. Follow the *prev* links from s_0 up to the element x_0
   such that $prev(x_0) = $ NIL;
return x_0;

This is clearly a correct Las Vegas algorithm, since no matter from which element we start, following the *prev* links from there will bring us to the overall smallest element x_0. It is only the question whether we shorten the path starting from the smallest element of the sample set. To answer this question, let's analyze the running time of the RAND-MIN-SCL algorithm.

The first step is easy to implement: generate $\lfloor \sqrt{n} \rfloor$ pseudorandom numbers from the interval $[1, \ldots, n]$ and take them as the index positions. Since the list is compact, we are guaranteed that the generated numbers reference some list elements. Thus, this step takes $\Theta(\sqrt{n})$ time.

The second step can be deterministically performed in time $\Theta(\sqrt{n})$, since the sample size is $|S| = \lfloor \sqrt{n} \rfloor$.

The running time of the third step clearly depends on the number of elements in X that are less than s_0. So, given a random sample S, if we define the set $A = \{x \in X : x < s_0\}$, then the third step takes time proportional to the A's size $|A|$. However, the size $|A|$ is a random variable that depends on the random sample S. This means that it doesn't make sense to talk about the exact value of $|A|$. Instead, we consider its expected value and show that $\mathbb{E}(|A|) = \Theta(\sqrt{n})$.

Therefore, the total expected running time of the RAND-MIN-SCL algorithm is $\Theta(\sqrt{n})$. As we can see, the randomized algorithm outperforms the best possible deterministic algorithm.

The hard part is to prove the claim $\mathbb{E}(|A|) = \Theta(\sqrt{n})$. We will prove a more general result that can be used in various settings. To this end, let X be a set of n elements and f be a function defined on the power set of X.[4] For any subset $Y \subseteq X$, define two associated sets:

$$U(Y) = \{y \in Y : f(Y \setminus \{y\}) \neq f(Y)\}$$
$$V(Y) = \{x \in X \setminus Y : f(Y \cup \{x\}) \neq f(Y)\}$$

The set $U(Y)$ is called the set of (inside) *extreme* elements in Y, while the set $V(Y)$ is called the set of (outside) *violators* of Y. It is easy to see that $x \notin Y$ violates Y if and only if x is extreme in $Y \cup \{x\}$.

---

[3]For convenience, throughout this section we identify an element and its key, i.e., we simply write x even in cases when it is more appropriate to write $key(x)$.
[4]The only purpose of the function f is to partition the power set 2^X into equivalence classes.

If S is a sample chosen uniformly at random from the set X, then $|U(S)|$ and $|V(S)|$ are random variables. Given the sample size m, $(0 < m < n)$, consider the random variables $U_m: S \rightarrow |U(S)|$ and $V_m: S \rightarrow |V(S)|$ that give the sizes of the corresponding sets of extreme elements and violators of a random sample S of size m. The following lemma relates the expected values of the random variables U_m and V_m.

LEMMA (SAMPLING LEMMA) *For any sample size m, $(0 < m < n)$,*

$$\mathbb{E}(V_m) = \mathbb{E}(U_{m+1}) \cdot \frac{n-m}{m+1}.$$

PROOF: Let $I\{\cdot\}$ denote the indicator variable for an event in braces. By the definition of the random variables U_m and V_m we have

$$\mathbb{E}(V_s) = \frac{\displaystyle\sum_{\substack{S \subseteq X \\ |S|=m}} \sum_{x \in X \backslash S} I\{x \text{ violates } S\}}{\dbinom{n}{m}}$$

$$= \frac{\displaystyle\sum_{\substack{S \subseteq X \\ |S|=m}} \sum_{x \in X \backslash S} I\{x \text{ is extreme in } S \cup \{x\}\}}{\dbinom{n}{m}}$$

$$= \frac{\displaystyle\sum_{\substack{R \subseteq X \\ |R|=m+1}} \sum_{x \in R} I\{x \text{ is extreme in } R\}}{\dbinom{n}{m}}$$

$$= \frac{\dbinom{n}{m+1} \cdot \mathbb{E}(U_{m+1})}{\dbinom{n}{m}}$$

$$= \mathbb{E}(U_{m+1}) \cdot \frac{n-m}{m+1}. \quad \blacksquare$$

Returning now to the third step of the RAND-MIN-SCL algorithm, we use the Sampling Lemma to argue that $\mathbb{E}(|A|) = \Theta(\sqrt{n})$. For any subset $Y \subseteq X$, define $f(Y) = \min(Y)$.[5] Then, for a random sample $S \subseteq X$ of size m, observe that $V(S) = \{x \in X: x < \min(S)\} = A$. Also, the only extreme element in S is $\min(S)$, for any sample size. Thus, $V_m = |A|$ and $U_{m+1} = 1$. Since in our case $m = \lfloor\sqrt{n}\rfloor$, by the Sampling Lemma we have

$$\mathbb{E}(|A|) = 1 \cdot \frac{n - \lfloor\sqrt{n}\rfloor}{\lfloor\sqrt{n}\rfloor + 1} = \Theta(\sqrt{n}),$$

as required.

Consider now the general case of searching a compact sorted list. Our task is to test whether an element with a given value k appears in a sorted compact list L of n distinct elements. Notice that now we can assume availability of the

---
[5]The empty set is mapped to some element outside X.

header element of the list, since it doesn't help to deterministically search for k in the list in time faster than $\Theta(n)$ on average. However, if we again apply the same randomized approach, we can do better in the general case as well.

RAND-SEARCH-SCL(L, k)
1. Take a random sample $S \subseteq X$ of size $\lfloor \sqrt{n} \rfloor$;
2. Find the smallest element $s_0 \in S$ in the sample such that $s_0 \geq k$;
3. Follow the *prev* links from s_0 up to some list element x_0
   such that either $x_0 = k$ or $x_0 < k$;
if $x_0 = k$ **then**
   return TRUE;
else if $x_0 < k$ **then**
   return FALSE;

It is clear that the RAND-SEARCH-SCL algorithm is a correct Las Vegas algorithm. Besides, the running time analysis of the RAND-SEARCH-SCL algorithm completely goes along the lines of the analysis of the RAND-MIN-SCL algorithm. In this case however, for any subset $Y \subseteq X$ we define $f(Y) = \min\{y \in Y : y \geq k\}$. Then, for a random sample $S \subseteq X$ of any size, if we denote $f(S) = s_0$, then $V(S) = \{x \in X : k \leq x < s_0\}$ and $U(S) = s_0$. Substituting $m = \lfloor \sqrt{n} \rfloor$ in the Sampling Lemma, we obtain again the expected number of violators of the random sample S to be

$$\mathbb{E}(V_m) = 1 \cdot \frac{n - \lfloor \sqrt{n} \rfloor}{\lfloor \sqrt{n} \rfloor + 1} = \Theta\left(\sqrt{n}\right).$$

3. Monte Carlo algorithms

Another paradigm of randomized algorithms is called **Monte Carlo algorithms**. In this case, a randomized algorithm may err and output an incorrect answer, but the probability of error is negligible. In return, a Monte Carlo algorithm is always fast. This prominent feature of these algorithms is informally stated as them being *probably correct and always fast*.

3.1. Finding a high-ranking element.
To convey the idea of a Monte Carlo algorithm, let's take a simple example. Given an array A of n numbers in arbitrary order, consider the problem of finding one of its elements that is larger than half of the array elements. More precisely, we look for an element of A that is greater than or equal to the median element of A. We call such an element the high-ranking element because its rank is greater than or equal to the rank of at least half of the elements of A.

To solve this problem deterministically, we may use the FINDMAX algorithm from Chapter 2 since a maximal element is certainly larger than half of the array elements. However, the problem with this solution is that the FINDMAX algorithm takes $\Theta(n)$ time, and we would like a much faster algorithm. We could get a speed up by a factor of two if we limit our search for a maximal element to the first $n/2$ elements. In this case we need to scan only the first half of the array, but the asymptotic time is still $\Theta(n)$.

If we try a randomized approach, we can pick randomly an element x of A and output that element x as a high-ranking element. If m is a median element of A, than the probability that the randomly chosen x is a high-ranking element is

$$\mathbb{P}\{x \geq m\} = \frac{\text{number of array elements} \geq m}{\text{total number of array elements}} \geq \frac{n/2}{n} = \frac{1}{2},$$

that is, the probability of error is

$$\mathbb{P}\{x < m\} = 1 - \mathbb{P}\{x \geq m\} \leq 1 - \frac{1}{2} = \frac{1}{2}.$$

This appears like flipping a coin to decide whether a picked element is a high-ranking element, and so it does not seem to be a prudent approach. However, what is promising is that the algorithm is very fast (it takes constant time) and the probability of its error is bounded by a constant.

This suggests that we can improve the algorithm by repeating our random choice. For example, we may randomly choose two elements from A, and output the larger of the two as a high-ranking element of A. Let x_1 and x_2 be the elements of A that are independently at random chosen one after another (with replacement). Since $\max\{x_1, x_2\} < m$ if and only if $x_1 < m$ and $x_2 < m$, the error probability of this approach is

$$\mathbb{P}\{\max\{x_1, x_2\} < m\} = \mathbb{P}\{x_1 < m \text{ and } x_2 < m\} = \mathbb{P}\{x_1 < m\} \cdot \mathbb{P}\{x_2 < m\} \leq \frac{1}{2} \cdot \frac{1}{2} = \frac{1}{2^2}.$$

Here we use the property that the probability of a finite intersection of independent events is equal to the product of the probabilities of the individual events. Thus, the error probability is now bounded by a respectable $1/4$. Clearly, it can be further reduced if we randomly choose $k > 2$ elements and output the largest of them as a high-ranking element of A. A randomized algorithm for finding a high-ranking element of the array A is therefore extremely simple.

RAND-HIGH-RANK(A)
    Choose $x_1, x_2, \ldots, x_k$ uniformly at random from A;
    return $\max\{x_1, x_2, \ldots, x_k\}$;

Since k is a fixed positive integer independent of the array size n, the RAND-HIGH-RANK algorithm takes a constant time $\Theta(k)$. Furthermore, its probability of returning an incorrect result is

$$\mathbb{P}\{\max\{x_1, x_2, \ldots, x_k\} < m\} = \prod_{i=1}^{k} \mathbb{P}\{x_i < m\} \leq \prod_{i=1}^{k} \frac{1}{2} = \frac{1}{2^k}.$$

To better grasp the quality of the RAND-HIGH-RANK algorithm, take for instance $k = 20$. The algorithm then returns the correct answer with probability of roughly 0.999999 or, equivalently, the error probability is about 0.000001. And if $k = 100$, the error probability is astronomically small and, in fact, it is less than the probability that a hardware error occurs while the algorithm is run. We have got therefore an algorithm that performs a constant number of 100 basic operations independently of the array size n. Moreover, it errs to produce a high-ranking element of the array with probability that is negligible in all practical terms.

3.2. Randomized counting. Suppose we have a huge file of unknown size and want to determine its size, say, in bytes. A simple approach is to sequentially read the file byte by byte and count the bytes until we get to the end of the file. This approach takes the time proportional to the file size, and so it may be too slow to be practical in some applications. For example, sequentially finding the size of a database may be unacceptable for some interactive software.

We will rephrase the file size problem in more abstract terms and turn it into the set cardinality counting problem. Given a set S of (unknown) size n, the sequential method to count all its elements takes $\Theta(n)$ time. However, if n is very large, the linear time $\Theta(n)$ is unacceptable and instead we need a more efficient algorithm. However, without any previous knowledge on the structure of the set S, it is not clear whether we can come up with a better deterministic algorithm at all. This is an example where the probability comes to our rescue and enables us to devise a faster randomized algorithm, provided we can tolerate a small inaccuracy of the result.

It turns out that the problem of counting large-cardinality sets is related to a well-known mathematical problem under the catchy name of *birthday paradox*. The birthday paradox can be stated in many equivalent ways, and we are going to use the balls-and-bins model. Suppose we have n different bins $b_1, b_2, \ldots, b_n$ and an infinite supply of identical balls. We conduct an experiment by repeatedly and independently throwing balls into bins with equal probability. More precisely, in step 1 we throw the first ball, in step 2 we independently throw the second ball, and so on. In each step we assume that the probability that a ball falls in any bin is equal among all bins. Each time we toss a ball, we check whether any bin contains two balls. If so, we end our experiment; otherwise, we continue with tossing the next ball. Thus, we seek the first time when some bin contains two balls or, in other words, when the first collision of two balls occurs.

Let X be the step when the first collision occurs, which is a random variable ranging between 2 and $n + 1$ (where the upper bound follows from the pigeonhole principle). For the set cardinality counting problem it is important, as we will see, to determine the asymptotic behavior of the expected value $\mathbb{E}(X)$. In fact, we only seek its crude estimate and show $\mathbb{E}(X) = \Theta(\sqrt{n})$. A more precise asymptotic formula

$$\mathbb{E}(X) = \sqrt{\frac{\pi n}{2}} + \frac{2}{3} + O(1/\sqrt{n}) \approx 1.25\sqrt{n}$$

requires relatively advanced mathematical machinery (see, for example, [**32**] and [**29**]). Even with this minimalistic goal in mind, finding an asymptotic estimate for $\mathbb{E}(X)$ gets rather involved, so we present only major steps without too many technical details.

First, rewriting the expectation sum for the random variable X, we get

$$\mathbb{E}(X) = \sum_{k=2}^{n+1} k \mathbb{P}\{X = k\}$$
$$= \sum_{k=2}^{n+1} \mathbb{P}\{X = k\} + \sum_{k=3}^{n+1} \mathbb{P}\{X = k\} + \cdots + \sum_{k=n+1}^{n+1} \mathbb{P}\{X = k\}$$
$$= \mathbb{P}\{X > 1\} + \mathbb{P}\{X > 2\} + \cdots + \mathbb{P}\{X > n\}$$

$$= \sum_{k=1}^{n} \mathbb{P}\{X > k\}.$$

To determine $\mathbb{P}\{X > k\}$ for every $k = 1, 2, \ldots, n$, denote by β_i the bin into which the ball was tossed in step i. The value of the random variable X is greater than k if and only if a collision has not yet occurred in step k, which further holds if and only if the sequence of bins $\beta_1, \ldots, \beta_k$ into which the balls ended up in steps $1, \ldots, k$ has no repetition. Equivalently, $\beta_1, \ldots, \beta_k$ is a k-permutation of a set of n elements. Since the number of all k-permutations of a set of n elements is $\binom{n}{k} k!$ and the number of all k-strings of a set of n elements is n^k, we have

$$\mathbb{P}\{X > k\} = \frac{\binom{n}{k} k!}{n^k} = \frac{n(n-1)\cdots(n-k+1)}{n^k}.$$

Therefore,

(23) $$\mathbb{E}(X) = \sum_{k=1}^{n} \frac{n(n-1)\cdots(n-k+1)}{n^k}.$$

Second, we are going to asymptotically estimate the fractional part of the sum on the right-hand side of the equation (23). For any $n = 2, 3, \ldots$ and $k = 1, 2, \ldots, n$, denote the fractional part of the sum by

$$\alpha(n, k) = \frac{n(n-1)\cdots(n-k+1)}{n^k}.$$

If we take the natural logarithm of both sides (where the base of the logarithm is the number $e = 2.71828\ldots$), then

$$\ln \alpha(n, k) = \ln \frac{n(n-1)\cdots(n-k+1)}{n^k}$$

$$= \ln\left(n(n-1)\cdots(n-k+1)\right) - \ln n^k$$

$$= \sum_{i=n-k+1}^{n} \ln i - k \ln n.$$

To obtain the lower and upper bounds for $\ln \alpha(n, k)$, we can use the fact that

$$\int_{n-k}^{n} \ln x \, dx \leq \sum_{i=n-k+1}^{n} \ln i \leq \int_{n-k+1}^{n+1} \ln x \, dx$$

and the fact that for $0 < x < 1$ we have

$$\ln(1-x) = -\sum_{i=1}^{\infty} \frac{x^i}{i},$$

to show that

(24) $$\ln \alpha(n, k) \leq -\frac{k^2}{2n}$$

and

(25) $$\ln \alpha(n, k) \geq -\frac{k^2}{2n} - \frac{k^3}{2n^2}.$$

Combining (24) and (25) we get

$$-\frac{k^2}{2n} - \frac{k^3}{2n^2} \leq \ln \alpha(n, k) \leq -\frac{k^2}{2n},$$

that is,

$$1 \le \frac{\ln \alpha(n,k)}{-\frac{k^2}{2n}} \le 1 + \frac{k}{n} \le 2.$$

This implies $\ln \alpha(n,k) = \Theta\left(-\frac{k^2}{2n}\right)$ or, equivalently, $\alpha(n,k) = \Theta\left(e^{-\frac{k^2}{2n}}\right)$. Therefore,

$$\mathbb{E}(X) = \sum_{k=1}^{n} \alpha(n,k) = \sum_{k=1}^{n} \Theta(e^{-\frac{k^2}{2n}}) = \Theta\left(\sum_{k=1}^{n} e^{-\frac{k^2}{2n}}\right).$$

Third, to asymptotically estimate the last sum, put $q = e^{-\frac{1}{2}} = \sqrt{1/e} \approx 0.61$. Also, represent the whole sum as a double sum of roughly $\sqrt{n}$ smaller sums, each with $\sqrt{n}$ summands. Then we can write

$$\mathbb{E}(X) = \Theta\left(\sum_{k=1}^{n} e^{-\frac{k^2}{2n}}\right) = \Theta\left(\sum_{k=1}^{n} q^{\frac{k^2}{n}}\right) = \Theta\left(\sum_{i=1}^{\sqrt{n}} \sqrt{n} \cdot q^{i^2}\right) = \Theta\left(\sqrt{n} \cdot \sum_{i=1}^{\sqrt{n}} q^{i^2}\right)$$
$$= \Theta\left(\sqrt{n}\right),$$

because $\sum_{i=1}^{\infty} q^{i^2}$ is clearly a convergent series.

Turning back to the problem of counting a large-cardinality set, note that throwing a ball uniformly at random into some bin b_i is equivalent to choosing uniformly at random an element x_i from the set $S = \{x_1, x_2, \ldots, x_n\}$. If elements are repeatedly selected from the set S, the number of times when we expect to choose the same element from S is equal to the expected number of steps before the first collision of two balls occurs. Since this expected number is $\Theta(\sqrt{n})$, we can devise a simple randomized algorithm for our problem at hand. Namely, to count the number of elements n of a large set S, we repeatedly pick an element uniformly at random from the set S until we get an element that was already picked. If that has happened in step k, we output $n = k^2$.

RAND-COUNT(S)
$R = \emptyset;$ ⟦ the set of sampled elements ⟧
$k = 1;$ ⟦ number of sampled elements ⟧
Choose $x \in S$ uniformly at random;
while $x \notin R$ **do**
$\quad R = R \cup \{x\};$
$\quad k = k + 1;$
$\quad$ Choose $x \in S$ uniformly at random;
return $k^2;$

Clearly, the RAND-COUNT algorithm runs in $\Theta(\sqrt{n})$ expected time. A statistical analysis of the random variable k^2 as a function of n shows that the error probability of the RAND-COUNT algorithm is small. However, details are formidable and beyond the scope of this book.

3.3. Verifying equality of strings. Consider the problem of testing two long strings a and b for equality. For simplicity, suppose that $a = a_1 a_2 \cdots a_n$ and $b = b_1 b_2 \cdots b_n$ are n-bit binary strings. Of course, if we want complete certainty about equality of a and b, it is not possible to do better than comparing all of a with all of b. The problem with this deterministic approach is the case when the strings are stored in different locations and, for the comparison to be made, one of them would have to be transmitted to the other location. If we take that transmitting one bit of information is a costly operation, we would like to minimize the overall transmission cost. This sort of complexity analysis is referred to as the *communication complexity*. It is often the chief performance measure of distributed algorithms.

A practical example where this is meaningful is comparing the consistency of two replicas of a large database. If the strings a and b represent two copies of a replicated database in remote locations, we would like to decide whether they are equal, but would rather not compare them directly since transmitting one string has communication complexity $\Theta(n)$.

Generally speaking, we are interested in obtaining a "fingerprint" from each string so that by comparing these fingerprints we can decide whether the strings are equal. Moreover, the length of the fingerprint should be much less than the length of the original strings.

Given a prime p, for a binary string $x = x_1 x_2 \cdots x_n$ define the fingerprint of x by

$$f_p(x) = \sum_{i=1}^{n} x_i 2^{i-1} \pmod{p}.$$

In other words, the fingerprint of a binary string x is obtained by taking the integer whose binary representation is the string x, and then computing that integer modulo prime p. Although efficiency of this operation is not our primary concern, we can certainly perform it in $O(n \log p)$ time using the "elementary school" division method.

The following is an algorithm to compare two n-bit binary strings a and b, where the constant $k > 2$ is to be determined later.

RAND-EQUAL(a, b)
    Choose a prime p uniformly at random from $[2, \ldots, n^k]$;
    $s = f_p(a)$;
    $t = f_p(b)$;
    if $s = t$ **then**
        return TRUE;
    else
        return FALSE;

Notice that in the distributed case when a and b are stored in different locations, only the pair p and s needs to be transmitted to the other location, where we then compute t to compare it with s.

If RAND-EQUAL(a, b) returns FALSE, then certainly $a \neq b$. However, if it returns TRUE, it may still be the case that $a \neq b$. Hence, a mistake is only made if we choose prime p such that $f_p(a) = f_p(b)$, provided that $a \neq b$. But the probability that this happens, which is the error probability of the RAND-EQUAL algorithm, is small, as the following theorem demonstrates.

THEOREM *If p is a prime number chosen uniformly at random from the interval $[2, \ldots, n^k]$, then*

$$\mathbb{P}\{f_p(a) = f_p(b) \mid a \neq b\} = O(1/n^{k-2}).$$

PROOF: Think of a and b as the integers whose binary representations are the strings $a_1 a_2 \cdots a_n$ and $b_1 b_2 \cdots b_n$, respectively. If $a \neq b$, then they are equal modulo p if and only if p divides their difference $d = |a - b|$ or, equivalently, p is a prime factor of d. Since $d \leq 2^n$ and every prime number is at least 2, if we write the prime factorization of d as $d = p_1^{i_1} \cdots p_m^{i_m}$, then we have

$$n \geq \log d = \sum_{j=1}^{m} i_j \cdot \log p_j \geq \sum_{j=1}^{m} 1 \cdot 1 = m.$$

Thus, the number m of different prime factors (dividers) of d is bounded by n. On the other hand, by the famous Prime Number Theorem we know that the total number of prime numbers $\pi(N)$ in the interval $[2, \ldots, N]$ is about $N/\ln N$. More precisely, there are positive constants c_1 and c_2 such that

$$c_1 \cdot \frac{N}{\ln N} \leq \pi(N) \leq c_2 \cdot \frac{N}{\ln N}.$$

Therefore,

$$\mathbb{P}\{f_p(a) = f_p(b) \mid a \neq b\} = \frac{\text{the number of primes that divide } |a - b|}{\text{the number of primes in } [2, \ldots, n^k]}$$

$$\leq \frac{n}{\pi(n^k)} \leq \frac{n}{c_1 \cdot \frac{n^k}{k \ln n}} = \frac{k}{c_1} \cdot \frac{\ln n}{n^{k-1}} = O(1/n^{k-2}). \blacksquare$$

In other words, the RAND-EQUAL algorithm correctly compares two strings a and b with high probability. Not only that, since $p \in [2, \ldots, n^k]$, the numbers p, s, and t require only $k \log n = O(\log n)$ bits to represent them. Thus, the communication complexity of the distributed version of the RAND-EQUAL algorithm is $O(\log n)$. Moreover, for a fixed n, the choice of larger k increases the number of bits needed to represent these numbers only linearly, while it decreases the error probability exponentially.

4. Skip lists

In this section we discuss a randomized data structure called *skip list* that can be used to implement the dictionary abstract data structure from Chapter 9. Skip lists have many of the desirable properties of the balanced binary search trees (for example, AVL-trees), but their organization is completely different. To maintain certain balance conditions and assure good performance, operations on a balanced tree deterministically self-adjust the tree as the tree operations are performed on it. Instead, skip lists are balanced probabilistically, which is often easier than explicitly maintaining the balance. Although skip lists have bad worst-case performance, no input consistently causes the worst-case behavior. (This is much like the quick sort when the pivot element is chosen randomly.)

To represent a set S of n (distinct) elements, the idea behind skip lists is to start with just a single sorted linked list containing these elements. If our goal initially is to do just searches and ignore updates (insertions/deletions), we obviously need to look at n elements in the worst case. To speed up this process, we can make a second-level sorted list that contains roughly half of the elements from the original list. Specifically, for each element in the original list, we duplicate it with probability $1/2$. We then link together all the duplicates into a second linked list and add a pointer from each duplicate back to its original. To simplify the search operation, we also add sentinel elements at the beginning and end of both lists. These sentinels contain special values $-\infty$ and $+\infty$, where $-\infty$ is smaller than every possible element of S and $+\infty$ is larger than every possible element of S. Figure 3 illustrates this structure for an example set $S = \{17, 23, 34, 42, 50, 56, 69, 72, 79, 86, 90, 96\}$.

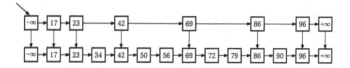

FIGURE 3. The set S represented by the original and shortcut lists.

Intuitively, the list with duplicates represents shortcuts we can use to quickly skip over some elements when searching for a given value. This is similar in nature to an address book with tabs, where the set S consists of names instead of numbers. To lookup up a name in such an address book, we first jump to the tab representing the first character of the desired name, and from there we then perform a sequential scan.

Thus, we can find a given value x in our augmented list structure using a two-phase algorithm. First, we scan for x in the shortcut list, starting from the $-\infty$ sentinel element, while the current element is less than or equal to x. If we find x, we are done. Otherwise, we reach some value v larger than x and we know that x is not further right in the sorted shortcut list. However, x may still be in the original list, but not to the left of the predecessor of v in the shortcut list. So, if u is the predecessor of v in the shortcut list, in the second phase of the search algorithm we go down u to move to the original list. Then we scan for x in the original list, starting from u, while the current element is less than or equal to x. Now, if we reach some value larger than x, we know for sure that x is not in the set S represented by the original sorted list.

For example, if we look for 56 in our example set S, in the first phase we stop at 69 in the shortcut list, drop down 42, and from there we sequentially reach 56 in the original list. This is illustrated in Figure 4.

Since each element appears in the shortcut list with probability $1/2$, the expected number of elements examined in the first phase is at most $n/2$. Observe also that at most one of the elements examined in the second phase has

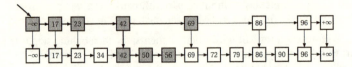

FIGURE 4. Searching for 56 in a two-list structure.

a duplicate. The probability that any element in the original list is followed by $k \geq 0$ elements without duplicates is $(1/2)^{k+1}$, so the expected number of elements examined in the second phase is at most $1 + \sum_{k=0}^{\infty} k(1/2)^{k+1} = 1 + 1 = 2$. By adding these random shortcuts therefore, we have reduced the worst-case cost of a search from n to $n/2 + 2$. This amounts to roughly a factor of two in savings.

Now we can make an obvious improvement to this approach and add another linked list that represents shortcuts to the shortcuts. That is, for each element in the shortcut list, we again duplicate it with probability $1/2$. Since the expected number of elements in the third list is $n/2^2$, we expect to cut the search time by a factor of 2^2. If we repeat this construction recursively, since each new linked list has about half the number of elements of the previous list, the total number of lists should be about $O(\log n)$. Moreover, each time we add another list of random shortcuts, we reduce the search time in half except for a constant overhead. So after $O(\log n)$ lists, we should have a search time of $O(\log n)$.

In fact, this is exactly how skip lists are constructed. For each element in the original sorted list, we repeatedly flip a coin until we get, say, tails. For each heads, we make a duplicate of the element. The duplicates are stacked up in levels, and the elements on the same level are linked together into sorted linked lists. An illustration of this randomized construction for our example set S is shown in Figure 5.

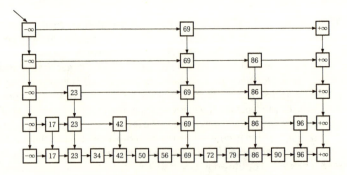

FIGURE 5. A skip list for the example set S.

As we can see from Figure 5, if we follow the links from the highest level, the structure of a skip list becomes similar to that of a binary search tree. Moreover, since each level contains roughly half of the elements from the previous level, we expect to get well-balanced binary tree and thus good performance on average.

The search operation for skip lists is very simple and, in fact, update operations are based on it. Starting from the leftmost element on the highest level, we scan through each level as far as we can without passing the target value, and then turn down to the next lower level. The search ends when we either find an element with the target value on some level, or fail to find the target value on the lowest level. Figure 6 shows the traversal of the data structure of Figure 5 when searching for the value 56.

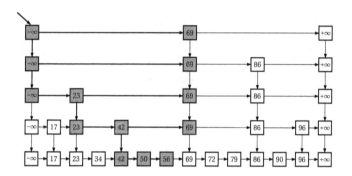

FIGURE 6. Searching for 56 in the example skip list.

4.1. The skip list data structure. Let us now consider an implementation of the skip list data structure and its operations in greater details. A *skip list* L for a dynamic set S consists of a sequence of sorted linked lists $L_0, L_1, \dots, L_m$. In addition, the lists in the skip list L satisfy the following:

1. The list L_0 contains every element of the set S, plus the two special elements with values $-\infty$ and $+\infty$.
2. For each $i = 1, \dots, m - 1$, the list L_i contains random duplicates of the elements in the list L_{i-1}, where each duplicate is included with probability $1/2$. Also, the list L_i always contains the special elements with values $-\infty$ and $+\infty$.
3. The list L_m contains only the special elements with values $-\infty$ and $+\infty$.
4. The skip list L is determined by a pointer (attribute object) L that points to the element $-\infty$ of the list L_m.

It is customary to visualize the skip list L with the list L_0 at the bottom and the lists $L_1, L_2, \dots, L_m$ above it. We view a skip list as a two-dimensional structure arranged horizontally into levels and vertically into columns. Each level i corresponds to the list L_i, and each column contains duplicates of the same element across consecutive lists. An example of the skip list data structure is shown in Figure 7.

Notice that we have pictured the skip list L in Figure 7 with links that have doubly pointed arrows. Although this is not strictly necessary, it makes implementation of the update operations easier. In this case, in addition to the *key* field for an element value, each element has four pointer fields *next, prev, up,*

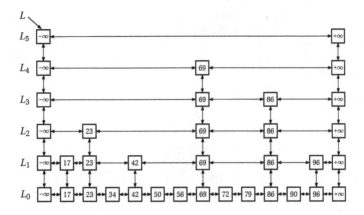

FIGURE 7. Representation of a skip list L.

and *down*. These pointers connect an element with the next and previous elements on the same level, as well as with the duplicates on immediately higher and lower levels.[6]

Since we already elaborated on the search operation, it is easy to give pseudocode for the procedure SL-SEARCH(L, x) that returns a pointer p to an element on the lowest level of the skip list L with the largest key (which could be $-\infty$) that is less than or equal to the given target x.

```
SL-SEARCH(L, x)
    p = L;
    while down(p) ≠ NIL do
        p = down(p);  [ move down ]
        while key(next(p)) ≤ x do
            p = next(p);  [ scan forward ]
    return p;
```

Notice that SL-SEARCH(L, x) returns p such that if $key(p) \neq x$, then the search is unsuccessful but p points to the proper element in L_0 after which x should appear. (Note the similarity with the result of a search in a binary search tree.)

The insert operation for skip lists adds a new element x to the skip list L. We begin with finding the right position p of the bottom-level list L_0 where x should be placed. For this, of course, we use the SL-SEARCH algorithm. Having inserted x in its right place on the list L_0, we "flip a coin." That is, we call the function $random(0, 1)$ that returns a real pseudorandom number between 0 and 1, and if that number is less than $1/2$, then we consider the coin flip to have come up heads, and otherwise tails. If it is tails, then we stop here. On the other hand, if it is heads, then we backtrack on the current level over the elements without duplicates, move to the next higher level and insert x in this level at the

---

[6]Elements of this type are usually called **quad-elements**.

appropriate position. We then again flip a coin: if it comes up tails, we stop; if it comes up heads, we go to the next higher level, insert x there at the appropriate position, and repeat. Thus, we continue to insert a duplicate of x on higher levels until we finally get a coin flip that comes up tails.

In addition to the SL-SEARCH procedure, our SL-INSERT algorithm uses a utility procedure SL-INSERTAAA (p, q, x) (for insert after-and-above) that inserts a new copy of the element x after the same-level element pointed to by p and immediately above the duplicate element of x on the previous lower level pointed to by q. SL-INSERTAAA appropriately adjusts relevant links and returns a pointer to the newly inserted copy. The details of this procedure are left as an exercise, and we just observe that it clearly takes a constant time.

```
SL-INSERT(L, x)
    p = SL-SEARCH(L, x);
    q = SL-INSERTAAA(p, NIL, x);
    while random(0, 1) < 1/2 do
        while up(p) = NIL do
            p = prev(p);   [ scan backward ]
        p = up(p);        [ move up ]
        q = SL-INSERTAAA(p, q, x);
    return q;
```

Figure 8 illustrates how the SL-INSERT algorithm works when an element is inserted into the skip list of Figure 7.

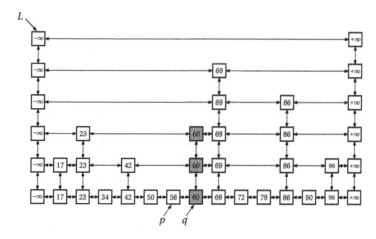

FIGURE 8. Inserting the element 60 into the skip list of Figure 7.

Finally, let us consider the delete operation. Since it is actually easier to delete an element than to insert a new element, we will only briefly explain the deletion routine for skip lists without going into the details of actual pseudocode.

Thus, if the target element is given by a pointer p to it on the bottom level (possibly by using the SL-SEARCH procedure to find it), we simply climb up the corresponding column using the *up* pointers and along the way we remove the column's elements adjusting the relevant links. Notice that before the end of deletion we may have to remove all but one list containing only the special values $-\infty$ and $+\infty$.

4.2. Analysis of skip lists. It is clear that the cost of all operations on a skip list depends on the number of levels of the skip list. That's why we begin our analysis by deriving a bound on the number of levels. Since skip lists are constructed in a randomized manner, the number of levels must be determined using probabilistic reasoning. However, instead of more common computation of the expected number of levels, we show a stronger result that holds with high probability.

LEMMA *With high probability, a skip list representing a set with n elements has $O(\log n)$ levels.*

PROOF: Let L be a skip list that represents a set S with n elements. Given $x \in S$, denote by $N(x)$ the random variable that gives the number of levels in L that contain x, not counting the bottom level. Each new copy of x is obtained with probability $1/2$ from the previous level, essentially by flipping a coin. Thus, x appears in at least $k \geq 0$ levels if and only if we have flipped at least k heads in a row. So,

$$\mathbb{P}\{N(x) \geq k\} = \sum_{i=1}^{\infty} (1/2)^{k+i} = (1/2)^k \sum_{i=1}^{\infty} (1/2)^i = 1/2^k.$$

In particular, for any constant $c > 1$,

$$\mathbb{P}\{N(x) \geq c\log n\} = \frac{1}{2^{c\log n}} = \frac{1}{n^c}.$$

Let N be the random variable that gives the number of levels of the skip list L. Since $N = \max_{x \in S} N(x)$ and $\max_{x \in S} N(x) \geq c\log n$ if and only if there is an element $x \in S$ such that $N(x) \geq c\log n$, we have

$$\mathbb{P}\{N \geq c\log n\} = \mathbb{P}\left\{\max_{x \in S} N(x) \geq c\log n\right\}$$

$$= \mathbb{P}\left\{\bigcup_{x \in S}\{N(x) \geq c\log n\}\right\}$$

$$\leq \sum_{x \in S} \mathbb{P}\{N(x) \geq c\log n\}$$

$$= n \cdot \frac{1}{n^c}$$

$$= \frac{1}{n^{c-1}}.$$

Therefore, given any constant $c > 1$, the number of levels is smaller than $c\log n$ with probability at least $1 - 1/n^{c-1}$. In other words, with high probability, a skip list representing a set with n elements has $O(\log n)$ levels. ∎

Once we have a good upper bound on the number of levels of a skip list L, we can easily get the expected running time of a search in L. In fact, we analyze the cost of a search in two ways. The first way is to recall that the search algorithm involves two nested **while** loops. The inner loop performs a scan forward on a level of L as long as the next element is no greater than the target. The outer loop moves down to the next lower level and repeats the scan-forward iteration. Since the number of levels of L is $O(\log n)$ with high probability, the number of iterations of the outer loop is therefore $O(\log n)$ with high probability.

To bound the number of iterations of the inner loop, i.e., the number of scan-forward steps that we make on any level, let x be the search target. Suppose next that u is the immediate higher level element from which we dropped down to the current level, and v is u's successor on the previous level from which we dropped down. Thus, u and v are the first two neighbor elements on the immediate higher level such that $u < x < v$ and with duplicates on the current level. Then, in the worst case, on the current level we examine all the elements without duplicates between u and v, plus the element v. This means it is enough to compute the expected number of successive elements without duplicates immediately following u on the current level.

But in our introductory discussion, when we argued that each new list of shortcuts reduces the search cost roughly by a factor of two, we basically showed that on any level we can expect that any element is followed by at most one element without duplicates. Thus, the expected number of scan-forward steps from the starting element on any level is at most 2. In other words, the expected time for the scan-forward iteration on any level is $O(1)$. In conclusion, since L has $O(\log n)$ levels with high probability, and on each level we spend $O(1)$ expected time, the search operation in L takes a total of $O(\log n)$ expected time.

The second, somewhat informal but stronger way to analyze the cost of a search is to imagine running the search algorithm "backwards." This means that we take the output from the SL-SEARCH algorithm as input and trace the search back through the skip list up to the upper left corner. If p is the pointer to some bottommost-list element in L that is returned by the SL-SEARCH algorithm, then tracing back a search is equivalent to the following code segment:

```
while p ≠ L do
    if up(p) ≠ NIL then
        p = up(p);
    else
        p = prev(p);
```

To analyze the running time of this code segment, we observe that for *every* element pointed to by p during the trace, $up(p) \neq$ NIL with probability $1/2$ by the construction of the skip list L. Thus, when we reach the upper left corner of L, the expected number of upward jumps is exactly the same as the expected number of leftward moves. Actually, it is not hard to see that these numbers of upward and leftward moves are the same with high probability. In other words, the running time of this code segment is proportional to twice the number of upward jumps, with high probability. Since the number of upward jumps is equal to the number of levels of L, and we have already shown that the number of levels is $O(\log n)$ with high probability, we can conclude that the search time is $O(\log n)$ with high probability.

A similar analysis for insertions and deletions shows that the expected running time of the insert and delete operations is $O(\log n)$.

Finally, consider the space used by a skip list L that represents a set of n elements. From our previous discussion, it follows that the expected number of elements on level i in L is $n/2^i$. Therefore, the expected total number of elements in L is

$$\sum_{i=0}^{m} n/2^i = n \sum_{i=0}^{m} 1/2^i < 2n.$$

This shows we can expect that the skip list L uses linear space $O(n)$ to represent n elements.

Exercises

1. In your favorite programming language implement

    (a) linear congruential generator;
    (b) Mersenne twister.

2. Describe how we can make use of a random number generator to implement the function $random(i, j)$ that returns a random number from the set of positive integers $\{i, i+1, \ldots, j\}$.

3. By a more careful estimate using the integration method, show that the expected number of comparisons in the RAND-QUICK-SORT algorithm is actually

$$\sum_{i=1}^{n-1} \sum_{j=i+1}^{n} \frac{2}{j-i+1} = \Theta(n \log n).$$

4. A randomized algorithm for selecting the k-th smallest element of a set S of n elements works as follows. We first pick a random element $x \in S$ and partition $S \setminus \{x\}$ into two sets S_1 and S_2 of elements smaller than or equal to x and larger than x, respectively. Then, if $|S_1| = k-1$, x is the desired element and we are done. Otherwise, if $|S_1| \geq k$, we recursively select k-th smallest element of the set S_1; and if $|S_1| > k$, we recursively select $(k-|S_1|-1)$-th smallest element of the set S_2. Show that the expected running time of the randomized selection algorithm is $O(n)$.

5. Show how we can implement the update operations on a skip list whose every element contains only two pointer fields *next* and *down*, in addition to the *key* field. What is the space requirement for such an implementation?

6. Assuming the implementation of a skip list with quad-elements, write a pseudocode for the utility procedure SL-INSERTAAA (p, q, x) used in the SL-INSERT algorithm. The procedure SL-INSERTAAA (p, q, x) inserts a new copy of the element x after the same-level element pointed to by p and immediately above the duplicate element of x on the previous lower level pointed to by q. The procedure should appropriately adjust relevant links and return a pointer to the newly inserted copy.

7. Assuming the implementation of a skip list with quad-elements, write a pseudocode for the procedure SL-DELETE(L, x) that deletes element x from skip list L.

12

NP-Completeness

In the previous chapters we have systematically studied the design and analysis of good algorithms. We hope that at this point you have a much better idea on how to go about finding an efficient algorithm when presented with a new algorithmic problem. You now know that you may try different design paradigms (divide-and-conquer, greedy method, dynamic programming, randomization, etc.), as well as that it is important to choose relevant data structures (trees, heaps, graphs, etc.) and their best representations (array or linked based trees, binary or binomial heaps, graph's adjacency lists or adjacency matrix, etc.). In addition, you have built up a solid repertoire of mathematical techniques to prove correctness and estimate the running time of your algorithm.

However, what if all of this does not help you discover an acceptably efficient algorithm for the problem at hand? For example, when you test one of your attempts, your algorithm runs on small problem instances reasonably efficiently, but on moderately large input it never terminates. Analyzing its running time, you realize that on input of size n the algorithm is running in "nonpolynomial" time, perhaps $n^{\log^2 n}$, or $2^{\sqrt{n}}$, or worse yet $n!$.

It is quite possible in practice that you try every trick in the book, and nothing seems to work. Does this mean that you are not smart enough, or that the problem is really computationally hard and actually there is no efficient algorithm to solve it? Should you give up on trying to come up with an efficient algorithm, or should you embark on a slippery road to prove that no efficient algorithmic solution exists? If you want to take this second path, the question is how to do this? Unfortunately, it is still very little known about the answer to the last question, because proofs of this kind are typically even harder to come by than efficient algorithms. Namely, to convince someone that a problem doesn't have an efficient algorithm, you have to prove that every algorithm solving it must be slow. In other words, you must prove that it is impossible to find algorithmically fast way to solve the problem. Notice that it is not enough to take a specific algorithm for the problem and lament its slowness—after all, *that* algorithm may be slow, but maybe there is a faster one.

All is not frustration, however. Although it's hard to prove that a problem doesn't admit an efficient algorithm, you can try to prove that your problem is at least as hard as some other known problem for which there seems to be no fast algorithmic solution. More precisely, there is a (growing) list of problems for which, although it is not provably known that they are computationally hard, it is known that they are *inherently* computationally hard. What this means is that, although there is no efficient algorithm for any of these problems on the list nor a proof to the contrary that any of them doesn't have an efficient algorithm, all of them are computationally equivalent in the sense that if any one of them can be solved efficiently, then all of them can be solved efficiently. In other words, this is not a mere list of seemingly difficult algorithmic problems, but it consists of strongly interrelated problems bound together by tight structure.

Thus, if you manage to add the new problem at hand to the list, you will show at least that your problem is not easy to solve efficiently. Otherwise, if it could, all the problems on the list could be solved efficiently, and no one has yet come up with an efficient algorithm for any of them despite years of coordinated effort and fruitless research. In fact, most computer scientists strongly believe that these problems cannot be solved efficiently, which you can take as a great relief when struggling with an efficient algorithm for the new problem.

This fascinating area is known as the theory of NP-completeness and we will be studying its basic concepts in this chapter. Notice that we are now taking up some sort of a negativistic approach—our goal is no longer to show that a problem can be solved efficiently by giving an efficient algorithm for it; instead, we will be trying to prove that a problem *cannot* be solved efficiently. This implies that our discussion regarding negative results must be more rigorous, because proving that something cannot be done efficiently requires that we have no ambiguities about any notion or claim. Otherwise, they would be open for different interpretations and soon would loose their value. This is radically different from the task of designing an efficient algorithm, where there's no harm of being rather informal with various concepts and rely on an intelligent reader to fill in any missing details.

1. Efficient and inefficient algorithms

Let's first clarify what is meant by an efficient algorithm for a problem. For this purpose we will take a broad perspective and make as general assumptions as possible. Up until now all the algorithms (well, almost all) we presented had the property that their worst-case running time is bounded above by some polynomial in the input size. For example, to sort n numbers, we know algorithms that can do that in $O(n^2)$ time. Likewise, we can find all-pairs shortest paths in a graph on n vertices in $O(n^3)$ time using Floyd's algorithm. Accordingly, we first define a **polynomial-time algorithm** to be an algorithm whose worst-case running time on an input of size n is $O(n^k)$, where k is some constant that is independent of the input size n. We then identify efficient algorithms with polynomial-time algorithms (although in practice this may be an exaggeration in case the running time is, say, $O(n^{100})$). Conversely, inefficient algorithms are regarded as those whose running time is not bounded above by any polynomial function of the input size.

We often use the words fast and slow as synonyms for efficient and inefficient, and thus refer to fast and slow algorithms. The corresponding terminology for problems is defined on the basis of efficiency of algorithms solving them. A problem is said to be **solvable in polynomial time** if there is a polynomial-time algorithm that solves it. We also use the phrase easy problem as synonym for a problem solvable in polynomial time. Similarly, a hard problem is problem that is not easy in this computational sense. Therefore, it is the polynomial time that makes the difference between efficient (fast) and inefficient (slow) algorithms, or easy and hard problems.

Some running-time functions of algorithms that don't look like polynomials (such as $n \log n$) are bounded above by polynomials (for instance, $n \log n = O(n^2)$). Hence the corresponding algorithms are efficient according to our definition. On the other hand, some running-time functions that do look like polynomials are not. For example, recall from Chapter 7 that the running time of the KNAPSACK algorithm was $O(nW)$, where n is the number of items to be packed in a knapsack with the weight limit W. Can we take that W is a constant, hence $nW = O(n)$, and thus consider KNAPSACK a polynomial-time (in fact, linear) algorithm? Well no, not really. The knapsack capacity W is a part of the input to the algorithm, so we could choose $W = 2^n$ and the running time of the KNAPSACK algorithm would not be bounded above by any polynomial in the input size (which is still in the order of n if W is represented as a binary number).

The last example also suggests that we need to be more precise about the notion of the input size of an algorithm. So far we have informally defined the input size as the number of "objects" constituting the algorithm's input: the number of array elements, the number of nodes in a tree, the number of vertices and edges in a graph, and so on. Clearly, on the real computer everything gets converted into bits (or bytes), so the input size reduces to the number of bits used to represent an input instance of a problem. For simplicity, we ignored this fact by implicitly taking for granted that the input is encoded by some reasonably efficient scheme that would not change the asymptotic running time. We did not elaborate much about what reasonably efficient encoding meant, so let's now outline why this is important by showing that different input encodings can actually turn an inefficient algorithm into efficient one.

For example, consider the following algorithmic problem: given an integer $n > 1$, is n prime? This simple problem just asks for a "yes/no" answer, and a straightforward algorithm that solves it can be informally described as follows. On input n, for each integer $k = 2, 3, \dots, n-1$ we compute the remainder of n divided by k to see if k evenly divides n. (For simplicity, we ignore the fact that it is actually enough to iterate k up to $\lfloor \sqrt{n} \rfloor$.) If none of k evenly divides n, we output "yes." Otherwise, the first time we encounter some k that evenly divides n, we output "no." Clearly, the algorithm in the worst case makes about n iterations, and each iteration takes a constant time dividing two integers and testing whether the remainder is zero. Thus, we might say that the worst-case running time of this algorithm is $\Theta(n)$, and so it seems as though this is a fast algorithm.

This could be true if we gave the input number n in unary notation as a string of n ones. For example, the decimal number 17 is expressed in unary notation as 11111111111111111 (the string of 17 digits of 1). Then computing the remainder of n divided by k, if both numbers are given in unary, reduces to splitting up the string of n ones into blocks each consisting of k ones and counting the length of the last block. Of course, the last block has exactly k ones if and only if the remainder is zero. Obviously this single step takes $O(n)$ time, and so the n iterations of the simple algorithm to decide whether n is prime take $O(n^2)$ time. Since the input size is really n, can we conclude then that the problem of testing the primality of a given integer n is easy? The answer is again negative, and the reason is that in this case we have used an input encoding that is not reasonably efficient. Specifically, we can represent the input number n in binary by a string with only about $\log n$ bits. This rules out the second version of our algorithm as an efficient candidate, because artificially increasing the input length to "speed up" an algorithm is unacceptable.

But neither the original version of our algorithm is efficient, if we represent the input number n in binary. To see why, denote input size by m so that roughly $m = \Theta(\log n)$. Then the worst-case running time of the original version is $\Theta(n) = \Theta(2^{\log n}) = \Theta(2^m)$, which is an exponential function of the input size m. That is, the algorithm is far from being efficient. Even the streamlined version that performs only about $\lfloor \sqrt{n} \rfloor$ iterations has the running time $\Theta(\sqrt{n}) = \Theta(2^{(\log n)/2}) = \Theta(2^{m/2})$, and that also grows much faster than any polynomial function of m.

Therefore, the algorithm that we have just discussed for testing the primality of a given integer is slow. In fact, it was a long-standing open question whether primality testing is an easy or seemingly/provably hard problem. Only recently the matter was settled by discovery of a fast algorithm to test for primality (see [1]). In other words, we now know that primality testing belongs to the class of easy problems.

Returning to the issue of input size, we thus formally define it as the number of bits it takes to represent an input instance using any reasonably efficient encoding scheme. Specifically, this means that there is not some significantly shorter way of describing the same input. For example, we saw that representing numbers in unary is not reasonably efficient, because binary notation provides an order of magnitude shorter description of the same information. Similarly, we could specify graphs in some highly inefficient way, such as by listing all of its paths, but this is also disallowed.

To make our discussion more explicit, we settle for one of many reasonably efficient encodings that we call the **standard binary encoding**. This a scheme in which

- characters are encoded in ASCII format (i.e., each character uses a constant number of bits), and strings are encoded as sequences of encoded characters;
- integer and real numbers are represented in binary;
- subsets of a finite universal set of n elements are represented by the strings of the form $\{i_1, i_2, \ldots, i_m\}$, where $i_j = k$ denotes the k-th element of the universal set represented by encoded integer k;

- graphs of n vertices are given using either adjacency matrices or adjacency lists, and their vertices are represented by the integers $1, 2, \ldots, n$ encoded in binary and their edges as pairs of encoded vertices;
- other compound objects are represented using a "natural" combination of the above simpler encoded objects.

Another notion that we should be more careful about is the worst-case running time of an algorithm. First of all, having formally defined the input size as the number of bits n to reasonably encode an input, we should formally define the worst-case running time of an algorithm A as a function of n to be the maximum time taken by A over all possible inputs whose encoding is n bits long. Second, we expressed the time taken by A in terms of the number of basic operations performed. This assumes that each basic operation takes a constant time, but from now on we should assume that any basic operation requires at least as much time as the number of bits used to represent its operands. For example, arithmetic operations on long integers cannot be performed in constant time.

However, don't get intimidated by these messy details. We ignored them in the previous chapters precisely because we would get bogged down in them and this would blur the essential facts we were trying to comprehend. Fortunately, it is not hard to check that if we are using the standard encoding scheme and we are not abusing the power of primitive operations, then the algorithms running in polynomial time and described in the informal way we are used to, will also run in polynomial time using the formal definitions of the input size and running time. Of course, the exponent of the polynomial obtained as a bound for the running time using formal definitions will be larger. However, what is important is that the exponent will be still a constant independent of the input size. Thus, an efficient informal algorithm is also efficient in the formal sense.

In conclusion, neither in this context is it necessary to be overly pedantic under the clear understanding that the formal definitions are important reference point. That is, in case of any doubt we must revert to them to resolve any ambiguities.

2. Decision problems

Proving that general problems cannot be solved efficiently is hard, so a natural approach to cope with this difficulty is to consider a restricted class of problems. The idea is first to pinpoint the essential source of the computational complexity in these simpler problems, and then hope to be able to extend the new insights to general problems.

The simpler problems on which we want to focus our attention are called decision problems. A ***decision problem*** is a problem whose result is a simple "yes" or "no" answer. In other words, the notion of a decision problem captures all problems with single output value of type $0/1$, TRUE/FALSE, accept/reject, or of any two-valued type. Since decision problems can have only two possible results, either the answer "yes" or the answer "no," they are traditionally stated in the informal form of a question having a "yes" or "no" answer. We follow this practice, although keep in mind that we are actually asking for an algorithm that solves a decision problem by returning 1 (TRUE) if presented by a yes-instance of the problem and 0 (FALSE) if presented by a no-instance of the problem.

We already met one decision problem called the *primality testing problem*: given an integer $n > 1$, is n prime? Many other interesting decision problems come from the graph theory. For example, recall that if $G = (V, E)$ is an undirected graph, then a Hamiltonian path (cycle) of G is a simple path (cycle) in G containing every vertex from V. Then the *Hamiltonian path problem* phrased in the form of a question with a "yes" or "no" answer is: given an undirected graph G, does G have a Hamiltonian path? Likewise, the *Hamiltonian cycle problem* is: given an undirected graph G, does G have a Hamiltonian cycle?

One of the reasons to consider decision problems is because many seemingly hard problems are phrased in the form of a question with a "yes" or "no" answer. Notice that for decision problems we require only the correct simple answer—"yes" or "no." In particular, it is not necessary to produce a factor of an integer if it's not a prime, or to exhibit a Hamiltonian path if a graph does have one.

Another reason to study decision problems is because a large class of other seemingly hard problems is the class of optimization problems, and we can often easily turn an optimization problem into a decision problem. Namely, since optimization problems ask for some optimal (maximal or minimal) value, we can introduce a new parameter k and ask if the optimal value is at least or at most k. An example of an optimization problem is the *traveling salesperson problem*: given a complete weighted graph with nonnegative edge weights, find a cycle that visits all vertices and has the minimal length. To turn this into a decision problem, we introduce a new input parameter k and rephrase the problem into the form of a question with a "yes" or "no" answer: given a complete weighted graph with nonnegative edge weights and given a number k, does the graph have a cycle that visits all vertices and whose length is at most k?

This may seem like a less interesting formulation of the problem. After all, it does not ask for the minimum-length cycle, nor even for the value of the minimal length. However, recall that our goal is to show that this kind of problems *cannot* be solved efficiently. Thus, if we show that a simpler decision problem cannot be solved efficiently, then the corresponding more general optimization problem certainly cannot be solved efficiently either.

In the previous example, if we succeeded to show that the traveling salesperson *decision* problem is hard, we could argue then by contradiction to show that the traveling salesperson *optimization* problem is hard. Namely, assuming the traveling salesperson optimization problem were easy, i.e., there were an efficient algorithm, say B, to solve it, we could also construct an efficient algorithm, say A, to solve the traveling salesperson decision problem. Since this would contradict the "proved" fact that the traveling salesperson decision problem is hard, it would follow that the traveling salesperson optimization problem is hard as well. Under the assumption that we have had B, we could design A in the following way. Given a complete weighted graph G with nonnegative edge weights and a number k, the algorithm A would first call B to find a minimum-length cycle in G. Then A would sum up the edge weights along the cycle to compute the length, say ℓ_{min}, of a minimum-length cycle in G. Finally, A would compare

ℓ_{min} and k, and if $\ell_{min} \le k$ it would output "yes" and otherwise "no." Clearly, A would take polynomial time to solve the traveling salesperson decision problem.

Another theoretical benefit of the decision problems is that they can be associated with sets in the mathematical sense, and so we can use all mathematical rigor to reason about them. Namely, a decision problem Π is entirely characterized by those problem instances that produce a yes-answer. If I_Π is the set of instances of Π, we can think of Π as the mapping $\Pi: I_\Pi \rightarrow \{yes, no\}$. Then Π can be identified with the set $\{x \in I_\Pi: \Pi(x) = yes\}$. For example, the primality testing problem, call it PRIME, is associated with the set of prime numbers

$$\{x \in I_{PRIME}: PRIME(x) = yes\} = \{n \in \mathbb{N}: n \text{ is prime}\} = \{2, 3, 5, 7, 11, 13, \dots\}.$$

Since we can identify a decision problem with its set of yes-instances, it is customary to use the same name for a decision problem and the associated set, e.g.,

$$PRIME = \{2, 3, 5, 7, 11, 13, \dots\}.$$

For the Hamiltonian path problem, call it HP, the corresponding set is

$$HP = \{G: G \text{ has a Hamiltonian path}\},$$

where G is a graph given by, say, its adjacency matrix. Similarly, for the Hamiltonian cycle problem, call it HC, we have

$$HC = \{G: G \text{ has a Hamiltonian cycle}\}.$$

Finally, for the traveling salesperson problem, call it TSP, the associated set is

$$TSP = \{(G, k): G \text{ has a cycle of length at most } k\}.$$

This set consists of ordered pairs (G, k) such that the first element is a wighted graph G and the second element is a number k.

Decision problems are useful on the practical side, too. Even though easiness of a decision problem does not immediately imply anything about the corresponding optimization problem—it could be easy as well as hard—with little extra work we can usually find that the corresponding optimization problem is also easy.

For example, consider the graph coloring optimization problem: given a graph, find the minimum number of colors to color vertices of the graph in such a way that no two neighboring vertices have the same color. The corresponding decision problem is: given a graph and an integer k, does the graph have a proper coloring of its vertices with at most k colors. Having a fast algorithm for this decision problem, we could use it to apply the "interval halving" technique in the design of a fast algorithm for the corresponding optimization problem. Informally, since the number of colors is between 1 and n, where n is the number of vertices, we would first solve the decision problem for $k = n/2$. If the answer is "yes," then the minimum number of colors is between 1 and $n/2$, and so we would next solve the decision problem for $k = n/4$. Otherwise, if the answer is "no," then the minimum number of colors is between $n/2$ and n, so we would next solve the decision problem for $k = 3n/4$. Clearly, continuing the procedure in this way and increasing the running time by only about a factor of $\log n$, we can efficiently solve the graph coloring optimization problem.

3. Complexity classes

We have seen that the question of how computationally hard or easy it is to solve a (decision) problem is answered on the basis of efficiency of algorithms that solve the problem. The first criterion we used to separate the class of easy problems from hard ones was existence of a polynomial-time algorithm for a problem. However, to get a finer structure on problems, we impose different conditions on the algorithms solving problems and group problems satisfying the criteria into *complexity classes*.

Since the conditions imposed on algorithms abound and are sometimes very subtle, we now need to be more careful about some notions related to algorithms. First of all, given a decision problem Π, an algorithm A **solves** (or **decides**) the problem Π if A halts on any input x and outputs 1 if and only if x is a yes-instance of Π. Such an algorithm A with the running-time function $T_A(n)$ is a **polynomial-time algorithm** if there exists a polynomial $p(n)$ such that $T_A(n) \le p(n)$ for every $n \in \mathbb{N}$.[1]

We define the complexity class P to be the set of all decision problems that can be solved in polynomial time, that is,

$$P = \{\Pi: \text{there is a polynomial-time algorithm } A \text{ that solves } \Pi\}.$$

The class name P stands for "polynomial," since this class contains all decision problems that can be solved (decided) in polynomial time of their input size. As already mentioned, we generally consider these problems as efficiently solvable.

Almost all problems that we have discussed so far in the book (turned into decision problems) belong to the class P. (A notable exception is the knapsack problem from Chapter 7.) As another example, consider the **minimum-path problem** (**MINPATH**): given a weighted graph G and its two vertices s and t, and given a number k, is there a path in G from s to t whose length is at most k? More formally,

$$\text{MINPATH} = \{(G, s, t, k): G \text{ has a path from } s \text{ to } t \text{ of length at most } k\}.$$

To show MINPATH $\in$ P, we present an algorithm A that decides MINPATH as follows. On input (G, s, t, k), A first runs Dijkstra's algorithm from s on G, and then compares its output value $d(t)$ with k. If $d(t) \le k$, A outputs 1; otherwise, A outputs 0. The running time $T_A(n)$ is clearly the same as Dijkstra's algorithm up to a constant, hence $T_A(n) = O(n^3)$.

Now we focus our attention on problems that are *not* known to be in P. It turns out that several problems that are not known to be in P have the following property: if x is a yes-instance of the problem, there is a "short" certificate y to the fact that x is a yes-instance. No such certificate may exist for no-instances (notice the asymmetry). "Short" here means that y's size is bounded by a polynomial in x's size. For example, suppose someone wants to convince you that a graph G with n vertices is a yes-instance of the Hamiltonian path problem (that is, $G \in$ HP). This means that the person claims that G contains a simple path that includes all vertices in G. To prove his/her claim, the person could give you a list of n vertices on a Hamiltonian path in G. This list represents a certificate

---

[1] Concisely and equivalently said, $T_A(n) = O(n^k)$ for some constant k.

to the alleged Hamiltonianicity of G. To get you convinced then is easy, since all you need to do is check whether each pair of consecutive vertices in the list is connected by an edge and whether each vertex of the graph appears exactly once in the list. You can do this efficiently, and if the given list passes the test there is no doubt that $G \in$ HP.

This idea of checking whether a potential certificate proves that an instance is indeed a yes-instance to a given problem is captured by the notion of a verifying algorithm. A **verifying algorithm** is a *two-argument* algorithm acting as a verifier for a decision problem, so we think of one argument as an instance of the problem and of the other argument as a certificate to the instance. Given a decision problem Π, a verifying algorithm $V_\Pi(x, y)$ **verifies** Π if for any yes-instance x of Π there exists a certificate y for which $V_\Pi(x, y) = 1$, and if for any no-instance of Π there exists no certificate y for which $V_\Pi(x, y) = 1$. More formally, expressing Π as the associated set of yes-instances we have

$$\Pi = \{x \in I_\Pi : \exists y, V_\Pi(x, y) = 1\}.$$

Notice that a verifying algorithm does not have to verify no-instances (it may even run forever for a no-instance and any certificate), but no fake certificate can fool the verifying algorithm.

Since we are interested in an efficient certificate-checking procedure, we further require that verifying algorithms run in time that is bounded by a polynomial in the size of yes-instances. This implies that without loss of generality we may assume that the size of a certificate is also bounded by a polynomial. More precisely, we say that $V_\Pi(x, y)$ is a **polynomial-time verifying algorithm** for a decision problem Π (or $V_\Pi(x, y)$ **verifies** Π **in polynomial time**) if there exists a polynomial $p(n)$ such that for any yes-instance x of Π with size n there exists a certificate y with size at most $p(n)$ for which $V_\Pi(x, y) = 1$ in time bounded by $p(n)$. In case of no-instances we impose no time constraints.

We define next the complexity class NP to be the set of all decision problems that can be verified in polynomial time, that is,

NP $= \{\Pi$: there is a polynomial-time verifying algorithm V_Π that verifies $\Pi\}$.

The class name NP does *not* mean "Not Polynomial time," but historically it meant "Nondeterministic Polynomial time." The original name has its roots in nondeterministic computational models, but for our purposes it is more intuitive to view the class from the perspective of verification.

As an example, our previous discussion on the motivation for introducing the concept of a verifying algorithm implies that HP $\in$ NP. An algorithm V_{HP} that verifies the HP problem takes a graph G and a list of its vertices as the certificate, and checks whether each pair of consecutive vertices in the list is an edge and whether each vertex of the graph appears exactly once in the list. The size of the certificate and the running time of the verifying algorithm are clearly polynomially related to the size of G, and so V_{HP} verifies HP in polynomial time.

Notice that not all decision problems have the property that they are easy to verify. For example, it is not clear that even the PRIME problem is such a problem.[2] Namely, what information should someone give us to certify that an integer n is a prime number? That person could give us a list of all remainders of n divided by the numbers between 2 and n, but it is not enough to check if none of them is 0. Thus we would anyway have to divide n with each number between 2 and n and compare our results with the given list. So, instead of comparing our remainders with 0 as we did in Section 1, we would compare them with the corresponding numbers in the list. This would not give us any improvement over the old algorithm, which we argued is not a polynomial-time algorithm.

Similarly, if we take the "complement" problem to HP as

$$\overline{\text{HP}} = \{G\colon G \text{ has no Hamiltonian path}\},$$

it is not clear again what certificate could be given to us to efficiently convince ourselves that a given graph G has no Hamiltonian path.

What can we say about the structure of the sets P and NP? Unfortunately, except for the obvious inclusion $P \subseteq NP$, not too much. This inclusion is clear because if $\Pi \in P$, that is, there is an algorithm A that decides the problem Π in polynomial time, we can use the same algorithm A to verify Π in polynomial time. If we simply add a dummy second argument to A, we do not have to use it at all as a certificate to get $A(x, y) = 1$ for any yes-instance x of Π in polynomial time. Thus, it follows that $\Pi \in NP$.

In fact, a humble-looking question like whether this inclusion is proper or the two sets are equal is probably a central open question in computer science: is $P = NP$? No one has yet found a decision problem that is in NP but not in P, nor proved $NP \subseteq P$ by showing that every problem in NP can be solved in polynomial time!

For example, we have shown $HP \in NP$, but no one knows whether $HP \in P$ or $HP \notin P$. That is, no one has come up with an algorithm that solves HP in polynomial time. The best algorithm that decides if an arbitrary graph G has a Hamiltonian path seems to be a simple brute-force approach that generates all $n!$ permutations of the n vertices in G and checks each permutation if it constitutes a Hamiltonian path in G. But in the worst case this takes $\Omega(n!)$ time, which is not a polynomially bounded time. On the other hand, neither has anyone proved that there can be no polynomial-time algorithm that solves HP. Thus, the question whether $HP \in P$ or $HP \notin P$ is still open.

The vast majority of researchers strongly believe that P is a proper subset of NP. This is justified for two reasons. First, there is a huge list of over a thousand of problems in NP, but it is still unknown a polynomial-time algorithm for a single one of them, despite coordinated effort and many years of research. Thus, it is very unlikely that $P = NP$. Second, intuitively seems harder to solve a problem in polynomial time than to verify the problem in polynomial time.[3] Thus, many

---

[2]But it *is* known that PRIME admits efficient verification and that, in fact, it is polynomially solvable.

[3]A nice mathematical analogy is that it is usually much harder to prove a theorem than to simply check whether a given proof to the theorem is correct.

believe that there must be at least one problem that is easy verifiable (in NP) but not easy solvable (not in P).

We hinted above at complement problems having a peculiar property that they might not be in NP even though original problems are in NP. Given a decision problem Π, the complement problem $\overline{\Pi}$ is associated with no-instances of Π, that is,

$$\overline{\Pi} = \{x \in I_\Pi : \Pi(x) = no\}.$$

In other words, for every instance $x \in I_\Pi$, $\overline{\Pi}(x) = yes$ if and only if $\Pi(x) = no$. The class of decision problems for which the corresponding complement problems belong to NP is called Co-NP:

$$\text{Co-NP} = \{\Pi : \overline{\Pi} \in \text{NP}\}.$$

Thus, a decision problem is in Co-NP if for every no-instance of it there is a short certificate to this fact that can be verified in polynomial time. Notice that the definitions of NP and Co-NP are not symmetric. Just because we can verify every yes-instance quickly may not imply that we are be able to check no-instances quickly, and vice versa. For example, we have seen that $\overline{HP} \in$ Co-NP, but it is not known whether $\overline{HP} \in$ NP. This raises another question: is NP = Co-NP? Again, we don't have a conclusive proof—everyone believes NP $\neq$ Co-NP, but nobody really knows. Figure 1 shows the known relationships between the three complexity classes mentioned so far.

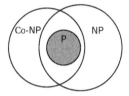

FIGURE 1. Hierarchy of complexity classes.

We do not know whether all of these complexity classes are distinct or whether they all collapse into the single class P. Put it another way, the following are some of the major unsolved questions:

- P = NP?
- NP = Co-NP?
- P = NP $\cap$ Co-NP?

4. Transformations (reductions) and NP-complete problems

Since most experts believe P $\neq$ NP, a natural approach in proving somehow this inequality is to isolate the "hardest" problems in NP and to try to show that they are not polynomial-time solvable. These NP problems should be hardest in a sense that every other problem in NP reduces to them. In the sense thus they should posses the largest intrinsic computational difficulty, so the best chance of showing nonpolynomial-time solvability is to try it on them. The hardest problems in NP are called NP-complete problems, and the subclass containing them is denoted NPC. Another nice property of the NP-complete problems is that if we can find a polynomial-time algorithm solving any one of them, then all problems in NP would be polynomial-time solvable or, equivalently, it would imply $P = NP$.

To clarify this discussion, we begin with the key notion of a *problem reduction*. A problem reduction is essentially a formalization of the programming concept of a subroutine. Informally, given two (not necessarily decision) problems Π_1 and Π_2, Π_1 reduces to Π_2 if an efficient algorithm for Π_2 implies that we can use it as a subroutine to efficiently solve Π_1.[4] For example, in showing MINPATH $\in$ P, we basically reduced the MINPATH problem to the SSSP problem[5] and used an efficient solution to the SSSP problem (Dijkstra's algorithm) as a subroutine to efficiently solve the MINPATH problem.

We take a somewhat more restricted view of reductions between two decision problems and think of them as efficient transformations of instances of one problem to instances of the other.[6] Moreover, some transformation of a decision problem Π_1 to a decision problem Π_2 satisfies that a yes-instance of Π_1 is transformed to a yes-instance of Π_2 and a no-instance of Π_1 is transformed to a no-instance of Π_2.

Formally, given two decision problems Π_1 and Π_2, we define a ***polynomial-time transformation*** of Π_1 to Π_2 to be a function $f\colon I_{\Pi_1} \rightarrow I_{\Pi_2}$ that maps instances of Π_1 to instances of Π_2 and satisfies the following two conditions:

1. For every instance $x \in I_{\Pi_1}$, x is a yes-instance of Π_1 if and only if $f(x)$ is a yes-instance of Π_2.
2. There is a polynomial-time algorithm A that computes f. That is, there exists a polynomial $p(n)$ such that on every input $x \in I_{\Pi_1}$ of size n, the algorithm A outputs $f(x) \in I_{\Pi_2}$ in time bounded by $p(n)$.

Roughly speaking, a polynomial-time transformation f of Π_1 to Π_2 is an algorithm A that quickly converts every instance x of Π_1 to an instance $f(x)$ of Π_2 such that x is a yes-instance of Π_1 if and only if $f(x)$ is a yes-instance of Π_2. Observe also that for an input x of size n, the size of the A's output $f(x)$ can be no grater than $p(n)$. This follows from the simple fact that an algorithm in each step can write at most one memory location.

If there is a polynomial-time transformation of Π_1 to Π_2, we write $\Pi_1 \leq \Pi_2$ (read "Π_1 transforms to Π_2"). Also, instead of the modifier "polynomial", we often use the term "efficient," or "fast," or simply drop it altogether but it should be implicitly understood. Figure 2 illustrates the definition of a polynomial-time transformation.

Before we consider the significance of problem transformations, it may be helpful to take a small example and carry out a transformation between two concrete problems. To this end we show HP $\leq$ HC, where HP is the Hamiltonian path problem and HC is the Hamiltonian cycle problem. Therefore, what we need to demonstrate is an algorithm that quickly converts an instance of HP,

---

[4]This type of reduction is sometimes called a *Turing* or a *Cook* reduction.

[5]The SSSP (single-source shortest paths) problem is: given a weighted graph G with nonnegative weights and a vertex s in G, output a list of shortest-path lengths from s to every other vertex in G.

[6]This type of reduction is sometimes called a *many-one*, a *Karp*, or a *mapping* reduction.

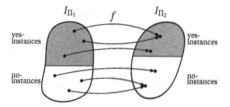

FIGURE 2. Reduction from Π_1 to Π_2 viewed as a transformation.

that is, an undirected graph G, to an instance of HC, that is, another undirected graph G', such that G has a Hamiltonian path if and only if G' has a Hamiltonian cycle. We first describe a function f that maps G to G', and then argue that the function can be algorithmically implemented in polynomial time in the size of G.

Now, given an undirected graph $G = (V, E)$ with n vertices and m edges, $f(G)$ is the graph G' that consists of all the vertices and edges of G plus one additional vertex not in V, say u, and n additional edges that connect u with each vertex from V, as illustrated in Figure 3. Formally $G' = (V', E')$, where $V' = V \cup \{u\}$ and $E' = E \cup \{(u, v) : v \in V\}$.

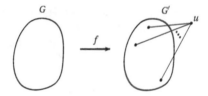

FIGURE 3. Transformation of an instance of HP to an instance of HC.

It is now easy to see that G is a yes-instance of HP if and only if G' is a yes-instance of HC. Namely, if G has a Hamiltonian path, say $s, \ldots, t$, containing every vertex from V exactly once, then clearly $u, s, \ldots, t, u$ is a Hamiltonian cycle in G'. Conversely, if G' has a Hamiltonian cycle, then the cycle must be in the form $u, s, \ldots, t, u$ with the portion $s, \ldots, t$ being a simple path in G that contains all its vertices. Hence G has a Hamiltonian path.

Given G, it remains to see that the construction of $G' = f(G)$ can be efficiently implemented. But this is trivial—(the representation of) the graph G' is obtained from (the representation of) the graph G simply by adding in one vertex and all edges from that vertex to the vertices of G. This can be clearly done by an algorithm whose input is the graph G, and the time it takes to produce G' is obviously proportional to the size of G'. Since the size of G' is $O(n + 1 + m + n) = O(n + m)$, it is the same order as the size of G, and so the algorithm runs in polynomial (in fact, linear) time in the size of G. This completes the argument that the HP problem is polynomial-time transformable to the HC problem.

Observe that if we had an efficient algorithm that solves the HC problem, then we could also solve the HP problem efficiently as follows. To decide whether an undirected graph G has a Hamiltonian path, we first use the above transformation to convert G to G', and then run the HC solver on G' returning its answer "yes" or "no" as the solution for G. More generally, the usefulness of transformations comes from the following lemma.

LEMMA 1 *If* $\Pi_1 \leq \Pi_2$ *then* $\Pi_2 \in P$ *implies* $\Pi_1 \in P$ *(or, equivalently,* $\Pi_1 \notin P$ *implies* $\Pi_2 \notin P$).

PROOF: Let $f\colon I_{\Pi_1} \to I_{\Pi_2}$ be the transformation of Π_1 to Π_2 that is computed by a polynomial-time algorithm A. Assuming $\Pi_2 \in P$, we have a polynomial-time algorithm B that decides Π_2. Then, a polynomial-time algorithm C that decides Π_1 can be constructed in the following way: on input $x \in I_{\Pi_1}$, C first calls $A(x)$ to compute $f(x) \in I_{\Pi_2}$, then calls $B(f(x))$, and finally returns the result of this call. In other words, $C(x) \equiv \mathbf{return}\ B(A(x))$. Clearly, the algorithm C correctly decides Π_1 because

x is a yes-instance of Π_1

$\Leftrightarrow$ $f(x)$ is a yes-instance of Π_2,	since A transforms Π_1 to Π_2
$\Leftrightarrow$ $B(f(x)) = 1$,	since B decides Π_2
$\Leftrightarrow$ $C(x) = 1$,	by construction.

Furthermore, the observation about the output size of a polynomial-time algorithm implies that if A's running-time function is $T_A(n) = O(n^{k_1})$ for some constant k_1, then on input x of size n the A's output $f(x)$ has size $O(n^{k_1})$. Thus, if B's running-time function is $T_B(n) = O(n^{k_2})$ for some constant k_2, then C's running-time function is

$$T_C(n) = O(n^{k_1}) + O\left((n^{k_1})^{k_2}\right) = O(n^{k_3}),$$

for some constant $k_3 = k_1 k_2$. ∎

This lemma justifies the notation "$\leq$" for a transformation of Π_1 to Π_2 because it indicates that Π_1 is an "easier" (or more precisely, "no harder") problem than Π_2. This is in the sense that if we could efficiently solve Π_2, then Π_1 would be readily solvable in polynomial time.

Another useful property of transformations between decision problems is that the transformation relation is transitive.

LEMMA 2 *If* $\Pi_1 \leq \Pi_2$ *and* $\Pi_2 \leq \Pi_3$, *then* $\Pi_1 \leq \Pi_3$.

PROOF: Let f and g be the transformations from Π_1 to Π_2 and from Π_2 to Π_3, respectively. Then their composition $h = g \circ f$ defines the desired transformation $h\colon I_{\Pi_1} \to I_{\Pi_3}$. Clearly,

x is a yes-instance of Π_1 $\Leftrightarrow$ $f(x)$ is a yes-instance of Π_2

$\Leftrightarrow$ $g(f(x))$ is a yes-instance of Π_3

$\Leftrightarrow$ $h(x)$ is a yes-instance of Π_3.

Also, a polynomial-time algorithm that computes h is constructed in an analogous way to that used in the proof of the previous lemma. ∎

We are now ready to give a formal definition of the NP-complete problems in terms of reducibility. A decision problem Π is **NP-complete** if

1. $\Pi \in NP$, and
2. for every $\Pi' \in NP$, $\Pi' \leq \Pi$.

In other words, a decision problem $\Pi \in NP$ is NP-complete if every other decision problem in NP is polynomial-time transformable to Π. This definition captures our intuition behind the NP-complete problems as the hardest problems in NP, because if we could solve any NP-complete problem in polynomial time, than all problems in NP would be solvable in polynomial time by Lemma 1, that is, P = NP. Conversely, if we could prove that any NP-complete problem cannot be solved in polynomial time, then no other NP-complete problem can be solved in polynomial time.

Sometimes it is not clear whether a decision problem belongs to NP, but it does satisfy the condition that every problem from NP transforms to it. Such problems are called NP-hard problems, that is, a decision problem Π is **NP-hard** if for every $\Pi' \in NP$, $\Pi' \leq \Pi$. We can therefore express NP-complete problems in terms of NP-hard problems by saying that a decision problem is NP-complete if (1) it belongs to the class NP and (2) it is NP-hard.

If we denote this subclass of all NP-complete problems by NPC, that is,

$$NPC = \{\Pi \in NP : \Pi \text{ is NP-complete}\},$$

then the tentative picture of the NP world in Figure 1 can now be refined as in Figure 4.

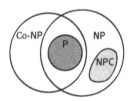

FIGURE 4. The (possible) structure of the NP world.

Consider now the question of how we go about proving that some problem is NP-complete. Suppose that we have a concrete decision problem Π at hand that we suspect is NP-complete. Presumably, we have tried hard to find an efficient algorithm for Π without success. However, an efficient verifying algorithm for Π is usually easier to come up with, so suppose that we know that our Π is in NP. So far, so good. Unfortunately, the second condition of the definition of an NP-complete problem asks that we show how to efficiently transform *every* problem in NP to our problem Π. There are potentially infinitely many problems in NP, so how can we ever hope to be able to do this? One way out of this difficulty is to use transitivity of the transformation relation and simplify the NP-hardness requirement. Namely, if we know one NP-complete problem Ψ, then it is enough to show that $\Psi \leq \Pi$, since every NP problem transforms to Ψ and so by transitivity also to our problem Π. This reasoning captures the essence of proving NP-completeness and proves the following lemma.

LEMMA 3   *A decision problem* Π *is NP-complete if*

1. Π ∈ NP, *and*
2. there is some NP-complete problem Ψ *such that* Ψ ≤ Π.

However, to start effectively using this lemma, we need to know at least one NP-complete problem. (Note that so far we even don't know whether NPC ≠ ∅.) To show then that a given problem in NP is NP-complete, it is enough to under-take the presumably simpler task of demonstrating a fast transformation of one of the known NP-complete problems to the given one. The first decision prob-lem shown to be NP-complete is called SAT (short for Boolean formula satisfi-ability).[7] Since then hundreds of NP-complete problems have been discovered using this technique of transforming one of the known NP-complete problems, with an ever-increasing list to choose from, to the problem in question.

Before specifying the SAT problem, we quickly recall some basic notions of the propositional logic. Given a set $\{x_1, x_2, \dots, x_n\}$ of Boolean variables that can have truth value 0 (false) or 1 (true), a **Boolean formula** is an expression con-structed over the variables $\{x_1, x_2, \dots, x_n\}$ and the Boolean operators ¬, ∧, and ∨ in such a way that the expression can be parsed only by applying the following two rules a finite number of times:

- the Boolean variables are Boolean formulas;
- if E and F are Boolean formulas, then the *negation* ¬(E), *conjunction* ($E \wedge F$), and *disjunction* ($E \vee F$) are also Boolean formulas.

For example,

$$(x_1 \wedge (x_2 \vee \neg(x_3))) \wedge (\neg(x_2 \vee x_3) \vee x_3)$$

is a Boolean formula F over the variables $\{x_1, x_2, x_3\}$.

The **size** of a Boolean formula is the number of *occurrences* of variables in the formula. Thus the size of the example formula F is 6.

The **truth value** of a Boolean formula is evaluated by assigning the truth value 0 or 1 to each variable and computing the truth values of the subformulas in the usual way: ¬(E) is 1 iff E is 0, ($E \wedge F$) is 1 iff both E and F are 1, and ($E \vee F$) is 1 iff E or F (or both) is 1. For example, the assignment $x_1 = 0, x_2 = 1, x_3 = 1$ results in the truth value 0 of the example formula.

We say that a Boolean formula is **satisfiable** if there is an assignment of val-ues to its variables such that the truth value of the formula evaluates to 1. For example, the assignment $x_1 = 1, x_2 = 0, x_3 = 1$ makes the example formula eval-uate to 1, so that formula is satisfiable.

Note that a Boolean formula is *not* satisfiable if no matter how we assign truth values to the variables, the truth value of the formula is always 0. For ex-ample, the formula

$$(x_1 \vee x_2) \wedge \neg(x_1 \vee x_2)$$

over $\{x_1, x_2\}$ is not satisfiable.

---

[7]This was proved by Stephen Cook in 1971 and independently by Leonid Levin in about the same time.

Under the standard precedence rules for the Boolean operators $\neg$, $\wedge$, and $\vee$, we adopt the usual convention of omitting unnecessary parenthesis in Boolean formulas. Also, some familiar facts from propositional logic (DeMorgan's laws, tautologies, and so on) are implicitly used when needed.

The ***satisfiability problem (SAT)*** is this simple question: given a finite set of Boolean variables and a Boolean formula over the variables, is the formula satisfiable? That is, is there an assignment of truth values to the variables so that the given formula evaluates to 1? SAT is clearly a decision problem, because every instance of it (a Boolean formula over some variables) has either "yes" or "no" answer (the formula is or isn't satisfiable). In the set notation, we have

$$\text{SAT} = \{F : F \text{ is a satisfiable Boolean formula}\}.$$

An obvious algorithm for testing the satisfiability of a formula is to systematically generate all combinations of truth values given to the variables and evaluate the formula for each combination, until the truth value of the formula is evaluated to be 1. Actually, this is more or less the best algorithm known for the general satisfiability problem. However, if the number of variables is n and the size of the formula is bounded by a polynomial of n so that the total input size of the problem is still an order of a polynomial of n, then the exhaustive search takes $\Omega(2^n)$ time in the worst-case, which is not polynomially related to the input size.

A natural question to ask then is whether there exists a polynomial-time algorithm for SAT at all? Unfortunately, our current knowledge is not strong enough to give a definite answer to this question. But at least we can prove that SAT is NP-complete, which strongly indicates that it is hard to solve efficiently.

The proof that SAT is NP-complete is involved and requires some background on formal-language theory and Turing machines. We present all of this in the next section that can be skipped on the first reading. The reader can merely take the fact that SAT is NP-complete for granted and move on to Section 6 to study some other interesting NP-complete problems.

5. The first NP-complete problem

We now study in more detail the satisfiability problem that has been a cornerstone of the NP-completeness theory. Since in this section we intend to prove that SAT is NP-complete, the concepts discussed previously must be understood without any ambiguity. In other words, we need more precise definitions of decision problems and algorithms. The theory of NP-completeness historically grew out of the automata and formal-language theory, hence the basic notions are usually formally presented in the formal-language framework.

5.1. Decision problems and formal languages. Decision problems are formalized in terms of languages in the formal-theoretic sense (sets of strings). To see how, we first review some basic notions from the formal-language theory.

An ***alphabet*** is a finite (nonempty) set of symbols, and a ***string*** over some alphabet is any finite sequence of symbols from the alphabet. A ***language*** over some alphabet is any set of strings made up of symbols from the alphabet. We denote alphabets by the Greek letters Σ, Γ, and so on. For example, $\Sigma = \{0, 1\}$

is the binary alphabet with only two symbols. We denote the empty string in a language by ϵ, and the empty language by $\emptyset$. The language of all strings over Σ is denoted by Σ^*. So, if $\Sigma = \{0,1\}$, then $\Sigma^* = \{\epsilon, 0, 1, 00, 01, 10, 11, 000, \dots\}$ is the set of all binary strings. Observe that every language L over Σ is just a subset of Σ^*, that is, $L \subseteq \Sigma^*$.

A decision problem is entirely characterized by those problem instances that produce a yes-answer. For example, the primality testing problem PRIME is associated with the set of prime numbers $\{2, 3, 5, 7, 11, 13, \dots\}$. Thus, using the standard binary encoding scheme to encode the problem instances, we can regard the problem PRIME as a language (set of strings) L_{PRIME} over $\Sigma = \{0,1\}$ such that

$$L_{\text{PRIME}} = \{\langle n \rangle : n \text{ is prime}\} = \{10, 11, 101, 111, 1011, 1101, \dots\},$$

where $\langle \cdot \rangle$ is used to denote the standard binary encoding. In the case of the Hamiltonian path problem HP, the corresponding language over binary alphabet is

$$L_{\text{HP}} = \{\langle G \rangle : G \text{ has a Hamiltonian path}\},$$

where $\langle G \rangle$ is the standard binary encoding of the graph G (e.g., as a string of concatenated rows of G's adjacency matrix). Finally, in the most complicated case of the traveling salesperson problem TSP the associated language is

$$L_{\text{TSP}} = \{\langle G, k \rangle : G \text{ has a cycle of length at most } k\}.$$

This set consists of encoded pairs $\langle G, k \rangle$, where the first element is the encoded weighted graph G (with the edge weights encoded in binary), followed by the number k encoded as a binary number.

In general, if Π is a decision problem, we can identify it with the language L_Π over $\Sigma = \{0,1\}$ such that

$$L_\Pi = \{\langle x \rangle \in \Sigma^* : x \text{ is a yes-instance of } \Pi\}.$$

We will thus use the terms "language" and "decision problem" interchangeably, since for our purposes they mean the same things. It is also sometimes convenient to use the same name for a decision problem and its associated language, e.g.,

$$\text{PRIME} = \{10, 11, 101, 111, 1011, 1101, \dots\}.$$

At first it may seem strange to identify problems with sets of binary strings, but obviously everything that is represented in a computer is broken down somehow into a string of zeros and ones. More importantly, this approach allows us to formalize the notion of algorithms for decision problems as "machines" that operate on languages. Intuitively, an algorithm for a decision problem Π answers "yes" if presented with a yes-instance of the problem Π on the input, and "no" otherwise. On the language level, that is, if Π is represented by its associated language L_Π, this amounts to simply checking if the input binary encoded instance belongs to the language L_Π. Thus, the notion of the algorithm boils down to the question of what it means to computationally recognize membership of a string in a language.

5.2. Turing machines. The notion of an algorithm is made precise by means of a special computational device called the Turing machine. The hardware of this device consists of a finite-state control unit, a read-write tape head, and a doubly infinite tape divided into cells that are numbered $\ldots, -2, -1, 0, 1, 2, \ldots$. The full name for this device is the ***deterministic one-tape Turing machine*** (or ***TM*** for short), because there are many variations on this basic hardware model.

A Turing machine is meant to be a primitive computational model of a general purpose computer, with the tape representing the (unlimited) memory, and the finite-state control unit representing the processor of a real computer. Don't get mislead by the primitiveness of Turing machines however—they can do everything that real computers can do, at least in principle and certainly more slowly. Figure 5 pictures schematically the hardware components of a Turing machine.

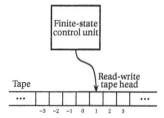

FIGURE 5. Deterministic one-tape Turing machine.

The following are capabilities of the hardware components of a Turing machine.

- Each cell of the doubly infinite tape can contain a single symbol from some alphabet, called the ***tape alphabet.***
- The tape head can read a single tape-alphabet symbol from a cell on the tape, write a single tape-alphabet symbol in a cell, or move its position one cell left or right relative to the current position.
- The control unit directs the functioning of the tape head and works in discrete time steps. In each step the control unit can be in exactly one of a finite number of states. Besides the regular states the unit can find itself in, there are additionally three special states: the ***start state*** q_0, in which the unit starts its operation, and two ***final states*** q_Y and q_N, in which the unit ceases its operation.

The way a Turing machine computes is straightforward. Initially the tape contains an input string whose individual symbols are stored in the consecutive cells numbered from 1 up to the length of the string, and blanks everywhere else. Symbols of the input string are drawn from a subset of the tape alphabet, called the ***input alphabet***. The control unit is put in the state q_0 and the tape head is positioned over the cell numbered 1. The machine then proceeds from this initial configuration in discrete time steps performing only one simple operation in each step: based on the current state and the symbol read in the current cell under the tape head, the control unit goes to another (possibly the same) state, the tape head writes another (possibly the same) symbol in the current cell, and the tape head moves one cell to the left or to the right. This basic operation is repeated continuously until the machine enters one of the two final states and then the computation halts. The final states q_Y and q_N in which the machine halts represent the output of the TM computation on the input string. If a TM computation ends up in the state q_Y, the output is "yes" (or 1), and we say that TM **accepts** the input string. If the final state is q_N, the output is "no" (or 0), and we say that TM **rejects** the input string. If the machine doesn't enter a final state, it will go on computing forever, never halting, and its computation doesn't produce an output for the input string.

Hardware of a Turing machine, like hardware of any device, is operated by software, so we next need to describe how we write programs for Turing machines. We have seen that the heart of a TM computation are simple operations that get the machine move from one step to the next. These transitions are made according to a table of instructions written in advance and stored in the control unit before the machine is activated in the initial configuration. This table is then looked up in each step to determine the instruction whose execution produces the next step based on the current state and the current symbol that the tape head is reading.

The machine instructions therefore should uniquely specify, for each possible state q the machine might be in (except q_Y and q_N) and each possible symbol a the tape head might be reading in the current cell, three pieces of information: what is the next machine state q', what is the symbol a' that is written in the current tape cell, and what is an increment ± 1 to the current head position (-1 meaning that the head moves one cell to the left, and $+1$ meaning that the head moves one cell to the right). These instructions thus associate pairs of the form (q, a) with triples of the form $(q', a', \pm 1)$, which is usually written as $(q, a) \rightarrow (q', a', \pm 1)$. In fact, since the order of the instructions is not important, the table of all instructions represents a function mapping the appropriate pairs to the appropriate triples. This function is called the **transition function** for a Turing machine.

Let's illustrate this by writing TM instructions (transition function) for a Turing machine that computes the parity of an input binary string. More precisely, the machine should determine whether an input binary string contains an odd number of 1's, and it should output "yes" (i.e., enter the state q_Y) if that is the case, or else output "no" (i.e., enter the state q_N). The idea is to scan the input string from left to right, keeping the machine in states q_0 or q_1 if an even or odd number of 1's, respectively, is scanned so far. The following is a transition function that does the job.

$$(q_0, 0) \rightarrow (q_0, 0, +1)$$
$$(q_0, 1) \rightarrow (q_1, 1, +1)$$
$$(q_0, \sqcup) \rightarrow (q_N, \sqcup, -1)$$
$$(q_1, 0) \rightarrow (q_1, 0, +1)$$
$$(q_1, 1) \rightarrow (q_0, 1, +1)$$
$$(q_1, \sqcup) \rightarrow (q_Y, \sqcup, -1)$$

For example, the step by step TM computation under this transition function on the input string 010011 is shown in Figure 6, where each computational step is represented by its state, the head position, and the contents of the relevant part of the tape.

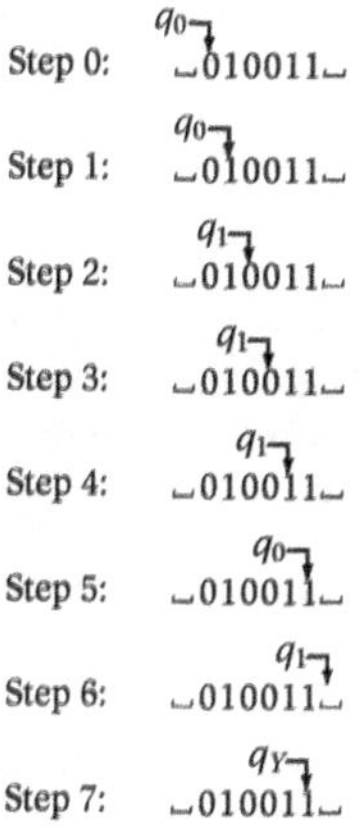

$$
\begin{array}{ll}
\text{Step 0:} & \overset{q_0 \rceil}{\underset{}{{}_{\sqcup}010011{}_{\sqcup}}} \\[6pt]
\text{Step 1:} & \overset{q_0 \rceil}{{}_{\sqcup}0\,10011{}_{\sqcup}} \\[6pt]
\text{Step 2:} & \overset{q_1 \rceil}{{}_{\sqcup}010011{}_{\sqcup}} \\[6pt]
\text{Step 3:} & \overset{q_1 \rceil}{{}_{\sqcup}010011{}_{\sqcup}} \\[6pt]
\text{Step 4:} & \overset{q_1 \rceil}{{}_{\sqcup}010011{}_{\sqcup}} \\[6pt]
\text{Step 5:} & \overset{q_0 \rceil}{{}_{\sqcup}010011{}_{\sqcup}} \\[6pt]
\text{Step 6:} & \overset{q_1 \rceil}{{}_{\sqcup}010011{}_{\sqcup}} \\[6pt]
\text{Step 7:} & \overset{q_Y \rceil}{{}_{\sqcup}010011{}_{\sqcup}}
\end{array}
$$

FIGURE 6. The step by step TM computation on input 010011.

In general, we see that a transition function together with the tape and input alphabets and the finite set of states constitutes a TM program for a Turing machine. It corresponds precisely to the notion of a program in some higher level programming language for the real computer. Thus, a TM program is specified by giving the machine's finite set of states, the tape alphabet, the input alphabet, and the transition function. Formally, we define (the syntax of) a **TM program** M as a seven-tuple $M = (Q, \Sigma, \Gamma, {}_{\sqcup}, \delta, q_Y, q_N)$, where

- Q is the finite set of regular states such that $q_0 \in Q$, $q_Y \notin Q$, and $q_N \notin Q$;
- Σ is the input alphabet such that $\Sigma \subseteq \Gamma$ and ${}_{\sqcup} \notin \Sigma$;
- Γ is the tape alphabet and ${}_{\sqcup} \in \Gamma$;
- ${}_{\sqcup}$ is the special blank symbol;
- $\delta : Q \times \Gamma \to (Q \cup \{q_Y, q_N\}) \times \Gamma \times \{-1, +1\}$ is the transition function;
- q_Y and q_N are the special final states.

For example, a syntactically correct program $P = (Q, \Gamma, \Sigma, {}_{\sqcup}, \delta, q_Y, q_N)$ that computes the parity of an input binary string is given by $Q = \{q_0, q_1\}$, $\Sigma = \{0, 1\}$, $\Gamma = \{0, 1, {}_{\sqcup}\}$, and δ such that

$$
\begin{aligned}
\delta(q_0, 0) &= (q_0, 0, +1) \\
\delta(q_0, 1) &= (q_1, 1, +1) \\
\delta(q_0, {}_{\sqcup}) &= (q_N, {}_{\sqcup}, -1) \\
\delta(q_1, 0) &= (q_1, 0, +1) \\
\delta(q_1, 1) &= (q_0, 1, +1) \\
\delta(q_1, {}_{\sqcup}) &= (q_Y, {}_{\sqcup}, -1).
\end{aligned}
$$

Now that we have the formal definition of a TM program, let's briefly reinforce the way it is executed on a Turing machine. A given TM program $M = (Q, \Gamma, \Sigma, {}_{\sqcup},$

$\delta, q_Y, q_N)$, which is stored beforehand in the control unit of the Turing machine, receives its input $x = x_1 x_2 \cdots x_n \in \Sigma^*$ in the tape cells numbered from 1 to n, and the rest of the tape is filled with blank symbol $\sqcup$ in each cell. The head starts positioned over the cell in location 1 scanning the first symbol of the input, and the machine is activated in the start state q_0. From then on the computation proceeds according to the rules prescribed by the transition function δ. The computation continues until the machine enters either of the two final states q_Y or q_N, at which moment it halts. If the machine never enters a final state, M goes on computing forever.

To see the Turing machine in action we don't have to build its hardware, instead we can simulate it on an ordinary computer. The following is an algorithm that mimics the execution of a TM program M on input x of length n, where we use an array t to represent the tape contents.

```
TM(M, x)
    for i = 1 to n do
        t(i) = x(i);              [ initial tape contents ]
    s = q_0; i = 1;               [ initial state and head position ]
    ℓ = 1; r = n;                 [ initial left and right tape boundaries ]
    repeat
        (s, t(i), d) = δ(s, t(i)); [ update the current state and cell contents ]
        i = i + d;                [ update the current head position ]
        if ¬(ℓ ≤ i ≤ r) then      [ update the tape boundaries ]
            t(i) = ⊔;
            if i < ℓ then ℓ = i; else r = i;
    until (s = q_Y) ∨ (s = q_N);
    if s = q_Y then
        return "yes";             [ accept x ]
    else
        return "no";              [ reject x ]
```

As a TM program computes, changes occur in the current state, the current tape contents, and the current head position of the Turing machine. Each step of the computation is completely described by the particular values of these three items, which is called a **configuration** of the Turing machine. For a state q and two strings y and z from Γ^*, we write yqz for the configuration in which the current state is q, the current relevant tape contents is yz (i.e., everything else are blanks), and the current head position is over the first symbol of the string z. For example, the configuration in step 2 of Figure 6 is conveniently represented by $01q_1 0011$.

The **initial configuration** of M on input $x \in \Sigma^*$ is the configuration $q_0 x$, which indicates that the machine is in the start state q_0 with its head positioned over the first symbol of x. In an **accepting** (or **rejecting**) **configuration**, the state of the configuration is q_Y (or q_N). The TM program M accepts a string $x \in \Sigma^*$ if there is an **accepting computation** of M on x, which is a (finite) sequence of configurations $C_0, C_1, \ldots, C_k$ such that

1. C_0 is the initial configuration of M on input x, and
2. for each $i = 1, \ldots, k$ the next configuration C_i legally follows from the previous one C_{i-1} in a single step according to the transition function δ, and
3. C_k is an accepting configuration.

We define analogously a **rejecting computation** of M on x if the last configuration C_k is a rejecting configuration. Notice that M might not halt on some input $x \in \Sigma^*$. These nonhalting computations together with rejecting computations are referred to collectively as **nonaccepting computations**.

The connection of TM programs with languages is provided by taking strings that the programs accept. The language $L(M)$ that is **recognized** by the TM program M is the set of all strings over M's input alphabet that M accepts, that is,

$$L(M) = \{x \in \Sigma^* : M \text{ accepts } x\}.$$

Observe that this definition of language recognition does not require that M halts for every input string $x \in \Sigma^*$, only for those in $L(M)$. If $x \in \Sigma^* \setminus L(M)$, then we know only that the computation of M on x is nonaccepting (i.e., rejecting or nonhalting).

The basic interpretation of the way a TM program is executed can be changed slightly to get other forms of computations. Firstly, a TM program that halts for all input strings can be used to compute functions if we change the meaning of the output of a program—instead of taking the state of the halting configuration as a result, we take the relevant contents of the tape. Specifically, a TM program $M = (Q, \Gamma, \Sigma, \_, \delta, q_Y, q_N)$ that halts for all input strings computes the function $f_M: \Sigma^* \to \Gamma^*$ so that, on input $x \in \Sigma^*$, $f_M(x)$ is the string of tape symbols in the last configuration between the leftmost and rightmost non-blank tape cells, inclusive. For example, the parity-computing TM program P may be also thought of as computing the function $f_P: \{0,1\}^* \to \{0,1,\_\}^*$, where $f_P(x) = x$ is the identity function.

Secondly, we can allow two-input TM programs if we initially place the second input string to the left of the first. Specifically, the initial configuration for two input strings x and y is $y^R\_q_0x$, where y^R is reverse of y. That is, initially $x = x_1 \cdots x_n$ occupies the tape cells $1, \ldots, n$ as before, $y = y_1 \cdots y_m$ occupies the tape cells $-1, \ldots, -m$, and the control unit is in the start state q_0 with the tape head positioned over the first symbol of x.

5.3. Algorithms = TM programs. In this section we look at the concepts discussed in earlier sections from the perspective of Turing machines. First of all, so far we have been using an intuitive notion of algorithm, which has been adequate for giving algorithms that solve particular problems. But proving that an algorithm does not exist requires a clear definition of algorithm. Now we *can* provide such a formalism and regard an algorithm simply as a TM program, by definition.[8]

This immediately clears up a somewhat fuzzy intuition we have had about the input size and running time of an algorithm, for example. The size of an

---

[8]This is the renown Church-Turing thesis, after Alonzo Church and Alan Turing.

input $x \in \Sigma^*$ for a TM program $M = (Q, \Sigma, \Gamma, \_, \delta, q_Y, q_N)$ is simply the length $|x|$ of the string x. The running time of M on input x is clearly the number of steps in the M's computation from the initial configuration $q_0 x$ of M on x. Accordingly, the running-time function $T_M(n)$ for M that halts for all inputs $x \in \Sigma^*$ is

$$T_M(n) = \max_{\substack{x \in \Sigma^*, \\ |x|=n}} \{k : M\text{'s computation on input } x \text{ takes } k \text{ steps}\}.$$

A TM program M is called a **polynomial-time TM program** if there is a polynomial $p(n)$ such that $T_M(n) \le p(n)$ for every $n \in \mathbb{N}$. Notice that a polynomial-time TM program implies that it halts for every input.

Formal counterpart to a verifying algorithm is a **verifying** two-input TM program M with one input $x \in \Sigma^*$ and the other $y \in \Gamma^*$. We say that M **verifies** $x \in \Sigma^*$ if there is some string $y \in \Gamma^*$ (called certificate) so that M's computation on input x and y halts in an accepting configuration. The language **verified** by M is the set of all first input strings verifiable by M with the help of second input strings (certificates), that is,

$L(M) = \{x \in \Sigma^* : \exists y \in \Gamma^*, M\text{'s computation on } x, y \text{ is an accepting computation}\}$

$\qquad = \{x \in \Sigma^* : M \text{ verifies } x\}.$

Given an $x \in L(M)$, there can be many certificates y for which M halts in an accepting configuration if started from the initial configuration with x and y. We thus define the time it takes a verifying TM program M to verify $x \in L(M)$ to be the minimum, over all accepting computations of M on fixed x and various y's, of the number of steps in these computations. Accordingly, the running-time function $T_M(n)$ for M is defined as

$$T_M(n) = \begin{cases} 1, & \text{no } x \in L(M) \text{ of length } n \text{ is verified by } M \\ \max_{\substack{x \in L(M), \\ |x|=n}} \{k : M \text{ verifies } x \text{ in } k \text{ steps}\}, & \text{otherwise} \end{cases}$$

$$= \begin{cases} 1, & \text{no } x \in L(M) \text{ of length } n \text{ is verified by } M \\ \max_{\substack{x \in L(M), \\ |x|=n}} \min_{y \in \Gamma^*} \{k : M\text{'s computation on } x, y \text{ takes } k \text{ steps}\}, & \text{otherwise}. \end{cases}$$

Notice that the running-time function $T_M(n)$ for a verifying TM program M depends only on the number of steps performed in accepting computations, so it is set to 1 by convention if no inputs of length n are verified by M. Also note that $T_M(n)$ does not depend on the length of certificates, but if $x \in L(M)$ there must a certificate of x with length at most $T_M(n)$.

Similarly to an ordinary TM program, a verifying TM program M is said to be a **polynomial-time verifying TM program** if there exists a polynomial $p(n)$ such that $T_M(n) \le p(n)$ for every $n \in \mathbb{N}$. Observe that we can assume without loss of generality that certificates for polynomial-time verifying TM programs have the length at most $p(n)$.

Since decision problems can be identified with languages, the correspondence between "solving (verifying) decision problems" and "recognizing (verifying) languages" is straightforward. Recall from Section 5.1 that a decision problem Π of the form "given an instance $x \in I_\Pi$, is x a yes-instance of Π?" translates to the question "given a string $\langle x \rangle \in \Sigma^*$, is $\langle x \rangle \in L_\Pi$?". Here, $\langle \cdot \rangle$ is the standard binary encoding scheme, $\Sigma = \{0, 1\}$, and $L_\Pi \subseteq \Sigma^*$ is the set of all encoded yes-instances of Π. Thus, an algorithm that solves Π is analogous to a TM program M that halts for all input strings and $L(M) = L_\Pi$. In other words, asking for an algorithm that solves (verifies) a decision problem Π is equivalent to asking for a TM program that recognizes (verifies) the language L_Π.

Once we have established the correspondence between the less formal world of decision problems and algorithms and the formal world of languages and TM programs, all concepts we introduced in the former can be easily phrased in the later. For example, the complexity classes P and NP become

P = $\{L \subseteq \Sigma^*$: there is a polynomial-time TM program that recognizes $L\}$,

NP = $\{L \subseteq \Sigma^*$: there is a polynomial-time verifying TM program that verifies $L\}$.

Also, a polynomial transformation of a language $L_1 \subseteq \Sigma^*$ to a language $L_2 \subseteq \Sigma^*$ is a function $f : \Sigma^* \to \Sigma^*$ such that

1. there is a polynomial-time TM program that computes f, and
2. for every $x \in \Sigma^*$, $x \in L_1 \Leftrightarrow f(x) \in L_2$.

The primitiveness of the Turing machine computational model is a doubly-sided sword. On the one hand, it has allowed us to get a firm formal ground for our discussion. On the other hand, even simple TM programs tend to be involved and require great care to specify them. Since the many details will only blur the big picture, we will be rather informal in our use of polynomial-time TM programs. We follow standard practice and describe them on algorithmic level of an ordinary computer. Moreover, we discuss them as if they are working directly on the components of an instance (the numbers, graphs, sets, etc.) rather than on their encoded counterparts. Thus our informal description of a TM program should be taken as a guide for anyone having enough patience to actually write the corresponding TM program. It should be also convincing to anyone having enough skills that an informal TM program runs in claimed polynomial time.

5.4. NP-completeness of the SAT problem. We are now ready to prove the Cook-Levin theorem.

THEOREM SAT *is NP-complete.*

PROOF: By definition, we need to prove two properties: SAT $\in$ NP and SAT is NP-hard.

Observe first that we are referring conveniently to SAT instead of more precisely to its associated language L_{SAT}. This and other languages should be taken over the alphabet $\Sigma = \{0, 1\}$, since we have fixed without loss of generality the standard binary encoding to map problems to languages.

To see the first property that SAT is in NP, we need to show that there is a TM program that verifies SAT in polynomial time. Since every satisfiable Boolean formula has a short certificate to this fact, which is the actual truth assignment to its variables that make the formula evaluate to 1, we can easily construct a TM program M that works as required. Namely, given an instance of SAT (a Boolean formula over some set of variables) and a certificate for the instance (a truth value assignment to the variables), M should only substitute the truth values given by the certificate for the variables in the given formula, evaluate the formula with this combination of truth values for the variables, and check the final result. If it is 1, M should accept. In all other cases (including when the certificate is incomplete or the formula is incorrectly specified, for example), M should reject. If the size of the input instance is n, it is clear that the size of the certificate is $O(n)$ and that an accepting computation takes time bounded by a polynomial of n. (This might take a minute of thought, but think of M as an ordinary program in a high-level programming language.) Therefore, M is a desired TM program that proves SAT $\in$ NP.

The second property that SAT is NP-hard requires that we show that every language in NP is polynomial-time transformable to SAT. That is, if $L \in$ NP, then there is a function $f : \Sigma^* \rightarrow \Sigma^*$ such that $x \in L \Leftrightarrow f(x) \in$ SAT and f can be computed by a polynomial-time TM program.

Now, if $L \in$ NP, then there is a TM program M that verifies strings in L of length n, if accompanied by a suitable certificate, in time bounded by some polynomial $p(n)$. Let $M = (Q, \Sigma, \Gamma, \sqcup, \delta, q_Y, q_N)$, where the total number of states (including the final states but excluding the start state) is s and the number of non-blank tape symbols is t. Thus, $Q = \{q_0, q_1, \ldots, q_{s-2}\}$, $q_{s-1} = q_N$, $q_s = q_Y$, and $\Gamma = \{a_0 = \sqcup, a_1, \ldots a_t\}$.

We first describe the function f by constructing for each string $x \in \Sigma^*$ an instance $f(x)$ of SAT so that $x \in L \Leftrightarrow f(x) \in$ SAT. Given $x = a_{i_1} a_{i_2} \cdots a_{i_n} \in \Sigma^*$ of length n, the instance $f(x)$ of SAT consists of a set of Boolean variables U and a Boolean formula F over U.

The idea for the construction of the instance $f(x)$ is that a truth assignment to the variables in U describes all possible configurations during any computation of M within $p(n)$ steps, while a satisfiable assignment for F additionally describes an accepting computation of M on x for some certificate y. If the input $x \in L = L(M)$, then there is a certificate $y = a_{i'_1} a_{i'_2} \cdots a_{i'_m} \in \Gamma^*$ and an accepting computation of M on x and y with the number of steps bounded by $p(n)$. Such a computation can involve only tape cells in locations $-p(n)$ through $p(n) + 1$, since the tape head begins at cell 1 and moves at most one cell left or right in any single step. Thus, the length m of the certificate y is at most $p(n)$ and there are at most $p(n) + 1$ single-step configurations that must be considered. Figure 7 illustrates such an accepting computation of M on x and y within $p(n)$ steps.

The variable set U contains variables of two types that describe the configuration in each step: one type gives the contents of each tape cell in the configuration, and the other determines the state and the head position in the configuration. Specifically, the set U contains Boolean variables $a_i^{\ell,k}$ and $q_j^{\ell,k}$ whose intended meaning is

- $a_i^{\ell,k}$ is 1 iff in the step k of M's computation, the tape cell in location ℓ contains the tape symbol a_i;
- $q_j^{\ell,k}$ is 1 iff in the step k of M's computation, the control unit is in the state q_j and its head is positioned over the tape cell in location ℓ.

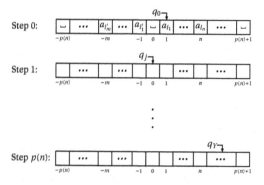

FIGURE 7. Accepting computation of M on x and y within $p(n)$ steps.

The indices i, j, ℓ, k of the variables $a_i^{\ell,k}$ and $q_j^{\ell,k}$ must run through those values so that we get all possible combinations of tape symbols, states, cell locations, and time steps during any computation of M within $p(n)$ steps. The index ranges are therefore $i = 0, \ldots, t$, $j = 0, \ldots, s$, $\ell = -p(n), \ldots, 0, \ldots, p(n) + 1$, and $k = 0, \ldots, p(n)$. This means that U is a relatively large set of $O(p(n)^2)$ variables, but this is still an order of a polynomial of n.

A computation of M induces a truth assignment on these variables in the obvious way, under the convention that if M halts in less than $p(n)$ steps, the configuration remains static in all later steps with the same final state, head position, and tape contents. On the other hand, an arbitrary truth assignment to these variables need not correspond at all to a TM computation, much less to an accepting computation of M. For example, $q_1^{5,17} = 1$ and $q_2^{10,17} = 1$ would mean that the machine in step 17 was in two different states q_1 and q_2 with tape positioned over two different cells 5 and 10. Similarly, $a_0^{1,17} = 1$, $a_1^{1,17} = 1$, and $a_2^{1,17} = 1$ would mean that in step 17 the cell 1 contains three different symbols a_0, a_1, and a_2. This is why we additionally need to restrict truth assignments to these variables so that they induce not only proper TM configurations but also an accepting computation of M. The formula F over these variables is intended for just this purpose, that is, a truth assignment to the variables must be a satisfying truth assignment for F if and only if the truth assignment corresponds to an accepting computation of M for the input x.

The formula F over U that simulates an accepting computation of M on x is constructed by further dividing F into four subformulas so that

$$F = F_I \wedge F_C \wedge F_D \wedge F_A,$$

where intended meaning of the subformulas is

- F_I is 1 iff M's computation starts in the initial configuration $q_0 x$;
- F_C is 1 iff M's computation consists of proper configurations;
- F_D is 1 iff M's computation legally follows the transition function δ;
- F_A is 1 iff M's computation halts in an accepting configuration.

We first introduce a shorthand notation for equivalent formulas that will simplify our description of the subformulas of F. If $\phi_1,\dots,\phi_n$ are Boolean formulas, we denote

$$\phi_1 \Rightarrow \phi_2 \equiv \neg\phi_1 \vee \phi_2$$

$$\phi_1 \Leftrightarrow \phi_2 \equiv (\neg\phi_1 \vee \phi_2) \wedge (\neg\phi_2 \vee \phi_1)$$

$$\bigwedge_{i=1}^{n} \phi_i \equiv \phi_1 \wedge \phi_2 \wedge \cdots \wedge \phi_n$$

$$\bigvee_{i=1}^{n} \phi_i \equiv \phi_1 \vee \phi_2 \vee \cdots \vee \phi_n$$

$$\bigYsum_{i=1}^{n} \phi_i \equiv \left(\bigvee_{i=1}^{n} \phi_i\right) \wedge \left(\bigwedge_{1 \le i < j \le n} \neg(\phi_i \wedge \phi_j)\right).$$

Observe that the last shorthand $\bigYsum_{i=1}^{n} \phi_i$ denotes the fact that exactly one of ϕ_i's is true: the first part in parenthesis says that at least one ϕ_i is true, while the second part says that no two (or more) ϕ_i's are true. Also observe that if each ϕ_i is a variable, then the size of the formula $\bigYsum_{i=1}^{n} x_i$ is $O(n^2)$.

Using these shorthands we now systematically build the subformulas F_I, F_C, F_D, and F_A. For future reference, we also estimate the size of each subformula along the way.

The initial configuration in step 0 of M's computation on x consists of

- the start state q_0 and the tape head positioned over the cell in location 1; and
- $x = a_{i_1} a_{i_2} \cdots a_{i_n}$ occupying cells in location 1 to n and all other cells containing blanks.

This can be expressed by the formula F_I such that

$$F_I = q_0^{1,0} \wedge \bigwedge_{\ell=1}^{n} a_{i_\ell}^{\ell,0} \wedge \bigwedge_{\ell=n+1}^{p(n)+1} a_0^{\ell,0} \wedge \bigwedge_{\ell=-p(n)}^{0} a_0^{\ell,0}.$$

It is easy to see that the size of the formula F_I is $O(p(n))$.

Proper configurations of M's computation are only those where

- in each step k, the cell in each location ℓ contains a unique tape symbol a_i; and
- in each step k, a unique cell in location ℓ is scanned in a unique state q_j.

These two conditions are expressed respectively by the formulas F_{C_1} and F_{C_2} such that

$$F_{C_1} = \bigwedge_{k=0}^{p(n)} \bigwedge_{\ell=-p(n)}^{p(n)+1} \left(\bigYsum_{i=0}^{t} a_i^{\ell,k}\right)$$

$$F_{C_2} = \bigwedge_{k=0}^{p(n)} \left(\bigYsum_{\ell=-p(n)}^{p(n)+1} \left(\bigYsum_{j=0}^{s} q_j^{\ell,k}\right)\right),$$

and so $F_C = F_{C_1} \wedge F_{C_2}$. The size of the formula F_C is $O(p(n)^3)$.

5. THE FIRST NP-COMPLETE PROBLEM

M's computation legally follows the transition function δ iff

- in each step k but the last, if a cell in location ℓ is not scanned in any of M's states q_j, the cell's symbol is not changed in the next step $k+1$; and
- in each step k but the last, if a cell in location ℓ with symbol a_i is scanned in state q_j and $\delta(q_j, a_i) = (q_{j'}, a_{i'}, d)$ is a rule of the transition function δ, then in the next step $k+1$ the cell in location ℓ contains $a_{i'}$, the state is changed to $q_{j'}$, and the head is moved over the cell $\ell + d$.

These two conditions together with our convention that the configurations don't change if M halts before step $p(n)$ can be expressed respectively by the formulas F_{D_1} and F_{D_2} such that

$$F_{D_1} = \bigwedge_{k=0}^{p(n)-1} \bigwedge_{\ell=-p(n)+1}^{p(n)} \bigwedge_{i=0}^{t} \left(\neg \left(\bigvee_{j=0}^{s} q_j^{\ell,k} \right) \Rightarrow \left(a_i^{\ell,k} \Leftrightarrow a_i^{\ell,k+1} \right) \right)$$

$$F_{D_2} = \bigwedge_{k=0}^{p(n)-1} \bigwedge_{\ell=-p(n)+1}^{p(n)} \bigwedge_{i=0}^{t} \bigwedge_{j=0}^{s} \left(\left(a_i^{\ell,k} \wedge q_j^{\ell,k} \right) \Rightarrow \left(a_{i'}^{\ell,k+1} \wedge q_{j'}^{\ell+d,k+1} \right) \right),$$

where in F_{D_2} for $j = s-1$ and $j = s$ we put $j' = j$, $i' = i$, $d = 0$. Thus $F_D = F_{D_1} \wedge F_{D_2}$, and the size of the formula F_D is $O(p(n)^2)$.

Finally, an accepting configuration of M's computation in step $p(n)$ is expressed by

$$F_A = \bigvee_{\ell=-p(n)}^{p(n)+1} q_s^{\ell,p(n)}.$$

The size of the formula F_A is obviously $O(p(n))$.

This completes the construction of $f(x)$ for every $x \in \Sigma^*$. It remains to show that $x = a_{i_1} a_{i_2} \cdots a_{i_n} \in L \Leftrightarrow f(x) \in \mathrm{SAT}$, as well as that $f(x)$ is computable by a polynomial-time TM program.

To see the first part of this claim, we argue as follows. If $x \in L$ with $|x| = n$, then there is a certificate $y = a_{i_1'} a_{i_2'} \cdots a_{i_m'} \in \Gamma^*$ of length $m \le p(n)$ such that an accepting computation of M on input x and y has no more than $p(n)$ steps. This computation induces the obvious truth assignment to the variables in U that reflects their intended meaning:

- the variables $a_i^{\ell,k}$ receive 0 or 1 according to the contents of each cell in each configuration of the M's accepting computation. For example, $a_{i_\ell}^{\ell,0} = 1$ for $\ell = 1, \ldots, n$ and $a_{i_\ell}^{-\ell,0} = 1$ for $\ell = 1, \ldots, m$;
- the variables $q_j^{\ell,k}$ receive 0 or 1 according to the actual position of the machine's head and its state in each configuration of the M's accepting computation. For example, $q_0^{1,0} = 1$ and $q_0^{\ell,0} = 0$ for $\ell \ne 1$

Given the interpretation of the subformulas of F, it is easy to see that they all evaluate to 1 under this assignment, and thus F is satisfied. In other words, $f(x) \in \mathrm{SAT}$.

Conversely, if $f(x) \in \mathrm{SAT}$, the construction of $f(x)$ is such that any truth assignment to the variables in U that satisfies the formula F determines two things. First, it gives in the obvious way a certificate $y \in \Gamma^*$ of length $p(n)$, that is, $y = a_{i'_1} a_{i'_2} \cdots a_{i'_{p(n)}}$ such that $a_{i'_\ell}^{-\ell,0} = 1$ for $\ell = 1, \ldots, p(n)$. Second, the satisfying truth assignment describes, step by step in minute detail, an accepting computation of M on input x and y of length $p(n)$ or less. Thus, $x \in L$.

Finally, given a TM program M and a polynomial $p(n)$ that bounds its running time, it is easy to see that $f(x)$ can be constructed from x algorithmically (i.e., by another TM program) in polynomial time. Informal justification for this is that the construction is highly repetitive without any complicated steps (for instance, a search or branching), and amounts to little more than straightforward filling in the blanks in a standard, though large, formula F. In other words, the construction takes time proportional to the size of the set U and the formula F, which is $O(p(n)^3)$. This estimate is actually low by a factor of $O(\log n)$, because each variable has indices that can range up to $p(n)$ and so may require $O(\log n)$ bits to represent in the standard encoding. However, this additional factor does not change the polynomiality of the result.

This concludes the proof that every $L \in$ NP is polynomial-time transformable to SAT (more precisely, L_{SAT}), which in turn completes the proof that SAT is NP-complete. ∎

6. Additional NP-complete problems

Now that we know that SAT is NP-complete, proving that other problems are NP-complete need not be as complicated as that for SAT. This task for a decision problem Π is greatly simplified by Lemma 3 from Section 4 and consists of three steps:

1. showing that $\Pi \in$ NP;
2. selecting an NP-complete problem Ψ;
3. proving that $\Psi \leq \Pi$.

Notice that if we have a problem in NP that we want to prove NP-complete, we need to show that a *known* NP-complete problem is transformable to our *candidate* NP-complete problem. For now we have only one known NP-complete problem—SAT, but as we build our repertoire of them we are free to choose any of them.

The first step of the above procedure for proving NP-completeness results is usually the simplest, since a certificate and a verifying algorithm for Π often arise naturally. The second step may seem easy (because now we don't have much of a choice), but confronted with an extensive list of hundreds of NP-complete problems, one can have hard time to select the best suited NP-complete problem Ψ as the basis for the desired proof. The third step is the most laborious because it includes constructing a function f from Ψ to Π and proving that f is a polynomial-time transformation.

To build our experience and intuition for such proofs, in this section we study some of the "basic" NP-complete problems. In doing this there is no need to be as formal as in the previous section, so we switch our discussion back to the less formal world of problems and algorithms instead of formal languages and Turing machines.

6.1. CNF-formula satisfiability. We have seen that the problem of whether a general Boolean formula is satisfiable is NP-complete. In this and the next section we show that even if we restrict the formula to a special form, the question remains NP-complete.

Before we can state the CNF-formula satisfiability problem, we need some more notions from the propositional logic. A *literal* is either a Boolean variable or the negation of a Boolean variable. If x is a Boolean variable, its negation $\neg(x)$ or just $\neg x$ in this context is very often denoted $\overline{x}$. A *clause* is a disjunction ($\vee$) of one or more literals. For example, $(x_1 \vee x_2 \vee \overline{x_3} \vee \overline{x_2})$ is a clause of four literals. A Boolean formula is in the *conjunctive normal form (CNF)* if it is a conjunction ($\wedge$) of one or more clauses. For example, the following formula is in the conjunctive normal form

$$\overline{x_1} \wedge (x_1 \vee x_2 \vee \overline{x_3}) \wedge x_3 \wedge (x_2 \vee \overline{x_4}) \wedge (x_3 \vee x_4),$$

where we have omitted unnecessary parenthesis around the single-literal clauses $\overline{x_1}$ and x_3. A Boolean formula in the conjunctive normal form is called a CNF-formula.

The *CNF-formula satisfiability problem (CNF-SAT)* is a restricted version of the SAT problem when the Boolean formula is restricted to be in the conjunctive normal form. That is,

CNF-SAT = $\{F: F$ is a satisfiable Boolean CNF-formula$\}$,

or, in the form of a question, CNF-SAT is: given a finite set of Boolean variables and a Boolean CNF-formula over the variables, is the CNF-formula satisfiable? In other words, is there an assignment of truth values to the variables so that all clauses of the formula are simultaneously satisfied (i.e., evaluate to 1)? For example, the above formula is satisfiable by the truth assignment $x_1 = 0, x_2 = 1, x_3 = 1, x_4 = 0$. Although the more restricted CNF-SAT seems easier than the general SAT, it is still NP-complete, as we now show.

First of all, it is clear that CNF-SAT is in NP because a certificate for a satisfiable CNF-formula is the same as for a general formula—it represents an actual truth assignment to the variables. Moreover, a verifying algorithm for CNF-SAT is essentially the same as for SAT. That is, given an instance of CNF-SAT (a Boolean CNF-formula over some set of variables) and a certificate for the formula (a truth value assignment to the variables), a verifying algorithm should only substitute the truth values given by the certificate for the variables in the given formula and evaluate each clause of the formula with this combination of truth values for the variables. If each clause evaluates to 1, the algorithm should return "yes," and otherwise "no." If the size (i.e., the number of literals) of the input CNF-formula is n, it is easy to see that the size of the certificate is $O(n)$ and that the verification procedure takes polynomial (in fact, linear) time in the input size n. Thus, CNF-SAT $\in$ NP.

Next, we show how SAT (well, what else?) can be transformed to CNF-SAT. To show the existence of a function f that maps SAT to CNF-SAT and proving that f is a polynomial-time transformation, we first describe f by constructing from a (general) Boolean formula F its image $G = f(F)$ so that G is a CNF-formula and F is satisfiable if and only if G is satisfiable.

Since any Boolean formula F can be put in the conjunctive normal form by using the distributivity of $\vee$ over $\wedge$ and DeMorgan's laws, one may be tempted to repeatedly apply these transformations until one gets equivalent CNF-formula F'. The problem with this approach is that the size of the generated CNF-formula F' might grow exponentially in the size of F. Thus, we would have no chance of obtaining a polynomial-time transformation from SAT to CNF-SAT in this way.

We therefore must be more careful in transforming a given formula F to a CNF-formula G. The key observation is that we don't have to construct an equivalent formula to F, only a CNF-formula G that is satisfiable if and only if F is satisfiable. The best way to describe the right transformation is to think of F in terms of its parse tree and demonstrate the construction on a small example.

Let X be a set of Boolean variables and let F be a (fully parenthesized) Boolean formula F over X. As the running example, let's take

$$F = \Big((\neg x_1 \vee x_2) \vee \neg\big((\neg x_1 \wedge x_3) \vee \neg x_4\big)\Big) \wedge \neg x_2.$$

We build the variable set $Y \supseteq X$ and the CNF-formula G over Y that correspond to X and F in several steps (see Figure 8):

1. Create the binary parse tree for F using the formula's literals as leaves. The result of this step on our running example is shown in (1) of Figure 8.
2. Using DeMorgan's laws push the negations as far down as possible in the parse tree for F so that they appear only at the leaves. Along the way, replace two or more consecutive negations with their equivalent meaning of no negation at all or just one negation. This way we obtain an equivalent formula F' containing only the operators $\wedge$ and $\vee$ and literals. The resulting parse tree of the equivalent formula F' in our example is pictured in (2) of Figure 8.
3. Assign a new variable y_i to the output of each internal node N_i of the parse tree for F'. At the same time, for each node N_i with inputs z_1 and z_2 write the subformula G_i in the form $G_i = y_i \Leftrightarrow z_1 \wedge z_2$ or $G_i = y_i \Leftrightarrow z_1 \vee z_2$ depending on whether N_i is an $\wedge$ or $\vee$ node, respectively. In fact, we formally represent each G_i in the equivalent conjunctive normal form over its three variables as

$$G_i = y_i \Leftrightarrow z_1 \wedge z_2 \equiv (\overline{y_i} \vee z_1) \wedge (\overline{y_i} \vee z_2) \wedge (y_i \vee \overline{z_1} \vee \overline{z_2})$$

or

$$G_i = y_i \Leftrightarrow z_1 \vee z_2 \equiv (y_i \vee \overline{z_1}) \wedge (y_i \vee \overline{z_2}) \wedge (\overline{y_i} \vee z_1 \vee z_2).$$

The reader can easily verify these equivalences using truth tables. The result of this step in our example is given in (3) of Figure 8.
4. Construct the CNF-formula $G = y_1 \wedge G_1 \wedge G_2 \wedge \cdots \wedge G_m$, where y_1 is the variable attached to the output of the root of the parse tree for F' and m is the number of its internal nodes. The final result is then the variable set $Y = X \cup \{y_1, \ldots, y_m\}$ and the CNF-formula G.

Now we must first show that F is satisfiable if and only if G is satisfiable. Since F' is an equivalent formula to F, it suffices to argue F' is satisfiable if and only if G is satisfiable. Well, if there is a truth assignment to the variables in X so that F' evaluates to 1, then this assignment induces an obvious truth assignment to the variables $y_1, \ldots, y_m$ as the output of each internal node. Moreover, under this assignment each subformula G_i is true and the truth value of y_1 must be

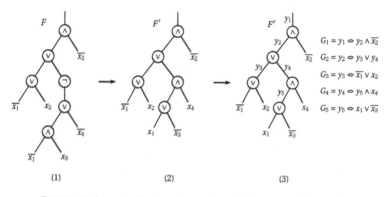

$$G_1 = y_1 \Leftrightarrow y_2 \wedge \overline{x_2}$$
$$G_2 = y_2 \Leftrightarrow y_3 \vee y_4$$
$$G_3 = y_3 \Leftrightarrow \overline{x_1} \vee x_2$$
$$G_4 = y_4 \Leftrightarrow y_5 \wedge x_4$$
$$G_5 = y_5 \Leftrightarrow x_1 \vee \overline{x_3}$$

(1)   (2)   (3)

FIGURE 8. Steps in the transformation of the example formula
F to the CNF-formula G.

1 since F' evaluates to 1. Thus, G is satisfiable. Conversely, observe that any satisfying truth assignment to the variables in Y for G must set $y_1 = 1$ and must satisfy each subformula G_i. But this implies that such an assignment restricted to the variables in X must also satisfy F' because $y_1 = 1$ is the result of F' under the restricted assignment. Thus, if G is satisfiable, then F' is satisfiable too.

We must also argue that we can generate G algorithmically in polynomial time in the size of the formula F. But this is clear taking into account the following observations about the construction of G. The first step of creating the parse tree for F isn't really necessary—it is only mentally helpful to follow the construction in this way. But even if we use the parse tree, it is easy to see that it can be created in linear time. The second step can also be done in linear time and it does not increase the size of the formula F' at all. In the third step we introduce m new variables y_i and m subformulas G_i, where m is the number of internal nodes of the parse tree for F'. Thus m is one less than the number of leaves in the parse tree, and the number of leaves is the size of the formula F'. Consequently, if the size of the formula F (or F') is n, then $m = n - 1$. Since the size of each G_i is 7, both the number of variables in Y and the size of the formula G is $O(n)$. In other words, given an instance of SAT, the construction of the desired instance of CNF-SAT can be accomplished by an algorithm in polynomial time in the size of the SAT instance. This completes the proof that SAT $\le$ CNF-SAT.

Therefore CNF-SAT is NP-hard and, being in NP, is NP-complete.

6.2. 3CNF-formula satisfiability. The 3CNF-formula satisfiability problem is another version of the Boolean formula satisfiability problem when we further restrict a given Boolean formula. We now require not only that a formula is in the conjunctive normal form, but also that each clause contains *exactly* three literals. Call a Boolean formula in the conjunctive normal form wherein each of its clauses contains exactly three literals a 3CNF-formula. In the form of a question, the **3CNF-formula satisfiability problem** (3CNF-SAT) is: given a finite set of Boolean variables and a Boolean 3CNF-formula over the variables, is

the 3CNF-formula satisfiable? That is,

$$3\text{CNF-SAT} = \{F : F \text{ is a satisfiable Boolean 3CNF-formula}\}.$$

The reason we care about yet another Boolean logic problem is that 3CNF-SAT is a much restricted version of SAT that we can use to prove other problems NP-complete. That is, 3CNF-SAT is easier to work with in transformations to other problems than the most general SAT or even less general CNF-SAT.

The proof that 3CNF-SAT is NP-complete is almost trivial and amounts to a simple modification of the proof that CNF-SAT is NP-complete. First, 3CNF-SAT is clearly in NP, since we can use the same verifying algorithm that we have used for CNF-SAT. Second, observe that the CNF-formula G constructed in the transformation from SAT to CNF-SAT contains *at most* three literals.[9] So, we transform SAT to 3CNF-SAT using the same construction as for CNF-SAT and slightly adapt G to get each of its clauses with exactly three literals. Specifically, given a Boolean formula F, we construct the CNF-formula $G = y_1 \wedge G_1 \wedge G_2 \wedge \cdots \wedge G_m$ as before, with the following modifications. Introducing two new variables, say, u_0 and v_0, the factor y_1 is replaced with its equivalent form

$$y_1 \equiv (y_1 \vee u_0 \vee v_0)) \wedge (y_1 \vee \overline{u_0} \vee v_0) \wedge (y_1 \vee u_0 \vee \overline{v_0}) \wedge (y_1 \vee \overline{u_0} \vee \overline{v_0}).$$

Similarly, we introduce two new variables u_i and v_i for each G_i, and replace the G_i's 2-literal clauses with equivalent 3-literal clauses using the tautology

$$a \vee b \equiv (a \vee b \vee c) \wedge (a \vee b \vee \overline{c}),$$

where c plays the role of u_i or v_i for the corresponding 2-literal clause.

Clearly, the new G is satisfiable if and only if the given formula F is satisfiable. Moreover, the number of new variables and the size of the new G are again linear in the size of F. Hence the construction of G can be done algorithmically in polynomial time in the size of F. This proves that 3CNF-SAT is NP-complete.

6.3. Clique. A *clique* in an undirected graph is a complete subgraph of the graph, that is, a subgraph in which any two distinct vertices are connected by an edge. A *k-clique* is a clique that has k vertices. For example, Figure 9 shows a graph with a 5-clique.

FIGURE 9. A graph with a 5-clique.

The *clique problem* (**CLIQUE**) is to determine whether a graph contains a clique of a given size. More formally,

$$\text{CLIQUE} = \{(G, k) : G \text{ is an undirected graph with a } k\text{-clique}\}$$

---

[9]3CNF-SAT is sometimes defined with the proviso that a given CNF-formula contains *at most* three literals, rather than *exactly* three literals. From the observation then it immediately follows that this version of 3CNF-SAT is NP-complete.

or, in the form of a question, CLIQUE is: given an undirected graph G and an integer k, does G have a k-clique?

To prove that CLIQUE is NP-complete, we show CLIQUE $\in$ NP and CLIQUE is NP-hard. The first property is easy to see, since a certificate that a graph G contains a k-clique is the actual subset of vertices in G that makes the clique. Thus, given a pair (G, k) and a subset C of vertices in G, a verifying algorithm should test whether C is a set of k vertices in G and whether G contains all edges connecting all pairs of vertices in C. If both tests pass, the algorithm returns "yes," and otherwise "no." The size of the certificate and the running time of the verifying algorithm are clearly polynomial in the size of G, and so CLIQUE is in NP.

To show that CLIQUE is NP-hard, we exhibit a transformation from 3CNF-SAT to CLIQUE. In general, the idea when constructing a polynomial-time transformation from 3CNF-SAT to another problem is to recognize some components in the nature of the target problem that can simulate variables and clauses in Boolean formulas. Such components are building blocks for the desired transformation and they are sometimes called **gadgets** or **widgets**. For example, in the transformation from 3CNF-SAT to CLIQUE, we have to look for graph-theoretic structures that can enforce certain Boolean-logic properties. As we will shortly see, for a given 3CNF-formula we construct a graph wherein vertices simulate literals and groups of three vertices simulate clauses. Moreover, a satisfying assignment is enforced by a clique of a particular size in the constructed graph.

So let's get back to our construction. Given a 3CNF-formula F with m clauses, we generate the pair (G, m) as an "equivalent" instance of CLIQUE, where G is an undirected graph constructed as follows.

First, for each occurrence of a literal in F, we make a vertex in G, labeled with the same literal name. For easier description, we organize the vertices into m groups of three vertices, called triples, associated with the clauses of F. In other words, each triple corresponds to one of the clauses of F, and each vertex in a triple corresponds to the same-name literal in the associated clause.

Second, we connect by edges all but two types of pairs of vertices in G—no two vertices in the same triple and no two "contradictory" vertices x_i and $\overline{x_i}$ from different triples are connected by an edge. Figure 10 illustrates this construction for $F = (x_1 \vee \overline{x_2} \vee \overline{x_3}) \wedge (\overline{x_1} \vee x_3 \vee \overline{x_3}) \wedge (x_2 \vee x_2 \vee x_3)$.

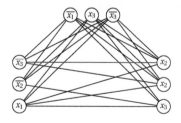

FIGURE 10. The graph G constructed for the example formula F.

Why this construction works, i.e., is F satisfiable if and only if G has an m-clique? Well, suppose first that F has a satisfying assignment. In that satisfying assignment, at least one literal must be 1 in every clause of F, and so in each triple of G we can select one vertex corresponding to such a literal. (If more than one literal is 1 in a particular clause, we choose one of the corresponding vertices arbitrarily.) We claim that the m vertices selected in this way form an m-clique in G. To see why, observe that each pair of selected vertices is not one of the edge-exception type in the construction. The pair could not be from the same triple because we selected only one vertex from a triple. Two vertices of the pair could not have contradictory labels because the corresponding literals were both 1 in the satisfying assignment. Each pair of selected vertices is therefore connected by an edge, hence G contains an m-clique.

Conversely, suppose that G has an m-clique. No two of the clique's vertices are from the same triple, because vertices in the same triple are not joined by edges. Thus, each of the m triples contains exactly one vertex of the m-clique. These vertices induce a truth assignment to the variables in F in the obvious way so that each literal labeling a clique vertex gets the truth value 1. This is a valid truth assignment, because two vertices labeled in a contradictory way are not connected by an edge and hence both cannot be in the m-clique. This truth assignment satisfies each clause of F, because each triple contains a clique vertex and so each clause contains a literal that gets 1 in the assignment. Therefore F is satisfiable.

It remains to see that the transformation from 3CNF-SAT to CLIQUE is polynomial. But this is clear, since the construction of the graph G from a 3CNF-formula F is straightforward. Moreover, the size of G is at most quadratic in the size of F. Therefore the transformation can be implemented by a polynomial-time algorithm. This concludes the proof that CLIQUE is NP-complete.

6.4. 3-colorability. The problem of determining whether the vertices of an undirected graph can be colored with three colors so that no edge has both endpoints of the same color is called the *3-colorability problem (3COL)*. More precisely,

3COL = {G: G is an undirected graph that is 3-colorable},

that is, 3COL asks the following question: given an undirected graph, can the vertices of the graph be colored with three colors so that no edge has both endpoints of the same color?

As usual, to prove that 3COL is NP-complete, we use the standard technique of showing that 3COL $\in$ NP and transforming a known NP-complete problem to 3COL.

It is not hard to see that 3COL belongs to NP. Given an an undirected graph G, a natural certificate that G is 3-colorable is an assignment of three colors to its vertices. A verifying algorithm can easily verify if such an assignment constitutes a proper 3-coloring of G by simply looking at each edge and checking that its endpoints have different colors. If every edge passes this test, the algorithm returns "yes;" otherwise, if some edge fails the test, the algorithm returns "no." Clearly, both the size of the certificate and the running time of the verifying algorithm are polynomial (in fact, linear) in the size of G.

Next we need to show that 3COL is NP-hard by transforming a known NP-complete problem to 3COL. A natural candidate for this is 3CNF-SAT, although the transformation itself is a little bit tricky.

Given an instance of 3CNF-SAT, i.e., a 3CNF-formula F over $X = \{x_1, \ldots, x_n\}$, we show how to transform it to an instance of 3COL, i.e., an undirected graph G, such that F is satisfiable if and only if G is 3-colorable. The graph G will be consisting of two types of "gadgets" that we call a variable-gadget and a clause-gadget. The variable-gadget is a part of the graph G made of one triangle with special vertices named T (true), F (false), and S (special). These names will also be used for the colors that are assigned to these vertices when the graph is 3-colored. The variable-gadget has additionally n triangles with vertices S, x_i, and $\overline{x_i}$ that correspond to the literals in the formula F. The variable-gadget is pictured in Figure 11.

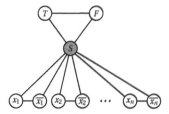

FIGURE 11. The variable-gadget.

Observe that any truth assignment to the variables in X corresponds to a proper 3-coloring of the variable-gadget by coloring each vertex x_i with T or F depending on whether the variable x_i gets 1 or 0 in the assignment, respectively, and by coloring the vertices T, F, and S with the same-name color. Conversely, any proper 3-coloring of the variable-gadget induces a valid truth assignment to the variables in X, since each vertex x_i must get either T or F color (i.e., different from S) and $\overline{x_i}$ gets the complementary color in the Boolean sense (i.e., different from S and the x_i's color).

The clause-gadget is a part of the graph G that is constructed for each 3-literal clause in F and looks like the one given in Figure 12.

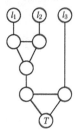

FIGURE 12. The clause-gadget associated with a clause $(l_1 \vee l_2 \vee l_3)$.

The clause-gadget is constructed so that its 3-colorability "computes" the truth value of a clause $(l_1 \vee l_2 \vee l_3)$. Namely, the reader can easily verify that the only way to color the clause-gadget with the three colors T, F, and S, without having an edge whose two endpoints have the same color, is to use the color T for at least one of the vertices l_1, l_2, and l_3. Conversely, if we assign colors T and F to the three vertices l_1, l_2, and l_3, then we can complete the coloring of the clause-gadget to a proper 3-coloring if and only if at least one of l_1, l_2, and l_3 is colored T.

Notice that the vertices l_1, l_2, l_3, and T of a clause-gadget are not new vertices, rather they represent the corresponding vertices of the variable-gadget. In other words, the resulting graph G has only one vertex T and one copy of the vertices (representing literals) $x_1, \overline{x_1} \dots, x_n, \overline{x_n}$ in the variable-gadget, and they are shared between all the clause-gadgets of G used for the clauses. For example, in Figure 13 is given the graph G that is obtained using this construction for the formula $F = (x_1 \vee \overline{x_2} \vee \overline{x_3}) \wedge (\overline{x_1} \vee x_3 \vee \overline{x_3})$.

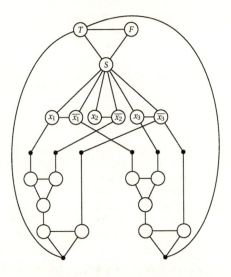

FIGURE 13. The graph G constructed for the example formula F.

Let us now prove that the 3CNF formula F over X is satisfiable if and only if the undirected graph G constructed in this way is 3-colorable. Suppose first that F is satisfiable, that is, we can assign the values 0 and 1 to the variables in X so that every clause evaluates to true. Clearly the truth assignment corresponds to a proper 3-coloring of the variable-gadget of G as mentioned before, and we have yet to see that this coloring can be extended to a proper 3-coloring of all the clause-gadgets of G used for clauses of F. But since every clause is satisfied by our choice of values for the variables, each clause has at least one of its literals equal to 1. This further implies that at least one of the three corresponding vertices of each clause-gadget is colored T. Thus, by the 3-colorability property of a clause-gadget, we can properly 3-color each clause-gadget extending the coloring of the variable-gadget of G.

Conversely, suppose that the graph G can be colored using 3 colors. Since a proper 3-coloring of the whole graph implies a proper 3-coloring of its variable-gadget, such a coloring induces a valid truth assignment to the variables of X as mentioned before. Moreover, a proper 3-coloring of the whole graph implies a proper 3-coloring of each of its clause-gadgets. By the 3-colorability property of a clause-gadget, this further implies that one or more of the three vertices corresponding to literals of each clause-gadget is colored T in a proper 3-coloring of G. But this means that every clause of F must be true under the induced truth assignment, hence F is satisfiable.

Since the construction of the graph G for a given 3CNF-formula F is highly repetitive and the size of G is linear in the size of F, the transformation can be clearly implemented by a polynomial-time algorithm. This shows 3CNF-SAT $\leq$ 3COL and completes the proof that 3COL is NP-complete.

6.5. Vertex cover. A *vertex cover* in an undirected graph is a subset of its vertices such that every edge has at least one endpoint in the subset. For example, the two grayed vertices in Figure 14 form a vertex cover for the given graph.

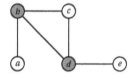

FIGURE 14. A graph and a vertex cover $\{b, d\}$.

The *vertex-cover problem* (VC) is: given an undirected graph G and an integer k, does G have a vertex cover of size k? In the set notation, this means

$$\text{VC} = \{(G, k): G \text{ is an undirected graph with a vertex cover of size } k\}.$$

It is easy to see that VC is in NP. Given a graph $G = (V, E)$ and a subset $V' \subseteq V$ as the certificate, a verifying algorithm checks that V' is a k-vertex subset and that every edge in E touches one of the vertices in V'. If both tests pass, the algorithm returns "yes," and otherwise "no." Clearly, the certificate and running time of the algorithm are polynomial in the size of G.

To show that VC is NP-hard, we give a transformation from CLIQUE to VC. The transformation uses the complement graph of a given graph. If $G = (V, E)$ is an undirected graph, its complement graph $\overline{G}$ is defined as $\overline{G} = (V, \overline{E})$, where

$$\overline{E} = \{(u, v) \in V \times V: u \neq v \text{ and } (u, v) \notin E\}.$$

In other words, $\overline{G}$ is the graph over the same set of vertices as G containing exactly those edges that are not in G. Figure 15 gives an example of a graph and its complement graph.

The reason for using the complement graph is because cliques in G and vertex covers in $\overline{G}$ are closely related, as the following lemma shows.

LEMMA *Let $G = (V, E)$ be an undirected graph and $\overline{G} = (V, \overline{E})$ be its complement graph. A subset $V' \subseteq V$ is a clique in G if and only if $V \setminus V'$ is a vertex cover in $\overline{G}$.*

G $\overline{G}$

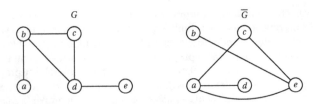

FIGURE 15. A graph and its complement graph.

PROOF: If $V' \subseteq V$ is a clique in G, then no edge in $\overline{E}$ has both endpoints in V'. Equivalently, if $(u, v) \in \overline{E}$ then at least one of u or v is not in V', which means at least one of u or v is in $V \setminus V'$. Thus each edge in $\overline{E}$ has at least one endpoint in $V \setminus V'$, implying $V \setminus V'$ is a vertex cover in $\overline{G}$.

Conversely, if $V \setminus V'$ is a vertex cover in $\overline{G}$, then every edge $(u, v) \in \overline{E}$ has at least one of u or v in $V \setminus V'$, which means at least one of u or v is not in V'. Contrapositively, no edge in $\overline{E}$ has both endpoints in V'. Thus every pair of vertices in V' is connected by an edge in E, implying V' is a clique in G. ∎

As the interested reader might suspect, the lemma hints at the transformation of CLIQUE to VC. It simply maps an instance (G, k) of CLIQUE, where G is a graph with n verices, to the instance $(\overline{G}, n - k)$ of VC. From the lemma it immediately follows that G has a k-clique if and only if $\overline{G}$ has a vertex cover of size $n - k$. In other words, (G, k) is a yes-instance of CLIQUE if and only if $(\overline{G}, n - k)$ is a yes-instance of VC. Moreover, given an undirected graph $G = (V, E)$ and an integer k, we can surely construct the graph $\overline{G}$ and integer $|V| - k$ in time that is polynomial in the size of the input instance. Therefore VC is NP-hard and, being in NP, is NP-complete.

6.6. Subset sum. Contrary to the previously discussed NP-complete problems coming from Boolean logic and graph theory, the next problem we consider belongs to a different problem domain—integer arithmetic. It is particularly useful for proving other problems NP-complete that involve numerical parameters, such as lengths, costs, times, etc.

In the ***subset-sum problem*** (SUBSET-SUM), we are given a finite set of integers $S = \{x_1, \dots, x_n\}$ and a target integer t. We want to determine whether there exists a subset $T \subseteq S$ whose elements add up exactly to t.[10] More formally,

$$\text{SUBSET-SUM} = \left\{ (S, t) : \begin{array}{l} S = \{x_1, \dots, x_n\} \subset \mathbb{N}, \ t \in \mathbb{N}, \text{ and there is } T \subseteq S \\ \text{such that } \sum_{x \in T} x = t \end{array} \right\}.$$

For example, if $S = \{5, 10, 11, 17, 23, 30\}$ and $t = 69$, the subset $T = \{5, 11, 23, 30\}$ of S sums to $5 + 11 + 23 + 30 = 69$, so this is a yes-instance of SUBSET-SUM.

Given S and t, the certificate for SUBSET-SUM is just a subset $T \subseteq S$. So, a verifying algorithm should check whether the certificate T forms a subset of S, then compute the sum of elements in T, and finally verify that this sum equals t. The reader should have no difficulty to see that this can be done in polynomial time. However, in this case as in any arithmetic problem, the input encoding is

---

[10]More generally, S and T may be multisets or lists and thus allow for the repetition of elements.

important and so one should be more careful about the input size and the running time. Nevertheless, recalling that the input integers are encoded in binary and that we can add two b-bit binary numbers in $O(b)$ time, it is not hard to convince ourselves that the verifying algorithm indeed runs in polynomial time. Thus SUBSET-SUM is in NP.

As usual, we next show that SUBSET-SUM is NP-hard by transforming a known NP-complete problem, in this case VC, to SUBSET-SUM. In other words, we describe a polynomial-time transformation that maps an instance of VC (a graph G and integer k) to an instance of SUBSET-SUM (a set of integers S and target integer t) such that G has a vertex cover of size k if and only if S has a subset T summing to t. The SUBSET-SUM problem instance that we construct contains numbers of large magnitude expressed in ordinary decimal notation.

At first sight, it is not quite clear how to transform VC to SUBSET-SUM, because they come from seemingly totally unrelated problem domains such as graph theory and integer arithmetic. However, observe that both problems involve selection of some objects in their respective domains. In the vertex-cover problem we are selecting a subset of k vertices in a graph that cover all the edges, and in the subset-sum problem we are selecting a subset of a set of numbers that sum to t. Thus it seems logical to try first to somehow associate the corresponding objects, and then worry how to enforce other constraints. This intuition leads to the conclusion that we should map vertices to numbers. Moreover, a particular-size subset of these vertices that cover all the edges should enforce the condition that the sum of the corresponding numbers is equal to the target value.

One way to map vertices to numbers so that we can easily relate other constraints as well is based on an incidence-matrix representation of a graph. Let $G = (V, E)$ be an undirected graph with n vertices and m edges. If we arbitrarily number the vertices and edges of G so that $V = \{v_1, v_2, \ldots, v_n\}$ and $E = \{e_1, e_2, \ldots, e_m\}$, the **incidence matrix** of G is an $n \times m$ binary matrix $B = (b_{ij})$ such that

$$b_{ij} = \begin{cases} 1, & \text{if } e_j \text{ is incident with } v_i \\ 0, & \text{otherwise.} \end{cases}$$

An example of a graph and its incidence matrix is shown in Figure 16.

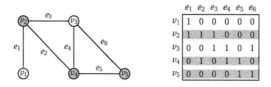

	e_1	e_2	e_3	e_4	e_5	e_6
v_1	1	0	0	0	0	0
v_2	1	1	1	0	0	0
v_3	0	0	1	1	0	1
v_4	0	1	0	1	1	0
v_5	0	0	0	0	1	1

FIGURE 16. A graph and its incidence matrix.

To map vertices to numbers now, we simply think of a vertex v_i as being represented by an m-digit number that forms v_i's row of the incidence matrix $B = (b_{ij})$ for G. That is, each $v_i \in V$ is mapped to the decimal number whose digits correspond to v_i's row of the incidence matrix for G. Formally, for $i =$

$1,\ldots,n$, the vertex v_i is mapped to the decimal number $x_i = \sum_{j=1}^{m} b_{ij} \cdot 10^{m-j}$. Actually, this is not yet our final association in the transformation, but gives an idea of how we are going to proceed.

Observe that the incidence matrix for G has exactly two 1's in any column, because each edge is incident with two distinct vertices.[11] Thus, if we take any subset of vertices and sum their corresponding numbers, the resulting sum is a number consisting of the digits 0, 1, or 2. That is, we will never generate a carry to the next column when summing the rows of the incidence matrix columnwise, since the rows represent the decimal digits of the numbers.

In addition, if selected subset of vertices is a vertex cover, then every edge will be covered by at least one of these vertices. This implies that the sum of the corresponding numbers will be a number consisting only of the digits 1 and 2. For example, the sum of the numbers corresponding to the grayed vertices that form a vertex cover in the graph of Figure 16 is 121121.

In other words, the resulting sum of the numbers corresponding to the vertices of a vertex cover is a number with no zero digits. Obviously, we would like the sum of these numbers to be our target value in the constructed instance of the subset-sum problem. However, we still do not have a unique target value, because this sum can be a number with any combination of 1's and 2's. To fix this problem, we take the target value t to be the number consisting of all 2's (m of them) and by construction add m additional *slack numbers*. For $j = 1,\ldots,m$, the j-th slack number y_j has all 0's except for a single 1 in the j-th position from the left, e.g., $y_4 = 000100$. Another way to think of the slack numbers is that they make an $m \times m$ identity matrix if they are taken in order from first to last as successive rows of a matrix. Figure 17 illustrates our transformation so far for the example graph of Figure 16.

	e_1	e_2	e_3	e_4	e_5	e_6	
x_1	1	0	0	0	0	0	
x_2	1	1	1	0	0	0	
x_3	0	0	1	1	0	1	vertex numbers
x_4	0	1	0	1	1	0	
x_5	0	0	0	0	1	1	
y_1	1	0	0	0	0	0	
y_2	0	1	0	0	0	0	
y_3	0	0	1	0	0	0	
y_4	0	0	0	1	0	0	slack numbers
y_5	0	0	0	0	1	0	
y_6	0	0	0	0	0	1	
t	2	2	2	2	2	2	} target number

FIGURE 17. Incomplete vertex-cover to subset-sum transformation.

Adding the slack numbers works because from the vertex numbers corresponding to a vertex cover we will get the sum consisting of 1's and 2's. Then, if necessary, for each position with a 1 we can supplement the final sum by adding

---

[11]Recall that we don't allow self-loops in undirected graphs. This restriction is not important here, since otherwise we would have at most two 1's in any column of the incidence matrix.

in the corresponding slack number. This way we can increase any number having digits 1 and 2 to the target number having all 2's. On the other hand, if there are any 0's in the sum of some vertex numbers, we won't have enough slack numbers to convert them into 2's.

The vertex mapping described so far may produce the same vertex numbers. For example, if a graph has an isolated edge (v, v'), the vertices v and v' are mapped to the same number. Since we need unique numbers in an instance of SUBSET-SUM, we add n additional columns behind the current m columns to provide for the uniqueness.

Now we show how the n additional columns are filled in. For $i = 1, \ldots, n$, each vertex number x_i has a single 1 in the i-th of these columns and all 0's otherwise. For $j = 1, \ldots, m$, each slack number y_j has all 0's in these columns. In other words, the blocks corresponding to the vertex and slack numbers in these columns comprise an $n \times n$ identity matrix and an $n \times n$ zero matrix, respectively. The target number t has all 3's in these columns, for reasons that will be apparent shortly.

This way we obviously get unique vertex (and slack) numbers, but now no new column can add up to a 3 in the target number. To rectify the construction, we add $2n$ different slack numbers $z_1, z_1', z_2, z_2', \ldots, z_n, z_n'$. The number z_i has a single 2 in the i-th of the last n columns and all 0's in the remaining columns, while the number z_i' has a single 3 in the i-th of the last n columns and all 0's in the remaining columns. This still incomplete construction is illustrated in Figure 18. Notice that in summing the vertex and slack numbers columnwise, we still have no carry from any of the $m + n$ columns, because the maximum sum columnwise is 6.

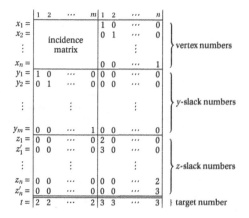

FIGURE 18. Still incomplete vertex-cover to subset-sum transformation.

The last issue we need to take care of is the parameter k that determines the size of the vertex cover in an instance of VC. We will control the number of vertices that go into the vertex cover by adding additional columns in front of the current $m + n$ columns. Let k be an ℓ-digit decimal number and $k = (k_1 \cdots k_\ell)_{10}$ be its decimal representation, that is, $k = \sum_{i=1}^{\ell} k_i \cdot 10^{\ell-i}$. We add ℓ additional columns in front of the $m + n$ columns so that each $(m + n)$-digit number constructed so far is prefixed by an ℓ-digit number as follows. For each vertex number we put a 1 in the ℓ-th column of the prefix part and a 0 in other prefix columns, if any. For each slack number we put a 0 in every prefix column. For the target number, we put the digits $k_1, \dots, k_\ell$ in this order in the prefix columns $1, \dots, \ell$. Thus to get the desired sum t, we must select exactly k of the vertex numbers, because there are no carries in the lower-order $m + n$ digits. This final modification is illustrated in Figure 19 with a partially filled table, where entries not specified should be taken to be 0 with obvious exceptions.

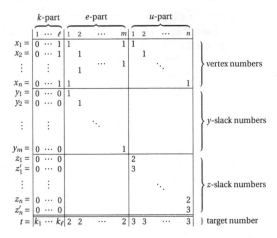

FIGURE 19. Complete vertex-cover to subset-sum transformation.

Let's recap complete transformation of the problem VC to the problem SUBSET-SUM. For an instance of VC consisting of a graph $G = (V, E)$ with n nodes and m edges and an ℓ-digit integer k, we construct n vertex numbers x_i, m slack numbers y_j, $2n$ slack numbers z_i, z_i', and the number t. These numbers comprise rows in a table presented in ordinary decimal notation. Each of these numbers is made of three parts: the prefix k-part has ℓ digits, the middle e-part has m digits, and the suffix u-part has n digits. In this way we get the set $S = \{x_1, \dots, x_n, y_1, \dots, y_m, z_1, z_1', \dots, z_n, z_n'\}$ and the target number t as the corresponding instance of SUBSET-SUM.

Observe that the transformation is a straightforward filling in the entries in a simple table, so it can be surely done algorithmically. Moreover, it can be done in polynomial time—in fact, the construction is proportional to the size of the table, which is $O((n + m + l)^2)$.

To prove that this transformation works, we need to show that G has a vertex cover of size k if and only if S has a subset that sums to t. First suppose that G has a vertex cover $V' = \{v_{i_1}, \dots, v_{i_k}\}$ with k nodes. We build the subset $T \subseteq S$ whose elements sum to t in the following way. We start with the vertex numbers corresponding to the vertices in V', that is, $T = \{x_{i_1}, \dots, x_{i_k}\}$. Then, for each edge that is covered only once in V', we take the corresponding y-slack number and add it in to T. Finally, to bring each of the last n digits of the sum up to 3, we add in to T exactly k nonprimed z-slack numbers $z_{i_1}, \ldots, z_{i_k}$, as well as $n - k$ primed z-slack numbers corresponding to the vertex numbers not in T (i.e., to the vertices not in V'). From our previous discussion it follows that the m-digit e-part and the n-digit u-part of the total sum of the T's numbers will consists of, respectively, all 2's and all 3's. In addition, because there are k vertex numbers in the subset T, the ℓ-digit k-part of the total sum will be exactly k. Thus the total sum of the T's numbers is equal to t.

For the reverse direction, if S has a subset T that sums to t, we first assert that T must contain exactly k vertex numbers $x_{i_1}, \ldots, x_{i_k}$. This follows from the fact that no carry is born in the e-part and the u-part of the summation, hence the k-part of the resulting sum comes solely from 1's in the k-part of the vertex numbers in T. In addition, for every $j = 1, \ldots, m$, there must be at least one of these numbers with the j-th digit in the e-part equal to 1. Otherwise, if for some j all the vertex numbers in T have had a 0 as the j-th digit in the e-part, then no matter which y-slack numbers there were in T, the j-th digit in the e-part of the resulting sum could not be a 2, because no carry is generated in summing up columnwise the e-part and u-part of the T's numbers. This means by construction that, if $V' = \{v_{l_1}, \ldots, v_{l_k}\} \subseteq V$ is the subset of vertices corresponding to the vertex numbers $x_{l_1}, \ldots, x_{l_k}$ in T, then every edge e_j for $j = 1, \ldots, m$ is covered by at least one vertex in V'. Thus V' is a vertex cover in G of size k.

This concludes the proof that SUBSET-SUM is NP-complete.

Exercises

1. The *compositeness problem* is complement to PRIME:

$$\text{COMPOSITE} = \{n \in \mathbb{N}: n \text{ is composite}\}$$

Argue that COMPOSITE is a polynomially verifiable problem.

2. Prove that if Π is NP-complete, then $\Pi \in P \Leftrightarrow P = NP$.

3. Prove that if Π is NP-hard, then $\Pi \in P \Rightarrow P = NP$.

4. Prove that the class of languages NP is closed under the following operations:

  (a) Union of two languages;
  (b) Intersection of two languages;
  (c) Concatenation of two languages. (*Hint:* Let # be a distinguished character not in the alphabet of the certificates. The following is an algorithm that verifies $x \in L_1 L_2$.)

$A(x, y)$
$[\![\, x = x_1 x_2 \cdots x_n \,]\!]$
if $y \neq y_1 \# y_2$ **then**
    return 0;
for $i = 1$ **to** n **do**
    if $(A_1(x_1 \cdots x_i, y_1) = 1) \wedge (A_2(x_{i+1} \cdots x_n, y_2) = 1)$ **then**
        return 1;
return 0;

5. Prove that $\Pi \leq \overline{\Pi}$ if and only if $\overline{\Pi} \leq \Pi$.

6. We say that an algorithm A runs in **quasi-linear** time if there are nonnegative constants c and k such that A's running-time function satisfies $T_A(n) \leq cn(\log n)^k$. Show that transformation in quasi-linear time is a transitive relation.

7. Give an example of a general Boolean formula so that one gets size-exponential blow up of its equivalent CNF-formula generated by successive use of the distributivity of $\vee$ over $\wedge$ and DeMorgan's laws.

8. Show that 2CNF-SAT is in P. That is, there is a polynomial-time algorithm that decides whether a Boolean CNF-formula with each clause having two literals is satisfiable. (*Hint:* Think of a clause $(a \vee b)$ as the implication $(\neg a \Rightarrow b)$ and make a directed graph with vertices being literals and edges being the implications derived from the clauses. Argue that the formula is not satisfiable if and only if any strongly connected component of this graph contains a variable and its negation.)

9. In the **double satisfiability problem** (**BI-SAT**) we are given a Boolean formula F over n variables and the question is whether there are two *distinct* satisfying assignments for F. Show that BI-SAT is NP-complete. (*Hint:* Show SAT $\leq$ BI-SAT using $F \rightarrow F \wedge (x_{n+1} \vee \neg x_{n+1})$.)

10. In the **distinct-formulas problem** (**DF**) we are given two Boolean formula F and G over the same set of variables and the question is whether there is a truth assignment for which $F \neq G$, that is, F and G evaluate to different truth values under the assignment. Show that DF is NP-complete. (*Hint:* Show SAT $\leq$ DF using $F \rightarrow F, G = x_1 \wedge \neg x_1$.)

11. A **triangle** in an undirected graph is a 3-clique. If

$$\text{TRIANGLE} = \{G: G \text{ is an undirected graph having a triangle}\},$$

show that TRIANGLE is in P.

12. An instance of the **two-cliques problem** (**2CLIQUE**) is an undirected graph G and two integers k_1 and k_2. The question is whether G has two *disjoint* cliques of size k_1 and k_2. Show that 2CLIQUE is NP-complete. (*Hint:* Transform CLIQUE to 2CLIQUE by converting an instance (G, k) of CLIQUE to an

instance of 2CLIQUE consisting of G augmented with an isolated vertex and $k_1 = k$ and $k_2 = 1$.)

13. A k-**coloring** of an undirected graph $G = (V, E)$ is an assignment of k colors to the vertices in G so that no two adjacent vertices are assigned the same color. Formally, a k-coloring is a function $c\colon V \to \{1, \dots, k\}$ such that $c(u) \neq c(v)$ for every edge $(u, v) \in E$. In the k-**colorability problem** *(k**COL***) we are given a graph G and an integer k, and the question is whether G has a k-coloring.

   (a) Show that 2COL $\in$ P.
   (b) Show that kCOL is NP-complete for $k \geq 3$. (*Hint:* Transform 3COL to kCOL by converting a graph G to G augmented with a k-complete graph.)

14. An **independent set** of an undirected graph $G = (V, E)$ is a subset $V' \subseteq V$ of vertices such that each edge in E has at most one vertex in V'. Equivalently, no pair of vertices in V' is connected by an edge in E. Given a graph G and an integer k, the **independent-set problem** *(IC)* asks whether G has an an independent set of size k. Prove that IC is NP-complete. (*Hint:* CLIQUE $\leq$ IC.)

15. Given two undirected graphs $G_1 = (V_1, E_1)$ and $G_2 = (V_2, E_2)$, we say that G_1 is **isomorphic** to G_2 if there is a one-to-one function $\pi\colon V_1 \to V_2$ such that $(u, v) \in E_1$ if and only if $(\pi(u), \pi(v)) \in E_2$. In other words, the graph G_2 is the same graph G_1 obtained by renumbering the vertices of G_1. In the **subgraph-isomorphism problem** *(SI)* we are given two undirected graphs G and H, and the question is whether H contains a subgraph H' that is isomorphic to G. Prove that SI is NP-complete. (*Hint:* Show CLIQUE $\leq$ SI using $(G, k) \to (K_k, G)$, where K_k is a k-complete graph.)

16. Give a dynamic-programming algorithm that solves the subset-sum problem in pseudo-polynomial time. That is, for a set $S = \{x_1, \dots, x_n\} \subset \mathbb{N}$ and a target number $t \in \mathbb{N}$, the algorithm should take $O(n \cdot t)$ time. (*Hint:* Build the i-th row of an $(n+1) \times (t+1)$ table T from its $(i-1)$-st row so that, for $i = 0, \dots, n$ and $j = 0, \dots, t$, $T(i, j) = 1$ if there is a subset of $\{x_1, \dots, x_i\}$ that sums to j, and $T(i, j) = 0$ otherwise.)

17. The **knapsack problem** *(KNAPSACK)* from Chapter 7 stated as a decision problem is the following. We have a finite set $S = \{(v_i, w_i) \in \mathbb{N} \times \mathbb{N}\colon i = 1, \dots, n\}$ of n integer pairs and a target pair $(V, W) \in \mathbb{N} \times \mathbb{N}$. The question is whether there is a subset $S' \subseteq S$ of the n pairs such that $\sum_{(v,w) \in S'} v \geq V$ and $\sum_{(v,w) \in S'} w \leq W$. Show that KNAPSACK is NP-complete. (*Hint:* Show SUBSET-SUM $\leq$ KNAPSACK using $(S, t) \to (S \times S, (t, t))$.)

18. In the **set-partition problem** *(SET-PARTITION)* we have a finite set $S \subset \mathbb{N}$ and the question is whether the set S can be partitioned into two disjoint sets S' and $S'' = S \setminus S'$ such that $\sum_{x \in S'} x = \sum_{x \in S''} x$. Show that SET-PARTITION is NP-complete. (*Hint:* Show SUBSET-SUM $\leq$ SET-PARTITION using $(S, t) \to S \cup \{\sum_{x \in S} x, 2t\}$.)

Bibliography

[1] Manindra Agrawal, Neeraj Kayal, and Nitin Saxena. PRIMES is in P. *Annals of Mathematics*, 160(2):781–793, 2004.

[2] Alfred V. Aho, John E. Hopcroft, and Jeffrey D. Ullman. *The Design and Analysis of Computer Algorithms*. Addison-Wesley, 1974.

[3] Alfred V. Aho, John E. Hopcroft, and Jeffrey D. Ullman. *Data Structures and Algorithms*. Addison-Wesley, 1987.

[4] Alfred V. Aho and Jeffrey D. Ullman. *Foundations of Computer Science*. W. H. Freeman, 1992.

[5] Michael O. Albertson and Joan P. Hutchinson. *Discrete Mathematics with Algorithms*. John Wiley, 1988.

[6] Sara Baase and Allen Van Gelder. *Computer Algorithms: Introduction to Design and Analysis*. Addison-Wesley, third edition, 1999.

[7] Jon Bentley. *Programming Pearls*. Addison-Wesley, second edition, 1999.

[8] Kenneth P. Bogart, Clifford Stein, and Robert Drysdale. *Discrete Mathematics for Computer Science*. Key College, 2005.

[9] Gilles Brassard and Paul Bratley. *Fundamentals of Algorithmics*. Prentice Hall, 1995.

[10] Alan Cobham. The intrinsic computational difficulty of functions. In *Proceedings of the 1964 Congress for Logic, Methodology, and the Philosophy of Science*, pages 24–30. North-Holland, 1964.

[11] Stephen A. Cook. The complexity of theorem-proving procedures. In *Proceedings of the Third Annual ACM Symposium on Theory of Computing*, pages 151–158. ACM Press, 1971.

[12] Thomas H. Cormen, Charles E. Leiserson, Ronald L. Rivest, and Clifford Stein. *Introduction to Algorithms*. The MIT Press, second edition, 2001.

[13] M. de Berg, M. van Kreveld, M. Overmars, and O. Schwarzkopf. *Computational Geometry: Algorithms and Applications*. Springer-Verlag, second edition, 2000.

[14] Marianne Durand and Philippe Flajolet. Loglog counting of large cardinalities. In *Proceedings of the 11-th Annual European Symposium on Algorithms*, volume 2832 of *Lecture Notes in Computer Science*, pages 605–617. Springer-Verlag, 2003.

[15] Jack Edmonds. Matroids and the greedy algorithm. *Mathematical Programming*, 1:126–136, 1971.

[16] William Feller. *An Introduction to Probability Theory and Its Applications*. John Wiley, third edition, 1968.

[17] William Ford and William Topp. *Data Structures with C++*. Prentice Hall, 1996.

[18] D. Gale and L. S. Shapley. College admissions and the stability of marriage. *American Mathematical Monthly*, 69:9–15, 1962.

[19] Michael R. Garey and David S. Johnson. *Computers and Intractability: A Guide to the Theory of NP-Completeness*. W. H. Freeman, 1979.

[20] G. H. Gonnet. *Handbook of Algorithms and Data Structures*. Addison-Wesley, 1984.

[21] Michael T. Goodrich and Roberto Tamassia. *Algorithm Design: Foundations, Analysis, and Internet Examples*. John Wiley, 2001.

[22] Ronald L. Graham, Donald E. Knuth, and Oren Patashnik. *Concrete Mathematics: A Foundation for Computer Science*. Addison-Wesley, second edition, 1994.

[23] Dan Gusfield and Robert W. Irving. *The Stable Marriage Problem: Structures and Algorithms*. The MIT Press, 1989.

[24] John E. Hopcroft and Jeffrey D. Ullman. *Introduction to Automata Theory, Languages, and Computation*. Addison-Wesley, 1979.

[25] Juraj Hromkovic. *Design and Analysis of Randomized Algorithms*. Springer, 2005.

[26] Richard Johnsonbaugh. *Discrete Mathematics*. Prentice Hall, sixth edition, 2004.

[27] Richard Johnsonbaugh and Marcus Schaefer. *Algorithms*. Prentice Hall, 2003.

[28] Richard M. Karp. Reducibility among combinatorial problems. In R. E. Miller and J. W. Thatcher, editors, *Complexity of Computer Computations*, Proc. Sympos. IBM Thomas J. Watson Research Center, pages 85–103. Plenum Press, 1972.

[29] M. S. Klamkin and D. J. Newman. Extensions of the birthday surprise. *Journal of Combinatorial Theory*, 3:279–282, 1967.

[30] Jon Kleinberg and Éva Tardos. *Algorithm Design*. Addison-Wesley, 2005.

[31] Donald E. Knuth. *Seminumerical Algorithms*, volume 2 of *The Art of Computer Programming*. Addison-Wesley, second edition, 1981.

[32] Donald E. Knuth. *Fundamental Algorithms*, volume 1 of *The Art of Computer Programming*. Addison-Wesley, third edition, 1997.

[33] Donald E. Knuth. *Stable marriage and its relation to other combinatorial problems: An introduction to the mathematical analysis of algorithms*, volume 10 of *CRM Proceedings & Lecture Notes*. American Mathematical Society, 1997.

[34] Donald E. Knuth. *Sorting and Searching*, volume 3 of *The Art of Computer Programming*. Addison-Wesley, second edition, 1998.

[35] Dexter C. Kozen. *The Design and Analysis of Algorithms*. Springer-Verlag, 1992.

[36] Anany Levitin. *Introduction to the Design and Analysis of Algorithms*. Addison-Wesley, second edition, 2006.

[37] Michael Machtey and Paul Young. *An Introduction to the General Theory of Algorithms*. Elsevier North Holland, 1978.

[38] Makoto Matsumoto and Takuji Nishimura. Mersenne twister: A 623-dimensionally equidistributed uniform pseudorandom number generator. *ACM Transactions on Modeling and Computer Simulation*, 8(1):3–30, 1998.

[39] Kurt Mehlhorn. *Data Structures and Efficient Algorithms, Vol. 1–3*. EATCS Monographs on Theoretical Computer Science. Springer-Verlag, 1984.

[40] Michael Mitzenmacher and Eli Upfal. *Probability and Computing: Randomized Algorithms and Probabilistic Analysis*. Cambridge University Press, 2005.

[41] Rajeev Motwani and Prabhakar Raghavan. *Randomized Algorithms*. Cambridge University Press, 1995.

[42] Joseph O'Rourke. *Computational Geometry in C*. Cambridge University Press, second edition, 1998.

[43] S. K. Park and K. W. Miller. Random number generators: Good ones are hard to find. *Communications of the ACM*, 31:1192–1201, 1988.

[44] William Pugh. Skip lists: A probablistic alternative to balanced trees. *Communications of the ACM*, 33(6):668–676, 1990.

[45] Kenneth H Rosen. *Discrete Mathematics and Its Applications*. McGraw-Hill, fifth edition, 2003.

[46] Robert Sedgewick and Philippe Flajolet. *An Introduction to the Analysis of Algorithms*. Addison-Wesley, 1995.

[47] Robert Sedgewick and Michael Schidlowsky. *Algorithms in Java*. Addison-Wesley, third edition, 2002.

[48] Michael Sipser. *Introduction to the Theory of Computation*. PWS Publishing Company, 1997.

[49] Steven S. Skiena. *The Algorithm Design Manual*. Springer-Verlag, 1998.

[50] James A. Storer. *An Introduction to Data Structures and Algorithms*. Birkhauser Boston, 2002.

[51] Robert E. Tarjan. *Data Structures and Network Algorithms*. CBMS-NSF Regional Conference Series in Applied Mathematics. Society for Industrial and Applied Mathematics, 1983.

[52] Gabriel Valiente. *Algorithms on Trees and Graphs*. Springer-Verlag, 2002.

[53] Mark Allen Weiss. *Algorithms, Data Structures, and Problem Solving with C++*. Addison-Wesley, 1996.

[54] Herbert S. Wilf. *Algorithms and Complexity*. A K Peters, second edition, 2003.

Index

inversion, 135
ITER-DFS(s), 197
iterative algorithms
  loop invariant, 8
  proving correctness, 8
iterative substitution method, 121–122

K

k-colorability problem (kCOL), 355
k-coloring, 355
KNAPSACK(v, w, W), 159
  running time, 160, 311
knapsack problem, 157–160, 355
  0-1 knapsack problem, 157
Kruskal's algorithm, 258–260

L

labeled trees, 76–77
language, 325
  recognized by TM program, 331
  verified by TM program, 332
Las Vegas algorithms, 284–292
  flipping a biased coin to simulate a fair coin, 288
  randomized quick sort, 285
  searching a sorted compact list, 289
LCG(s), 281
LCS(x, y), 157
leftmost-child, right-sibling representation of trees, 81
length of a path, 166, 262
limit rule, 40, 50
linear congruential generator, 279–281
  full-period, 279
  multiplicative, 279
  period, 279
  seed, 279
linear probing, 231–232
linearity of expectation, 95, 119, 229, 287
linked list, 53–62
  deleting from, 55
  doubly linked, 57
  implementing, 57
    multiple-array C implementation, 59
    multiple-array representation, 58
    single-array representation, 58
  initializing, 54
  inserting into, 55
  searching, 56
LIST-BREAK(L, L_1, L_2), 112
LIST-DELETE(L, p), 56
LIST-HEAD-DELETE(L), 56
LIST-HEAD-INSERT(L, x), 55
LIST-INSERT(L, x, p), 55
LIST-MAKE(L), 55
LIST-MERGE(L_1, L_2, L), 111

list-of-children representation of trees, 80–81
LIST-SEARCH(L, k), 56
literal, 339
log-cost model, 4
longest common subsequence, 154–157
  length of, 155
    dynamic programming technique, 156
    recursive definition, 156
loop invariant, 8
LR-ROTATE(x), 218

M

MAKE-CHANGE(n), 138
MAKE-CHANGE(n, C), 154
map coloring, 145
master method, 126–130
master theorem, 127
matching
  perfect, 32
  stable, 33
maximum contiguous subsequence sum problem, 27–32
  divide-and-conquer algorithm, 135
  dynamic-programming algorithm, 162
  linear algorithm, 30
  quadratic algorithm, 28
maximum element of an array, 18–19
maximum-size independent-set problem on trees, 145
maximum-weight independent-set problem on trees, 163
MCSS1(A), 29, 30
MCSS2(A), 32, 50
merge sort, 110–114
  analysis of algorithm, 112
MERGE-SORT(L), 112
Mersenne twister, 282–284
  period parameters, 283
  tempering parameters, 283
minimal spanning trees, 256–262
  Kruskal's algorithm, 258
  MST property, 257
  Prim's algorithm, 260
  spanning tree, 256
minimum distance, 263
minimum-path problem, 316
Monte Carlo algorithms, 292–298
  finding a high-ranking element, 292
  randomized counting, 294
  verifying equality of strings, 297
MST property, 257
MST-KRUSKAL(G), 259
MST-PRIM(G), 261
MT($\mathbf{x}(0), \mathbf{x}(1), \ldots, \mathbf{x}(n-1)$), 284
multiplication method, 225
multiplication of large integers, 130–133

www.ingramcontent.com/pod-product-compliance
Lightning Source LLC
LaVergne TN
LVHW042331060326
832902LV00006B/99

* 9 7 8 3 6 3 9 1 3 5 0 4 6 *